New Trends in the Chemistry of Nitrogen Fixation

Publisher's Note

This work has been put together as part of the 200th anniversary celebrations of the Portuguese Academy of Sciences. A special limited edition is to be published by the Academy in their series "Frontiers of Knowledge". We are grateful to the Academy for their agreement to the publication of this general edition and congratulate them on their anniversary.

The book is dedicated to the memory of the late General Luiz da Câmara Pina who, as Treasurer, played such an active role in the affairs of the Academy and was the inspiration of the "Frontiers of Knowledge" series.

Insignia of the Academia das Ciências de Lisboa

NEW TRENDS IN THE CHEMISTRY OF NITROGEN FIXATION

Edited by

J. Chatt
A.R.C. Unit of Nitrogen Fixation, University of Sussex, England

General L. M. da Câmara Pina
Academia das Ciências de Lisboa, Lisbon, Portugal

R. L. Richards
A.R.C. Unit of Nitrogen Fixation, University of Sussex, England

1980

ACADEMIC PRESS
A Subsidiary of Harcourt Brace Jovanovich, Publishers
London · New York · Toronto · Sydney · San Francisco

ACADEMIC PRESS INC. (LONDON) LTD
24/28 Oval Road
London NW1

U.S. Edition published by
ACADEMIC PRESS INC.
111 Fifth Avenue
New York
New York 10003

British Library Cataloguing in Publication Data

New trends in the chemistry of nitrogen fixation.
1. Nitrogen–Fixation
I. Chatt, J
II. Richards, R L
III. Camara Pina, L M da
546′.711′59 QR89.7 80-40236

ISBN 0-12-169450-X

Printed in Great Britain by
Spottiswoode Ballantyne Ltd.
Colchester and London

List of Contributors

J. Chatt

ARC Unit of Nitrogen Fixation, University of Sussex, Falmer, Brighton BN1 9RQ, England

G. J. Leigh

ARC Unit of Nitrogen Fixation, University of Sussex, Falmer, Brighton BN1 9RQ, England

A. J. L. Pombeiro

Complexo Interdisciplinar, Instituto Superior Técnico, Lisboa-1, Portugal

R. L. Richards

ARC Unit of Nitrogen Fixation, University of Sussex, Falmer, Brighton BN1 9RQ, England

G. N. Schrauzer

Department of Chemistry, Revelle College, University of California, P.O. Box 109, La Jolla CA 92037, USA

A. E. Shilov

Institute of Chemical Physics, USSR Academy of Sciences, Chernogolovka, Moscow 142432, USSR

V. B. Schur

Institute of Organometallic Compounds, USSR Academy of Sciences, Vavilova 28, Moscow 117334, USSR

K. Tamaru

Department of Chemistry, University of Tokyo, Hongo, Bunkyo-ku, Tokyo 113, Japan

J. H. Teuben

Laboratorium voor Anorganische Chemie, Rijksuniversiteit Groningen, The Netherlands

M. E. Vol'pin

Institute of Organometallic Compounds, USSR Academy of Sciences, Vavilova 28, Moscow 117334, USSR

Preface

This volume is part of a series "Frontiers of Knowledge" written to mark the bicentenary celebration of the foundation in 1779 of the Academia das Ciências de Lisboa at the suggestion of its Treasurer, General Luiz Maria da Câmara Pina. The authors wish to express their hopes for the continuing health and well-being of the Academy. They are sure that they are joined in this wish by all scientists from all over the world.

The authors also wish to express their deep sympathy to Mrs. da Câmara Pina and to the Academy on the death of General da Câmara Pina, which occurred while this book was in the late stages of preparation.

Contents

1

Introduction

JOSEPH CHATT

A.R.C. Unit of Nitrogen Fixation, University of Sussex, Brighton, England

THIS book has been published as part of a series entitled "Frontiers of Knowledge" produced in celebration of the bicentenary of the Academia das Ciências de Lisboa which was founded on 24th December 1779. It was a very happy thought of its Treasurer, General Luiz Maria da Câmara Pina, to produce this series of contributions from experts in many countries; and in this way to illustrate the international interdependence of science in scientific areas undergoing the most rapid development at the present time; and also in those of potentially greatest benefit to mankind. The papers collected here formed the basis of a symposium held in Lisbon in the rooms of the Academy from 30th July to 4th August 1979 at which the various authors discussed their contributions before a Portuguese audience.

The following chapters contain some very recently discovered and some controversial chemistry. Each chapter with its individual author is more or less self-contained, leading inevitably to a certain amount of repetition. The book is not a textbook nor a series of reviews. It is the history of fifteen years of great activity in an important area of chemistry described by some of those who have made the most significant contributions to it.

Over the past fifteen years nitrogen fixation has become very topical. This has arisen from the isolation in 1960 of active nitrogenase from nitrogen-fixing bacteria and from the realization that the present high level of agricultural production is absolutely dependent upon an adequate supply of nitrogenous fertilizer which is at present provided industrially by the catalytic reduction of elementary nitrogen to ammonia by hydrogen. The industrial process known as the Haber or Haber–Bosch process consumes a considerable quantity of energy yet ammonia is by any standards a cheap chemical and it will be difficult indeed to better the Haber process.

It is estimated that at present the total nitrogen fixed on a global basis is about $2{\cdot}4 \times 10^8$ tonnes. Some 65% of this is supplied by nitrogen-fixing microorganisms and the remainder by two chemical processes; the first industrial and the second incidental. The first, which is the Haber process, accounts for some 25% of the total and the second, accounting for the remaining 10%, is the oxidation of nitrogen in the air. This is brought about by ultraviolet radiation, fires, lightning, and now, in significant quantity, by the internal combustion engine.

The importance of the industrially fixed nitrogen is much greater than the figure of 25% would indicate because that nitrogen is used to fertilize agricultural land where it is needed. Its use is essential to sustain the present level of the world's population (4×10^9) and some four times the present production will be needed to provide the food for the projected world population (6 or 7×10^9) by the end of this century.[1]

Ammonia is manufactured from molecular nitrogen (correctly named dinitrogen) and dihydrogen on a promoted (K_2O, BaO, MoO_3, etc.) iron–alumina catalyst at around 350–1000 atmospheres pressure and 300–400°C. The dihydrogen needed for this reaction is at present obtained from the reaction of natural gas with water. When natural gas becomes exhausted oil can take its place, followed by coal, or any other form of energy which can be used to liberate dihydrogen from water. But all of this involves greatly increasing costs.

Relative to the high temperatures and pressures of the Haber process the natural biological process appears to be effortless. Some very primitive microorganisms using water and nothing more reducing than carbohydrate are able to convert molecular nitrogen to ammonia. It is this knowledge which has spurred chemists to attempt to produce ammonia more cheaply than by the Haber process. However it was the isolation in 1960[2] of the enzyme nitrogenase, on which the biological production of ammonia occurs, that sparked the present phase of interest in the biochemistry and chemistry of nitrogen fixation. The series of "oil crises" over the past five years have added even greater emphasis to the need to find an alternative source of fertilizer nitrogen. Work in this area is now concentrated on two general goals. The first aim is to find by a study of the natural process a more economical chemical process for fixing nitrogen specifically for agricultural or horticultural purposes. The second depends on the rapid development of microbial genetics over the past ten years. This opened up the possibility of transferring the genes which specify nitrogen fixation and those which regulate nitrogen fixation from nitrogen-fixing microorganisms into other microbes or perhaps even plants. This latter would be an ideal solution; solar power collected through the leaves of the plants would provide the energy for nitrogen fixation within the plant; but this is not a subject for us here.

This book is concerned only with recent trends in the chemistry of nitrogen fixation. The subject has two aspects, an attempt to prepare improved catalysts for a Haber-type process and the search for a new reaction of molecular nitrogen to form the basis of an industrial process.

Heterogeneous Catalysts for Nitrogen Fixation

The Haber process, which uses a heterogeneous catalyst at high temperatures and pressures, is the only viable industrial process at the present time. If the equilibrium between dinitrogen, dihydrogen and ammonia could be established at low temperatures it would lie well over to the side of ammonia (98·5% at 27°C) at atmospheric pressure (Reaction (**1**)), but at 327°C the equilibrium mixture contains only 8·7% of ammonia and at 627° only 0·21%

$$N_2 + 3H_2 \rightarrow 2NH_3 + 11{\cdot}0\ \text{kcal} \quad (\text{at } 20^\circ\text{C}) \qquad (\mathbf{1})$$

There are no catalysts sufficiently active to give a reasonable rate of reaction at temperatures under 350°C and at such temperatures high pressures are also necessary to make the process viable. More efficient catalysts would save energy by working at lower temperatures and pressures. The process is now carried out on such a scale that minor increases in catalyst efficiency can have highly significant economic consequences. Professor K. Tamaru describes the current progress in this area of endeavour in Chapter 2.

The search for new reactions of molecular nitrogen has commonly been based on a very rudimentary knowledge of the chemistry of the bacterial process. During the past fifteen years it has led to the discovery of a variety of chemical reactions undergone by molecular nitrogen at ordinary or moderate temperatures and pressures. All recent discoveries involve the reaction of transition metal compounds with molecular nitrogen under reducing conditions. The most interesting have involved the early transition metals especially titanium, zirconium, vanadium, molybdenum and tungsten: with very strong reducing agents, such as metallic sodium or its amalgam, magnesium, vanadium(II), sodium borohydride, Grignard and similar reagents. Other chemical connections between the various reactions are so tenuous that it is best to take them in the chronological order of their main periods of development as follows.

Nitriding Reactions of Molecular Nitrogen

This topic developed completely independently of the nitrogenase reaction. Its two earliest and independent contributors[3,4] were both interested in carbenes and attempted to prepare titanium analogues. They subjected titanium compounds to strong reduction under dinitrogen as an "inert atmosphere" and obtained products which on hydrolysis produced ammonia. The development of this area is described in Chapter 3 by Professor M. E. Vol'pin of Moscow.

Fixation of Nitrogen in Aqueous Solution

There have been many attempts to achieve the reduction of dinitrogen to ammonia in aqueous solution and a great number of claims of success over the past fifty years. Few of these have been substantiated, and indeed this is one of the most controversial areas of recent work. In the two quite different approaches of Part 2 the editors have made no attempt to disguise the controversy. Indeed debate and controversy is an important stimulus to research and only further work can resolve the matter. Nevertheless, one can confidently conclude that molecular nitrogen can be reduced by suitable treatment in methanolic or aqueous solutions.

Stable Dinitrogen Complexes as Intermediates in Nitrogen Fixation

Perhaps the most important recent development was the discovery in 1965 that dinitrogen can be bound as a ligand to transition metals in coordination compounds, also known as complex compounds or complexes. Although the first well characterized example of a dinitrogen complex $[Ru(NH_3)_5(N_2)]Cl_2$ was prepared indirectly from hydrazine,[5] it was very soon discovered that dinitrogen itself could combine directly and rapidly with transition metal salts in a suitable ligand environment under strongly reducing conditions even at temperatures as low as $-80°C$.[6,7] This was then followed by the discovery that dinitrogen could attach itself to a transition metal complex in aqueous solution at ordinary temperature and pressure (Reaction (**2**)).[8] At higher pressures dinitrogen entered the same ruthenium complex as a terminal ligand (Reaction (**3**)).

$$2[Ru(NH_3)_5(H_2O)]^{2+} \xrightarrow[(H_2O)\ 20°C]{N_2,\ 1\ atm.} [(NH_3)_5Ru{-}N{\equiv}N{-}Ru(NH_3)_5]^{4+} \quad \textbf{(2)}$$

$$[Ru(NH_3)_5(H_2O)]^{2+} \xrightarrow[(H_2O)\ 20°C]{N_2,\ 100\ atm.} [Ru(NH_3)_5(N_2)]^{2+} \quad \textbf{(3)}$$

There seems no doubt that a reaction such as (**2**) or (**3**) on molybdenum centres in the enzyme is the initial step in the biological reduction of molecular nitrogen to ammonia. The developments in this area leading to a very satisfying mechanism for the reduction of dinitrogen on a single metal site are the subject of Part 3. Understanding of the reactions of dinitrogen ligating metal sites could also lead to a wider range of reactions than formation of nitrogen hydrides. The formation of nitrogen carbon bonds is such a newly developing area and is also reviewed in Part 3.

Reactions of Ligands Analogous to Dinitrogen

Nitrogenase is not specific for the catalysis of the reduction of dinitrogen. Many other small molecules such as carbon monoxide, dihydrogen, hydrogen cyanide, acetylene, organic isocyanides, etc., either inhibit the nitrogenase reaction or are reduced on nitrogenase. The reactions of complexes analogous to dinitrogen complexes or partially reduced dinitrogen complexes have been examined for analogies with the nitrogenase reaction (G. N. Schrauzer in Chapter 4 and A. J. L. Pombeiro in Chapter 6).

Some investigators in all but the heterogeneous area of catalysis have drawn proposals as to the possible chemical mechanism of fixation on the enzyme from their results. These mechanisms are often in conflict with each other and no attempt has been made to set out the pros and cons of each. With the development of more sensitive spectrometers, for instance in ^{15}N NMR spectroscopy, and the accumulation of spectroscopic data from the models of the various artificial systems of Chapters 3 to 9, it should be possible within the next decade to reach a reasonably firm conclusion as to the chemistry of the natural process.

Since rudimentary knowledge of the nitrogenase reaction has led to such an extensive investigation of the reactions of molecular nitrogen with transition metal salts, it is useful for the understanding of the book to provide a summary of the present state of biochemistry relevant to the chemistry of the natural process.

Nitrogenase is a brown complex molybdo–iron enzyme which catalyses the reduction of dinitrogen to ammonia in some very primitive microorganisms. After extraction from nitrogen-fixing microorganisms it is highly sensitive to dioxygen and is destroyed by it. The conversion of dinitrogen to ammonia occurs on it in aqueous solution at around neutral pH. It requires a reducing agent, magnesium ion, and a source of ATP. During the reaction the ATP is reduced to ADP which is an inhibitor of the reaction. Sodium dithionite is the most convenient artificial reducing agent.

The biochemistry of the nitrogenase reaction is now well understood at the descriptive level, but there is still very little definite information about its chemistry. The enzyme is not specific and besides the reduction of dinitrogen it catalyses the reduction of dinitrogen oxide, azide ion, cyanide ion, acetylene, methyl isocyanide, cyclopropene and some of its derivatives, and, much less efficiently, methyl cyanide. Cyanide ion is reduced mainly to methane and ammonia, but acetylene is reduced only to ethylene. Dihydrogen is a competitive inhibitor of nitrogen fixation and carbon monoxide a strong inhibitor.[9] All of these are substances which bind strongly to transition metals in low oxidation state complexes, hence the relevance of the chemistry of Chapters 3 to 10.

The reduction of dinitrogen on nitrogenase *in vivo* and *in vitro* is accompanied by the energetically wasteful production of dihydrogen which is countered in some aerobes, e.g. Azotobacter, which have a hydrogenase to reabsorb the dihydrogen before it escapes.[10] When nitrogenase functions *in vitro* under argon in place of dinitrogen there is greatly increased evolution of dihydrogen and under acetylene dihydrogen evolution can be almost completely suppressed. C_2D_2 is reduced to *cis*-1,2-$C_2H_2D_2$ and under a mixture of D_2 and N_2, HD is produced, but only when the enzyme is fixing nitrogen. It is not produced when the enzyme is confined under a mixture of Ar and D_2—then the enzyme expends all its reducing capacity in producing dihydrogen.[9,10]

Nitrogenase can be separated on a "Sephadex" column into a brown protein of molecular weight around 220 000 and a yellow protein of molecular weight around 68 000. The molecular weights vary slightly depending on the microorganisms from which the enzyme has been extracted. The larger of the two proteins contains ideally two atoms of molybdenum and some 24 to 34 atoms of iron. The iron content may vary according to the source of the protein, but now it is generally thought that the number of iron atoms is nearer 34 than 24. The small protein contains four atoms of iron and no molybdenum. The iron atoms are accompanied by sulphide ions in about equal number. In the smaller protein the iron is known to be present as one Fe_4S_4 ferredoxin-type cluster and in the larger protein at least half of the iron appears to be combined in that manner. Such clusters are bonded through four cysteine sulphur atoms into the protein. One cysteine sulphur binds to each iron atom. The nitrogenase proteins are sensitive to air. The smaller is immediately destroyed on the briefest exposure to air and the larger one is rather less sensitive.[9]

The reduction of dinitrogen to ammonia appears to occur on the large protein. The small protein acts as a specific electron carrier or electron pump which conveys electrons from the reducing agent to the large protein. It can only do this in the presence of the monomagnesium salt of ATP which is hydrolysed to ADP and inorganic phosphate, one molecule for each electron transferred. The Fe_4S_4 clusters in the large protein probably provide the electron storage system from which electrons pass to the active site and thence into a dinitrogen molecule which simultaneously picks up protons from its environment until it has been completely reduced to ammonia. The ammonia, presumably, is then displaced by dinitrogen and the cycle repeats.[9]

The molybdenum atoms are believed to provide the active site or sites for the reduction of dinitrogen. Until 1978 all attempts to determine their environment in the enzyme by spectroscopic means were unsuccessful. The molybdenum concentration is too low for any but the most sensitive methods. However, recent studies of the extended X-ray absorption fine structure (EXAFS) spectrum of the large protein now confirms the long entrenched belief that the molybdenum, like the iron, is associated with sulphide ions. These studies[11] indicate that the molybdenum is surrounded by three or four sulphur atoms at $2{\cdot}35 \pm 0{\cdot}03$ Å, one or two sulphur atoms at $2{\cdot}49 \pm 0{\cdot}35$ Å, and two or three iron atoms at $2{\cdot}72 \pm 0{\cdot}05$ Å. Two types of sulphur-bridged structure containing iron and molybdenum were proposed as satisfying the spectroscopic criteria, but when the measurements were made no such structures were known. One proposition[11] was that the molybdenum was contained in an $MoFe_3S_4$ cluster analogous to the ferredoxin Fe_4S_4 type cluster but having one iron atom replaced by molybdenum (Fig. 1). Now two compounds each containing two such clusters have been described, $[Et_4N]_3[\{(FeSEt)_3\text{-}MoS_4\}_2S(SEt)_2]$[12] and $[Bu^n{}_4N]_3[\{(FeSPh)_3MoS_4\}_2(SPh)_3]$.[13]

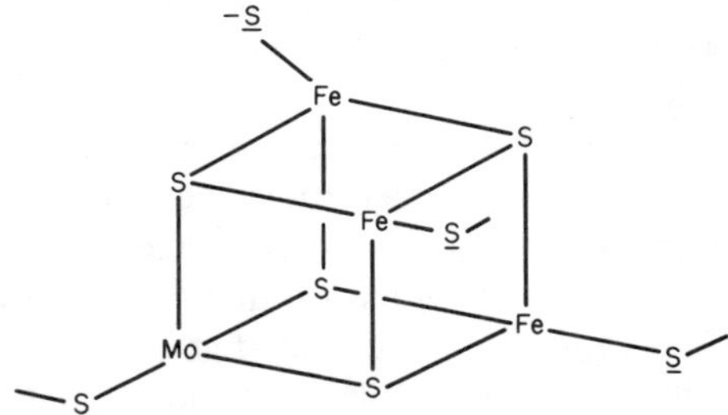

Fig. 1. Proposed $MoFe_3S_4$ cluster bonded through cysteine sulphur S̲ into the protein.

None of these compounds have so far been shown to catalyse the reduction of dinitrogen to ammonia.

A very unstable iron–molybdenum cofactor has been obtained from the large protein by treatment with citric acid and extraction with dimethyl formamide. This cofactor activates the large protein obtained in inactive form from mutants of nitrogen-fixing bacteria which produce nitrogenase containing no molybdenum.[14] It is the only molybdenum-containing substance which is known to do so. It is not certain how closely it is related to the molybdenum site in nitrogenase. The cofactor is very unstable in aqueous solution; it contains almost no protein or amino acids, apparently has a low molecular weight, and contains molybdenum, iron and sulphur probably in the ratio of 1:8:6.[14]

The enzyme shows a number of ESR absorptions which can be attributed to the iron–sulphur clusters. The changes in these absorptions induced by oxidation or reduction of the proteins have been extensively investigated. However there is no ESR absorption which can be attributed to molybdenum.[9]

Whether the molybdenum atoms in the enzyme react with dinitrogen as a pair or independently is important. There is no direct evidence on this question, but circumstantial evidence suggests that they may behave independently. This evidence is that there is probably only one molybdenum atom in the iron–molybdenum cofactor and that the large protein can be separated into two pairs of identical sub-units. One of these pairs must contain the molybdenum atoms, one in each of its two matching sub-units. This suggests that the molybdenum atoms may be far apart, although they could be paired on the boundary between the matching sub-units. As the enzyme is usually isolated from some bacterium, it is deficient in molybdenum and its activity is more nearly linearly related to the concentration of molybdenum than to the square or some other such functions.[15] This suggests that the molybdenum atoms behave independently or that they are inserted into the enzyme in pairs.

Three quite different outline proposals concerning the chemical mechanism of the reduction in the enzyme stand before chemists at the present time. The first, described in Chapter 4 supposes that the dinitrogen molecule is bonded side-on to one molybdenum atom and the reduction occurs through the stages of side-on bonded diazene and hydrazine to ammonia. The second, described in Chapter 5 supposes that the dinitrogen molecule bonds between two metal atoms and is reduced somewhat as organic azo-compounds. The third, described in Chapters 7 and 8 supposes that the dinitrogen molecule is end-bonded to one molybdenum atom and that protonation and reduction occurs from the free end of the molecule taking away first the outer nitrogen atom as ammonia and then the one adjacent to the metal.

Until recently no intermediate between dinitrogen and ammonia had been found on the enzyme. However, when the functioning enzyme is destroyed by addition of either acid or alkali, free hydrazine is produced. The quantity is the same whether an acid or alkali quench of the nitrogenase reaction is used and the amount of hydrazine is proportional to the rate of production of ammonia.[16] Since all of the above proposals involve going through hydrazine or a derivative of hydrazine the detection of hydrazine in the quenched reaction is not diagnostic for any particular mechanism. The next ten years should at least see the establishment of a firm outline mechanism for the nitrogenase reaction and the demise of at least two of the above proposals.

References

1. C. C. Delwiche, *Ambio* (*Stockholm*) (1977) **6**, 106; R. W. F. Hardy and U. D. Havelka, *Science N.Y.* (1975) **188**, 633.
2. J. E. Carnahan, L.E. Mortenson, H. F. Mower and J. E. Castle, *Biochim. Biophys. Acta* (1960) **38**, 188.

3. M. E. Vol'pin and V. B. Shur, *Nature* (*London*) (1966) **209**, 1236, and references therein.
4. E. E. van Tamelen, G. Boche and R. Greeley, *J. Amer. Chem. Soc.* (1968) **90**, 1677, and references therein.
5. A. D. Allen and C. V. Senoff, *Chem. Commun.* (1965) 621.
6. A. E. Shilov, A. K. Shilova, E. F. Kvashina and T. A. Vorontsova, *Chem. Commun.* (1971) 1570, and references therein.
7. A. Sacco and M. Rossi, *Chem. Commun.* (1967) 316.
8. D. F. Harrison, E. Weissberger and H. Taube, *Science, N.Y.* (1968) **159**, 320.
9. R. R. Eady and B. E. Smith, *In* "Dinitrogen Fixation" (Ed. R. W. F. Hardy) Vol. 11, 399. Wiley–Interscience, New York (1978).; L. E. Mortenson and R. N. F. Thorneley, *Ann. Rev. Biochem.* (1979) **48**, 387; W. H. Orme-Johnson, L. C. Davis, M. T. Henzl, B. A. Averill, N. R. Orme-Johnson, E. Munck and R. Zimmerman, *In* "Recent Developments in Nitrogen Fixation" (Ed. W. Newton, J. R. Postgate and C. Rodriguez-Barrueco), 131. Academic Press, New York and London (1977).
10. C. C. Walker and M. G. Yates, *Biochemie* (1978) **60**, 225.
11. S. P. Cramer, K. O. Hodgson, W. O. Gillum and L. E. Mortenson, *J. Amer. Chem. Soc.* (1978) **100**, 3398; S. P. Cramer, W. O. Gillum, K. O. Hodgson, L. E. Mortenson, E. I. Stiefel, S. R. Chisnell, W. J. Brill and V. K. Shah, *J. Amer. Chem. Soc.* (1978) **100**, 3814.
12. T. E. Wolff, J M. Berg, C. Warrick, K. O. Hodgson, R. H. Holm and R. B. Frankel, *J. Amer. Chem. Soc.* (1978) **100**, 4630.
13. G. Christou, C. D. Garner, F. E. Mabbs and T. J. King, *J.C.S. Chem. Commun.* (1978) 740.
14. V. K. Shah and W. J. Brill, *Proc. Natl. Acad. Sci., U.S.A.* (1977) **74**, 3249; B. E. Smith, *Proc. Internat. Symposium on Molybdenum Chemistry of Biological Significance, Shiga, Japan, 1979.*
15. R. R. Eady, B. E. Smith, K. A. Cook and J. R. Postgate, *Biochem. J.* (1972) **178**, 655; B. E. Smith, R. N. F. Thorneley, M. G. Yates, R. R. Eady and J. R. Postgate, "Proc. 1st Internat. Symposium on Nitrogen Fixation" (Ed. W. E. Newton and C. J. Nyman) **1**, 150. Washington State University Press (1976).
16. R. N. F. Thorneley, R. R. Eady and D. J. Lowe, *Nature* (*London*) (1978) **272**, 557.

Part 1

Nitrogen Fixation Involving Nitride and Related Intermediates

2

Developments in Ammonia Synthesis and Decomposition on Metals

KENZI TAMARU

Department of Chemistry, The University of Tokyo, Japan

In the famous book "Catalysis in Theory and Practice" by Rideal and Taylor[1] my topic is described as follows: "The technical production of pure ammonia by the catalytic combination of nitrogen and hydrogen must be considered as one of the greatest triumphs of modern physical and engineering chemistry." Since most of the natural nitrogen resources for production of fertilizer and nitric acid depended upon Chile saltpetre in the beginning of the present century, the direct synthesis of ammonia from its elements has been considered a revolutionary innovation essential for the survival of mankind.

In the last century there were many patents and reports which claimed that ammonia could be synthesized from nitrogen and hydrogen and also that high pressure would favour the synthesis. At the beginning of this century there were repeated discussions of the equilibrium between ammonia, nitrogen and hydrogen, and whether ammonia could possibly be synthesized from its elements. It was Haber and his co-workers[2] who successfully demonstrated thermodynamically and kinetically that ammonia can be synthesized under high pressures and preferably at lower temperatures with the aid of active catalysts, and also that the synthesis and decomposition proceed reversibly. There resulted an elegant example of the thermodynamic and kinetic approaches systematically combining to achieve the greatest triumph both of science and technology.

The mechanism of the synthesis and decomposition of ammonia has been studied by many investigators, and almost every time a new tool appeared in the field of catalysis it was applied to the ammonia system to assess its effectiveness: adsorption measurements,[3] kinetic approaches,[4] isotope techniques,[5] spectrophotometric methods such as infrared techniques,[6] electron spectroscopy,[7] work-function measurements,[8] dynamic approaches of chemisorbed species[9] and so forth were all examined. In this respect the history of research into ammonia synthesis and decomposition parallels that of heterogeneous catalysis in general.

As is generally the case in the field of heterogeneous catalysis, many results reported so far have frequently exhibited poor reproducibility and sometimes even conflict with each other. Many conclusions derived from these indirect methods have been considered as speculative and are based on a series of assumptions. However, in recent years many powerful new physical tools such as the branches of electron spectroscopy (XPS, AES, UPS, etc.) have become available and increasingly effective techniques are arising for looking at the catalyst surfaces directly in studying adsorption and catalysis—sometimes on well-defined surfaces—which will certainly help to elucidate the mechanism of catalysis in more detail and more precisely.

At the present moment, however, we still have many serious problems left unsolved regarding ammonia synthesis and decomposition. Most of these are qualitatively the general problems of heterogeneous catalysis; for instance, why is iron a good ammonia catalyst, while other metals such as platinum are not? Why do promoters such as K_2O and Al_2O_3 accelerate the rate of the reaction over iron catalysts? What is the real reaction mechanism for the overall synthesis or decomposition of ammonia?

I. Decomposition of Ammonia on Metals

It is generally accepted that catalysts do not change the equilibrium of reactions, and if they catalyse a forward reaction they should also promote its backward reaction to the corresponding extent as well, especially in the neighbourhood of the equilibrium, according to the law of microscopic reversibility. If a reaction is far from equilibrium, the catalyst surface in its working state may be markedly different in the forward and backward reactions, but the extrapolation toward equilibrium should always give information associated with the backward reaction.

Under normal reaction conditions ammonia decomposition is easier to carry out than ammonia synthesis, since the equilibrium is normally shifted far to the side of dinitrogen and dihydrogen, rather than ammonia, under normal pressures.

Clean tungsten surfaces are comparatively easily prepared by flashing under high vacuum and the decomposition of ammonia, in this sense, has been one of the few catalytic reactions where reproducible experimental information has been obtained by many investigators over many years. The decomposition of ammonia on a tungsten filament was studied by Hinshelwood and Burk,[10] who reported that the reaction is zero-order and the rate is independent of the pressures of ammonia, dihydrogen and dinitrogen, as shown in Fig. 1.

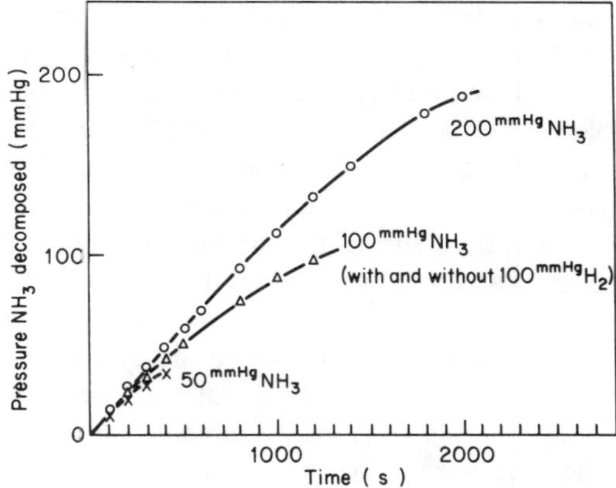

Fig. 1. Pressure–time curves for the decomposition of ammonia on tungsten at three different initial pressures. ×, 50 mmHg NH_3; △, 100 mmHg NH_3 (with and without 100 mmHg NH_3; ○, 200 mmHg NH_3.[10]

However, the reaction is not strictly zero-order, as the decomposition curves in Fig. 1 at different ammonia pressures are not superposable and the rate of the reaction changes as the decomposition proceeds, although the initial rates looked the same for different pressures of ammonia.

The zero-order kinetics were interpreted as indicating that the catalyst surface is saturated with ammonia molecules adsorbed during the course of the reaction. Frankenburger and Holder,[11] however, observed that when ammonia is brought into contact with a tungsten surface, dihydrogen formation takes place even at 150°C, and at 250°C more than one molecule of dihydrogen is released per adsorbed ammonia molecule, but no nitrogen molecules are formed:

$$NH_3 \rightarrow N(a) + \tfrac{3}{2}H_2(g).$$

At the reaction temperature of ammonia decomposition which is normally above 600°C, accordingly, it is improbable that the tungsten surface is saturated with ammonia molecules. It is more plausible to assume that the tungsten surface is fully covered by chemisorbed nitrogen, all the hydrogen in the ammonia molecule being desorbed as molecular hydrogen before nitrogen desorption takes place, and the rate-determining step is nitrogen desorption from the saturated surface, which is expected to remain constant being independent of the ambient partial pressures of ammonia, dihydrogen and dinitrogen.

It is, however, of great interest to note that a kinetic isotope effect for hydrogen in ammonia was observed by Jungers and Taylor,[12] and Barrer.[13] NH_3 decomposes 1·6 times as fast as ND_3 at 1000 K over tungsten filaments.

If the desorption of nitrogen from the surface saturated with chemisorbed nitrogen is the rate-determining step of the overall reaction, NH_3 and ND_3 should decompose at the same rate, as all the hydrogen would be split off to form hydrogen molecules prior to the nitrogen desorption process. This kinetic isotope effect must be explained by some other reaction mechanism.

In these circumstances, Tamaru[14] proposed a new approach involving direct monitoring of the adsorption during the course of the reaction in order to identify the surface species which fully covers the surface independently of the ambient pressures, and also the dynamic behaviour of the adsorbed species under the reaction conditions. He measured adsorption by a volumetric method under the reaction conditions using a closed circulating system and a tungsten powder catalyst with a large surface area which was carefully reduced by dihydrogen. When ammonia was introduced onto the tungsten catalyst, hydrogen was only produced below 723 K and a negligible amount of molecular nitrogen was formed. The rate of dihydrogen production was initially independent of the ammonia pressure, but became pressure-dependent at later times.

At reaction temperatures above 773 K, considerable nitrogen formation was observed and a negligible amount of hydrogen was adsorbed on the catalyst surface in any form. On the other hand, a large amount of nitrogen was adsorbed, several times that for monolayer adsorption under normal ammonia pressures, which indicates that several layers of surface nitride are formed under reaction conditions. Molecular nitrogen and a tungsten surface do not form nitride layers under normal pressures, which demonstrates that the chemical potential of the nitrogen sorbed in the tungsten surface in its working state is much higher than that of the ambient nitrogen gas.

The behaviour of the nitrogen in the tungsten surface was studied under reaction conditions. The rate of nitride formation decreased rapidly as the thickness of the layer increased, whereas the rate of nitrogen desorption or that of decomposition of the surface nitride increased rapidly as the layer became thicker. Accordingly, the thickness of the surface nitride layer during the course of the reaction is determined by the balance between the rate of supply of chemisorbed nitrogen from the decomposing ammonia and its rate of desorption. The kinetic isotope effect would be derived from the different rate of supply of the chemisorbed nitrogen, the thickness of the nitride layer in the steady state being larger for NH_3 than for ND_3.

The rate of decomposition of the surface nitride layer was separately measured in the absence of ambient ammonia as a function of its thickness and was compared with that of nitrogen production in the overall rate at equal degrees of nitrogen sorption, from which it was demonstrated that both rates are in reasonable agreement. These experimental results demonstrate that nitrogen desorption is one of the rate-determining steps of the overall reaction, and also that molecular nitrogen is formed by the desorption of nitride layers and not by a process in which ambient gas, such as $NH_3(g) + N(a)$, participates, or through the decomposition of surface species such as N_xH_y which would exist to a small amount in the surface:

$$N_xH_y(a) \rightarrow \frac{x}{2} N_2(g) + \frac{y}{2} H_2(g).$$

Matsushita and Hansen[15] studied ammonia decomposition on polycrystalline tungsten in an ultrahigh vacuum apparatus and estimated the amount of nitrogen uptake by tungsten in the course of the reaction using flash-filament desorption spectroscopy. The amount of nitrogen chemisorbed through the decomposition of ammonia, which they called *x*-nitrogen, was estimated by the peak area of the flash-desorption spectrum. Nitrogen desorption started at 810 K with a peak around 1100 K, while the corresponding temperatures are 1200 K and 1530 K, respectively, for nitrogen adsorbed from molecular nitrogen (β-nitrogen), as shown in Fig. 2. The experimental results they obtained are in reasonable agreement with those obtained by Tamaru over tungsten powder.

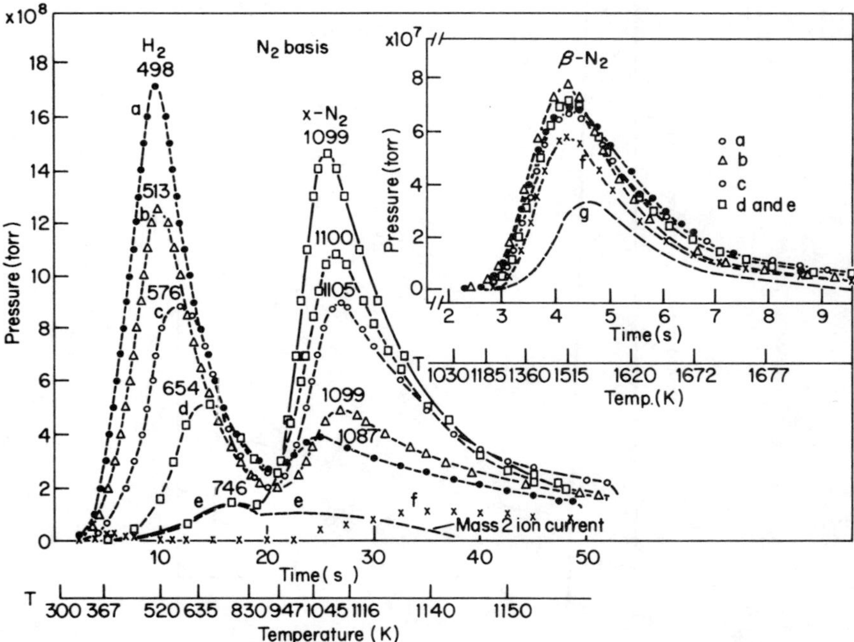

Fig. 2. Flash desorption spectra showing the separation of dihydrogen, *x*-, and β-nitrogen upon decomposition of ammonia.[15]

McAllister and Hansen[16] used different faces, (100), (110), and (111), of single-crystal tungsten for ammonia decomposition. The rate (R) of the reaction in the range of ammonia pressures which they studied is given by:

$$R = A + BP_{NH_3}^{2/3},$$

(see in Fig. 3) and is independent of the partial pressures of dinitrogen and dihydrogen. The constants A and B varied with the crystal face, the value of B over the (111) face being the largest. In the decomposition of NH_3 and ND_3 on the (111) face at 860 K, $A_{NH_3} = A_{ND_3}$ and $B_{NH_3} = 1{\cdot}47B_{ND_3}$. These results may be interpreted in various ways, since the reaction mechanism in general cannot be uniquely determined on the basis of the kinetic equation.

Recently the adsorption and the behaviour of the surface species under the reaction conditions have been examined by Shindo and others[17] employing a tungsten foil. The cleanliness of its surface was constantly checked by Auger electron spectroscopy. The amount of nitrogen chemisorbed was also studied by flash-desorption as well as Auger techniques. When ammonia is introduced onto the tungsten surface, the ambient ammonia decreases at the rate V_{in} given

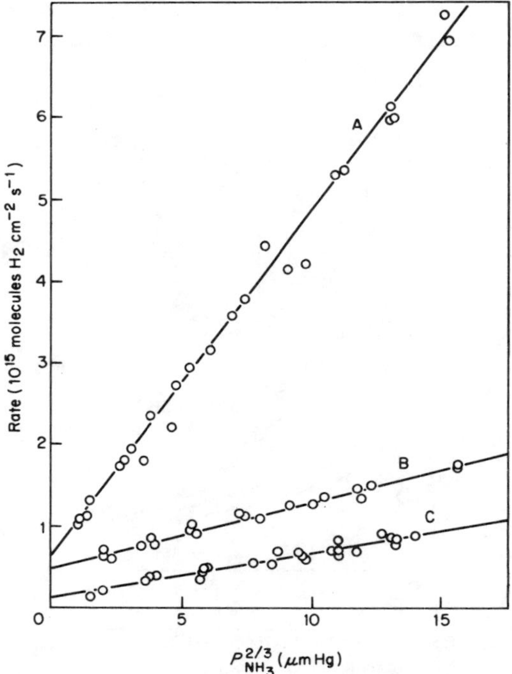

Fig. 3. Comparison of ammonia decomposition rates on W(111) face (curve A, T = 848 K), W(100) face (curve B, T = 863 K), W(100) face (curve C, T = 845 K).[16]

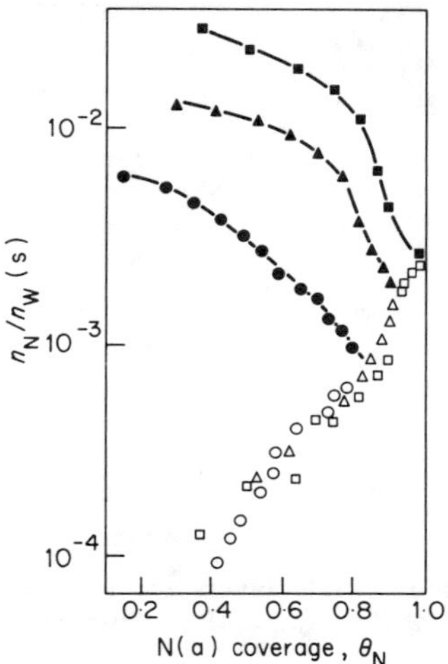

Fig. 4. Rates of ammonia uptake and dinitrogen desorption plotted against (N(a) coverage on tungsten. $T = 1073$ K. P_{NH_3}: ●, ○, $1{\cdot}7 \times 10^{-6}$ Pa; ▲, △, $5{\cdot}1 \times 10^{-6}$ Pa. ■, □, $9{\cdot}8 \times 10^{-6}$ pa. ●, ▲, ■: $V_{in} = -dn_{NH_3(g)}/dt$. ○, △, □: $V_{out} = 2dn_{N_2(g)}/dt$. $V_{in} = dn_{N(a)}/dt + V_{out}$.[17]

in Fig. 4. It is dependent upon the ammonia pressure, and slows down as the nitrogen coverage increases. On the other hand, the production of nitrogen molecules through the desorption of chemisorbed nitrogen proceeds at the rate V_{out} which increases with the nitrogen coverage, being independent of the ammonia pressure.

As has previously been observed by Tamaru[14], the nitrogen uptake during the course of the reaction reaches a steady state of dynamic balance between supply and consumption. Such balancing behaviour of nitrogen uptake was also examined using NH_3 and ND_3, which demonstrated that NH_3 gives more nitrogen uptake than does ND_3 (Fig. 5).[18] especially at higher pressures of ammonia. Although practically no hydrogen is being adsorbed on the catalyst surface, and the addition of ambient molecular hydrogen affects neither the nitrogen uptake nor its desorption rate, the kinetic isotope effect (NH_3, ND_3) appears at the step of nitrogen supply from ammonia onto the catalyst surface, which results in different nitrogen adsorption as well as the rate of nitrogen production.

The amounts of nitrogen chemisorbed during the course of ammonia decomposition are certainly much more than those at adsorption equilibrium with ambient molecular nitrogen, and the rate of nitrogen desorption is equal to that of nitrogen production during the decomposition of ammonia provided the amount of nitrogen uptake is the same. These experimental results obtained for

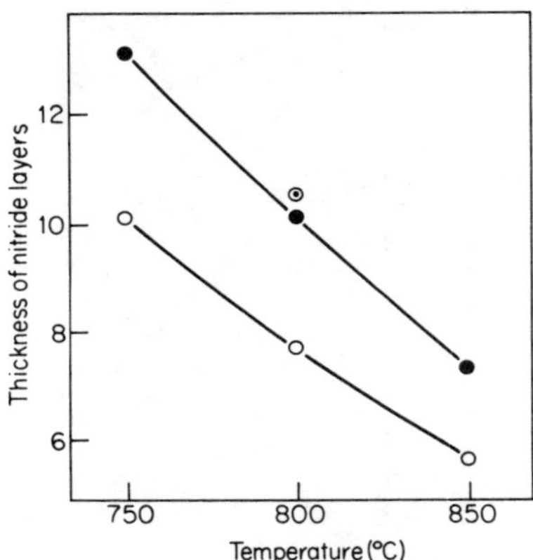

Fig. 5. Average thickness of nitride layers formed during NH_3 and ND_3 decomposition on tungsten, 15 min, $P_{NH_3,ND_3} = 0{\cdot}23$ Torr. ○, ND_3; ●, NH_3; ⊙, NH_3 with $P_{H_2} = 0{\cdot}2$ Torr.[18]

a clean tungsten surface reasonably support the reaction mechanism of "dynamic balance", which Tamaru proposed previously on the basis of the results obtained with tungsten powder, although the turnover frequency per surface tungsten atom is smaller in the former case, presumably because of the unavoidable surface contamination of the tungsten surface in the conventional reaction vessel used. Actually when the effect of contamination by carbon on the reaction rate was studied by Egawa and others,[19] it was revealed that the contamination just corresponds to the decrease in the surface area of the catalyst.

The state of nitrogen in ammonia adsorbed on a molybdenum surface was also examined by means of high-resolution Auger electron spectroscopy (AES).[20] The shape and the position of the AES peaks reflect the environment of the atom, in particular when the valence electrons are involved in the Auger process. At lower temperatures such as −78°C, nitrogen KLL Auger electron spectroscopy of adsorbed ammonia exhibits a spectrum identical to that of molecular ammonia gas. When ammonia is introduced onto a molybdenum surface at 450°C, nitride formation can be demonstrated as shown in Fig. 6, whereas ammonia adsorption at room temperature shows a decrease in NH bonding which suggests dissociative adsorption (Fig. 6).

The interaction of ammonia with metal surfaces has been studied by many investigators, although their results are not necessarily directly associated with the mechanism of the overall reactions.[21] When ammonia is adsorbed on a polycrystalline iron surface at 85 K, according to Kishi and Roberts,[22] photoelectron spectroscopy gave a broad N(1s) peak centred at about 400 eV, which split to two peaks, 397 and 400 eV, on warming to room temperature. The peak at 397·2 eV is assigned to dissociated species (which is identical with

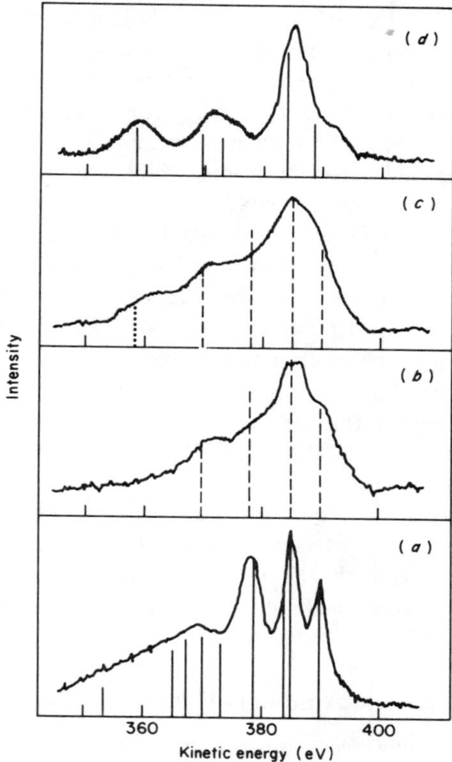

Fig. 6. High-resolution, non-derivative nitrogen KLL Auger spectra of (*a*) gaseous NH_3, adsorbed NH_3 on (*b*) a sulphur-segregated Mo—S surface at room temperature, (*c*) a clean Mo surface at about room temperature, and (*d*) a clean Mo surface at 450°C. The vertical solid lines in (*a*) and (*d*) represent the energies calculated in the equivalent core approximaton. The vertical broken lines in (*b*) and (*c*) represent observed energies and intensities for the peaks of gaseous NH_3. The dotted line in (*c*) shows the observed value of Mo_2N which is at the lowest energy in (*d*).[20]

that which follows sorption of dinitrogen), and the latter to molecular adsorption. A similar observation was reported by Gay *et al.* over the Fe(111) plane. The peak corresponding to the dissociated species is unaffected in intensity when the adlayer is heated up to 500 K and gradually decreases at temperature of 600 K or higher. Similar experimental evidence demonstrates that at lower temperatures NH_3 adsorbs without dissociation and transforms through a surface intermediate into atomic nitrogen and hydrogen.[24]

The interaction of ammonia with a Fe(110) surface was recently studied by Weiss, Ertl and Nitschke[25] by means of LEED, AES, UPS, work-function and thermal desorption measurements. Below room temperature ammonia is non-dissociatively chemisorbed through coupling of the N lone electron pair to the metal, which decreases the maximum work function by 2·4 eV. At 350 K a stable intermediate characterized by ionization potentials of 5·2 and 8·4 eV below the Fermi level and a 2 × 2 LEED pattern is formed which is identical with NH(ads). Above 400 K complete dissociation and desorption of H_2 takes place. The remaining adsorbed N atoms, which are easily identified from the UPS data, recombine and desorb as N_2 only above 850 K.

Such successive dissociation of adsorbed ammonia to adsorbed N atoms through NH(ads) was also demonstrated using secondary ion mass spectroscopy (SIMS) by G. Ertl, M. Weiss, M. Drechsler, H. Hoinkes, H. Kaarmans and H. Wilsch.[26] At lower temperatures such as 130 K $Fe_n(NH_3)_m^+$ ions appeared, which indicates molecular adsorption without dissociation, but at room temperature the NH^+ ion, but no NH_2^+, started to appear. It increased with temperature, passed through a maximum and then finally decreased to disappearance at 400 K, indicating that NH(ads) is a stable intermediate in the dissociation process.

When ammonia is adsorbed on Fe(111) and Fe(100) surfaces,[27] the consecutive dissociation of adsorbed ammonia to nitrogen atom takes place in a similar manner, although the behaviour in detail, such as the rate of dissociation, and isotope exchange between NH_3(ads) and D_2 are a little different.

II. Kinetics of Ammonia Decomposition and Synthesis

The kinetics of ammonia decomposition and synthesis over doubly promoted iron catalysts has been studied by Temkin and Pyzhev[4] through the adsorption behaviour of dinitrogen. The rate of nitrogen chemisorption (r_a) and desorption (r_d) on the catalyst are expressed by the following equations, as a function of nitrogen coverage (θ_N):

$$r_a = k_a P_N \exp(-g\theta_N), \qquad r_d = k_d \exp(h\theta_N).$$

In the presence of dihydrogen, hydrogen accelerates the rate of adsorption but similar rate equations may still be applied,[28] although the constants k_a and k_d, and g and h are different. If we assume that the nitrogen adsorption and desorption steps are rate-determining for ammonia synthesis and decomposition, respectively, the rate of the overall reaction should be determined by these equations, provided θ_N is the nitrogen coverage during the course of the reaction (θ_N^r).

The nitrogen adsorbed during the course of the reaction is not in adsorption equilibrium with the ambient dinitrogen gas, but is in equilibrium with dihydrogen and ammonia in the gas phase;

$$3H_2(g) + N_2(a) = 2NH_3(g),$$

where (a) and (g) represent the adsorbed and gaseous states, respectively, and N_2(a) does not imply the adsorbed state is in the molecular form.

If we designate the pressure of dinitrogen which would be in equilibrium with ambient dihydrogen and ammonia by $P^*_{N_2}$,

$$P^*_{N_2} = P^2_{NH_3}/(KP^3_{H_2}).$$

This is the pressure of dinitrogen which would be in adsorption equilibrium with the nitrogen adsorbed during the course of the reaction.

The dependence of θ_N upon dinitrogen pressure in adsorption equilibrium is obtained by equating r_a and r_d as follows:

$$k_a P^e_N \exp(-g\theta^e_N) = k_d \exp(h\theta^e_N)$$

$$\theta^e_N = \frac{1}{g+h} \ln\left(\frac{k_a}{k_d} P^e_{N_2}\right)$$

where θ^e_N and P^e_N represent the respective values of θ_N and P_N in adsorption equilibrium. Since θ_N during the course of the reaction, θ^r_N, is in adsorption equilibrium with $P^*_{N_2}$,

$$\theta^r_N = \frac{1}{g+h} \ln \frac{k_a}{k_d} P^*_{N_2} = \frac{1}{g+h} \ln\left(\frac{k_a}{k_d} \frac{P^2_{NH_3}}{KP^3_{H_2}}\right).$$

Accordingly, as the rate of adsorption or desorption at θ^r_N is the rate of ammonia synthesis or decomposition, we get the following rate equations for ammonia synthesis (V_+)

$$\begin{aligned} V_+ &= k_a P_{N_2} \exp(-g\theta^r_N) \\ &= k_a P_{N_2} \exp\left[-\frac{g}{g+h} \ln\left(\frac{k_a P^2_{NH_3}}{k_d KP^3_{H_2}}\right)\right] \\ &= kP_{N_2}(P^3_{H_2}/P^2_{NH_3})^\alpha \end{aligned}$$

and for ammonia decomposition (V_-)

$$\begin{aligned} V_- &= k_d \exp(h\theta^r_N) \\ &= k_d(k_a/k_d)^{h/(g+h)} K^{-h/(g+h)} (P^2_{NH_3}/P^3_{H_2})^{h/(g+h)} \\ &= k(P^2_{NH_3}/P^3_{H_2})^{1-\alpha} \end{aligned}$$

where $\alpha = g/(g + h)$. This rate equation is in agreement not only with those obtained over the double promoted iron catalyst where $1 - \alpha$ is 0·3, but also with those obtained over other metal catalysts, as shown in Tables 1 and 2.[29–43]

Logan and Kemball[34a] studied the decomposition of ammonia on various evaporated metals. Their experimental results and those obtained by other

Table 1. Rate of ammonia synthesis:
$V = k_a P_{N_2}(P^3_{H_2}/P^2_{NH_3})^{\alpha} - k_d(P^2_{NHe}/P^3_{H_2})^{1-\alpha}$

Metal	T (°C)	α	Ref.
Fe	300 ~ 500	0·5 ~ 0·7	(29)
Ru	608 ~ 740	0·5	(30)
Mo	447 ~ 550	0·5	(31)
W	582 ~ 678	0·5	(32)
Os	400 ~ 450	0	(33)
	550 ~ 600	0·5	

Table 2. Kinetics of ammonia decomposition:
$V = kP^x_{NH_3}/P^y_{H_2}$

Metal	T (°C)	x	y	$-y/x$	Ref.
Pt	490 ~ 1350	1·4 ~ 2·0	2·2 ~ 3·0	1·5 ~ 1·7	(34)
		1	1 (low P_{H_2})	1	(34b, 35)
		1	0 (high P_{H_2})	0	(34b, 35b)
Re	380 ~ 570	0·53 ~ 0·7	0·89 ~ 1·4	1·68 ~ 2·0	(34a)
Rh	420 ~ 500	1·35	2·45	1·81 ~	(34a, 37)
Ru	270 ~ 735	0·6 ~ 1·2	0·9 ~ 2·0	1·5 ~ 1·75	(34a, 38)
Cu	495 ~ 910	1	1 ~ 1·5	1 ~ 1·5	(39)
Ni	300 ~ 500	0·96 ~ 1·0	1·5 ~ 1·53	1·5 ~ 1·59	(34a, 40)
Co	370 ~ 480	0·85	1·42	1·67	(34a)
Fe	335 ~ 524	0·5 ~ 1·0	0·7 ~ 1·5	1·4 ~ 1·67	(41)
V	400 ~ 480	1·0	1·5 (high P_{H_2})	1·5	(42)
		0·5	0 (low P_{H_2})	0	
W	275 ~ 1250	0	0		(34a, 43)
Mo	800 ~ 1000	0			(43e)

workers are plotted against the parameter ($-\Delta H^0_0$) as will be explained later), which is a measure of the strength of chemisorption of gases on various metals as shown in Figs. 7 and 8,[44] where x, y and δ are obtained from the following equations:

$$\text{rate} = kP^x_{NH_3}P^y_{H_2},$$

$$\tfrac{1}{2}x - \tfrac{1}{3}y = 2\delta = 2(1-\alpha).$$

It is shown in the figures that ($-\Delta H^0_0$) can also be correlated with the catalytic behaviour of metals. The magnitude of x, y and δ decreases down to zero as ($-\Delta H^0_0$) increases. During the decomposition of ammonia on such metals as tungsten, vanadium and tantalum, all of which have high values of ($-\Delta H^0_0$), the formation of nitride has been observed by means of X-ray

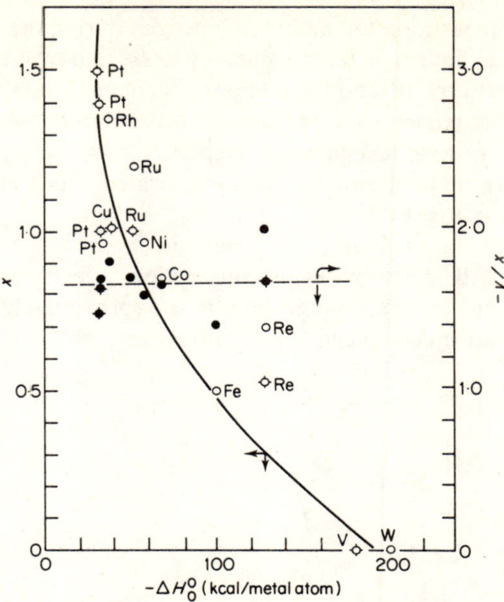

Fig. 7. The dependence of x and $-y/x$ upon $(-\Delta H_0^0)$, the heat of formation of the highest oxide per metal atom, where x and y are the orders of the ammonia decomposition reaction on various metal catalysts with respect to ammonia and dihydrogen: ○, ●, results obtained by Logan and Kemball, -○- and -●- by other workers.[44]

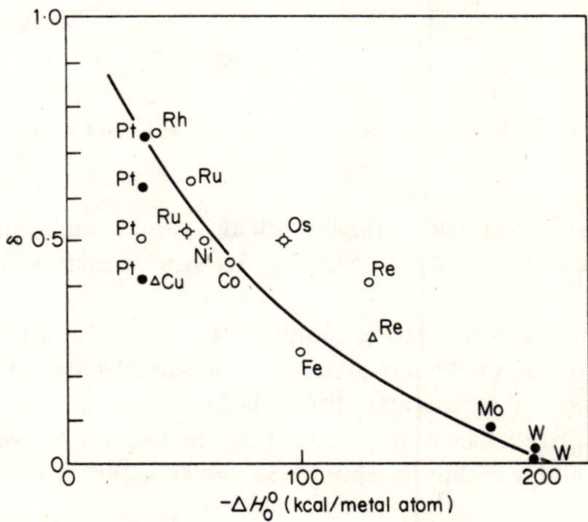

Fig. 8. The dependence of δ upon $(-\Delta H_0^0)$ in the decomposition of ammonia on various metal catalysts: ●, wire; ○, evaporated film; △, powder; and -○-, deposited on SiO_2.[44]

diffraction or by measuring the uptake of nitrogen during the reaction. Metals such as iron, which have a medium value of $(-\Delta H_0^0)$, show nitriding to various extents according to the reaction conditions. The higher the values of $(-\Delta H_0^0)$ become, the greater tendency of nitriding observed during the reaction. It is to be noted that this general tendency can be observed on various forms of metal catalysts such as metal powder, wire, evaporated film, and deposited on carriers as shown in Fig. 8.

When nitriding takes place, the reaction approaches zero order (δ approaches zero). If the nitrogen chemisorption is extremely weak, or its coverage is low, on the other hand, the rate is approximately proportional to $P^*_{N_2}$ and δ becomes unity. When the empirical rule proposed by Evans and

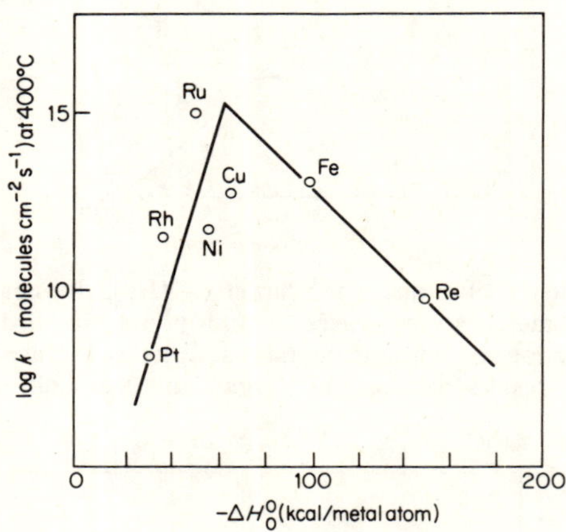

Fig. 9. The activity of metals for ammonia decomposition as a function of $(-\Delta H_0^0)$.[44]

Polanyi[45] is taken into consideration, a volcano shaped curve may be expected, as is actually the case when the catalytic activity of metals is plotted against $(-\Delta H_0^0)$ (Fig. 9).

Frankenburg suggested that the requisite for a good ammonia catalyst is a limited stability of the intermediary complex between catalyst and dinitrogen. That is, if it is too stable, a stable nitride is formed, resulting in poor activity, while, if it is so unstable as to have negligible interaction between catalyst and dinitrogen, the catalytic activity should be low. This general trend is explicitly shown in Fig. 9.

In the steady state of the reaction, the rate of the overall reaction is the difference between the forward and backward rates of each step divided by its "stoichiometric number" (s), which is the number of times that the step must be repeated in order to obtain, by summation over all the steps, the overall stoichiometric equation for the reaction as written:

$$V = (1/s_1)(v_{+1} - v_{-1}) = (1/s_2)(v_{+2} - v_{-2}) = \ldots.$$
$$= (1/s_n)(v_{+n} - v_{-n}).$$

If ammonia synthesis, $N_2 + 3H_2 = 2NH_3$, took place in the following manner,

		s
(1)	$N_2(g) \rightarrow 2N(a)$	1
(2)	$H_2(g) \rightarrow 2H(a)$	3
(3)	$N(a) + H(a) \rightarrow NH(a)$	2
(4)	$NH(a) + H(a) \rightarrow NH_2(a)$	2
(5)	$NH_2(a) + H(a) \rightarrow NH_3(g)$	2

the stoichiometric number of the individual steps would be that shown on the right.

In an elementary reaction, A $\rightleftarrows$ B, the free energy drop, G, which accompanies the reaction is given in general by the following equation:

$$-\Delta G = -\Delta G_0 + RT \ln \frac{[A]}{[B]}$$

where ΔG_0 is the standard free energy increase in the reaction under the standard conditions, [A] = [B] = 1. Since

$$K = k_+/k_- = [B]_e/[A]_e = \exp(-\Delta G_0/RT),$$

where K, k_+ and k_- are respectively the equilibrium constant, and the rate constants for forward and backward rates,

$$-\Delta G = -\Delta G_0 + RT\ln \frac{k_+[A]}{k_-[B]} \frac{1}{K}$$
$$= -\Delta G_0 + RT\ln(v_+/v_-) - RT\ln K$$
$$= RT\ln(v_+/v_-).$$

In this way the free energy drop, ΔG, is generally related to the ratio of the forward (v_+) and backward (v_-) rates.

Accordingly,

$$\bar{V} = \frac{v_{+1}}{s_1}\left(1 - \frac{v_{-1}}{v_{+1}}\right) = \frac{1 - \exp(\Delta G_1/RT)}{s_1/v_{+1}} = \ldots$$
$$= \frac{1 - \exp(\Delta G_i/RT)}{s_i/v_{+i}} = \ldots = \frac{1 - \exp(\Delta G_n/RT)}{s_n/v_{+n}},$$

where ΔG_1, ΔG_2, ..., ΔG_n are the free energy differences for each of the steps. If both v_+ and v_- are large enough and v_{+1}/v_{-1} is nearly unity, ΔG_i approaches zero, which means that the ith step is practically at equilibrium.

The free energy change for the overall reaction, ΔG is the sum of the free energy changes which accompanies each of the steps, as follows:

$$\Delta G = \sum_{1}^{n} s_i \, \Delta G_i.$$

If the rate of one of the steps (rth step) is considerably slower than any of the others (if the rth step is the rate-determining step), i.e. if

$$s_r/v_{+r} \gg s_i/v_{+i} \qquad (i \neq r)$$

all the steps except the rth step should be practically at equilibrium. In such a case

$$\Delta G = s_r \Delta G_r \quad \text{and} \quad V = (v_{4r}/s_r)(1 - \exp(\Delta G/s_r RT))$$

or

$$s_r = \frac{-\Delta G}{RT \ln\left(\dfrac{v_{+r}}{v_{-r}}\right)} = \frac{\Delta G}{RT \ln\left(\dfrac{v_+}{v_-}\right)}.$$

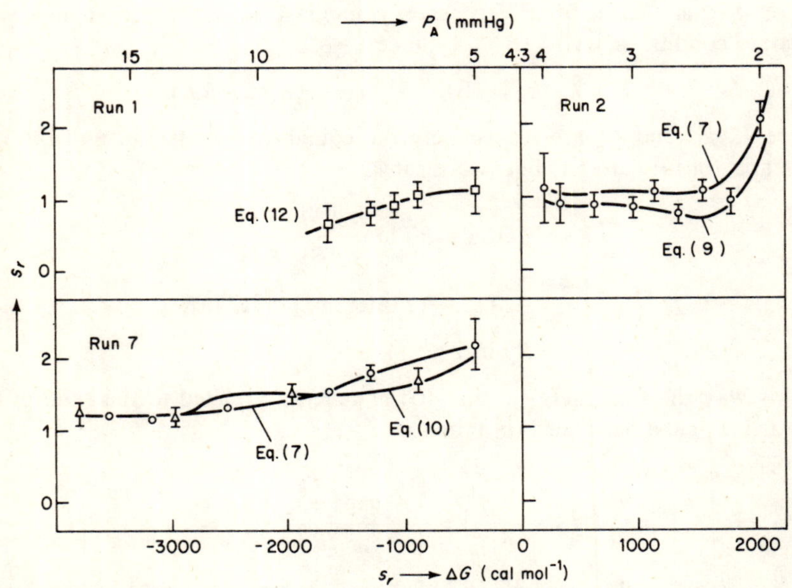

Fig. 10. Observed s_r values and their probable errors as a function of chemical affinity ($T = 390°C$).[47]

The value of ΔG corresponds to the distance from equilibrium and may be determined in the case of ammonia synthesis by

$$-\Delta G = RT \ln \frac{P_{NH_3}^2}{P_{N_2} P_{H_2}^3} \frac{1}{K}$$

and the ratio of the overall forward and backward rates may be estimated by following the behaviour of isotopic nitrogen in the reaction product during the course of the reaction. In this manner we can determine the stoichiometric number of the rate-determining step of the overall reaction.

In the case of ammonia synthesis the stoichiometric numbers of the elementary steps have different values ranging from 1 to 3 as mentioned above, and by determining the value of s_r we can obtain information on the rate-determining step of the overall reaction.

Such a kinetic approach, which was first proposed by Horiuti,[46] leads to the conclusion that $s_r = 1$ under normal reaction conditions, as given in Fig. 10,[47] which suggests that any steps of which the stoichiometric number is other than unity cannot be rate-determining, while the steps such as

$$N_2(g) \rightarrow N_2(a) \quad \text{or} \quad 2N(a)$$

or

$$N_2(a) + H_2 \rightarrow 2N_2H_2$$

can be rate-determining.[47]

III. Chemisorption of Dinitrogen and Dihydrogen on Metals

Dinitrogen is chemisorbed over clean surfaces of some metals even at room temperature. The metals which according to Trapnell[48] with later slight modifications by Bond,[49] chemisorb dinitrogen, dioxygen, carbon monoxide and hydrocarbons below room temperature are listed in Tables 3a,b.[50] In these experiments evaporated films of metals were employed to obtain clean surfaces. It is demonstrated in the table that the behaviour of various gases toward metal surfaces is very specific and platinum, for instance, does not chemisorb dinitrogen whereas iron does. In most cases transition metals are more reactive toward various gases than non-transition metals. As is shown in the table, nitrogen is one of the most inert gases and reacts with the least number of metal surfaces.

The inert behaviour of dinitrogen toward various metals may be associated with its thermodynamic properties as well as kinetic ones, and at higher temperatures more gases are generally chemisorbed, a phenomenon called "activated adsorption" in which appreciable activation energies are required

Table 3. Classification of the metals and semi-metals according to the chemical reactivity of their surfaces

(*a*)

Li	Be											B	C
Na	Mg							**D**				Al	Si
K	Ca	(Sc)	Ti	(V)	Cr	Mn	Fe	Co	Ni	Cu	Zn	Ga	Ge
Rb	Sr	(Y)	Zr	Nb	Mo	(Tc)	(Ru)	Rh	Pd	Ag	Cd	In	Sn
Cs	Ba	La	(Hf)	Ta	W	Re	(Os)	Ir	Pt	Au	Hg	Tl	Pb
	C		**A**			**B**		**E**					

( ) There are insufficient data to make a definite classification but the metal is presumed to have properties similar to its neighbours.

(*b*)

Metals	Reacting gases						
	O_2	C_2H_2	C_2H_4	CO	H_2	CO_2	N_2
Group A	3	3	3	3	3	3	3
Group B	3	3	3	3	3†	3	2
Group C	3	3	3	3	2 or 3	3	2
Group D	3	3	3	3	3	3	1
Group E	3	3	3	3	3	1 or 0	1 or 0
Cu	3	3	3	1	0	?	0
Ag	2 or 3	1	1	0	0	?	0
Au	0	3	3	1	0	?	0
B	3 or 2	?	?	3 or 2	3 or 2	?	3 or 2
Al	3	2	2	3	0	?	0
Si, Ge	3	?	?	0	2	2 or 3	0
K	3	3	0	0	0	?	0
Other metals	3	0	0	0	0	?	0

† The adsorption of H_2 on Mn is activated at 300 K.

3: A rapid and probably non-activated uptake of gas detectable at 300 K and 10^{-4} Torr pressure. 2: A slower and sometimes activated uptake of gas detectable at 300 K and 10^{-4} Torr pressure. 1: A detectable uptake of gas at 195 K but not at 300 K. 0: No detectable uptake of gas at 195 K or 300 K and 10^{-4} Torr pressure. ?: No data.

for the chemisorption. If we express the rate of adsorption of dinitrogen by its sticking probability, the metals which belong to group A exhibit values larger than 0·2 at room temperature, whereas Cr, Mn, Fe, Sr, and Ba in groups B and C, show in many cases, values less than 10^{-4}.

Tanaka and Tamaru[51] proposed a general rule connecting heats of adsorption of gases with the heats of formation of the corresponding metal

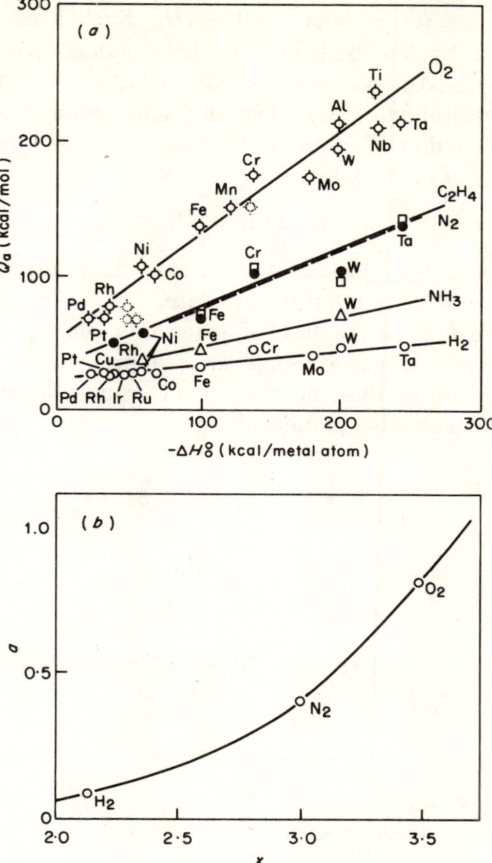

Fig. 11. (*a*) The correlation between the initial heat of adsorption of various gases (Q_{ads}) and the standard heat of formation of the highest oxide per metal atom ($-\Delta H_0^0$). (*b*) The values of α for oxygen, nitrogen, and hydrogen plotted against their electronegativity.[51]

oxides, as shown in Fig. 11. The initial heats of adsorption of gases on various metal surfaces are empirically expressed by the following equation:

$$Q_0 = a[(-H_0^0) + 37] + 20 \text{ (kcal mol}^{-1}\text{)}$$

where Q_0 is the initial heat of chemisorption, a is a constant which depends upon the electronegativityof gases adsorbed (see Fig. 11(*b*)) and $-\Delta H_0^0$ is the heat of formation per metal atom of the highest oxides. The initial heats of chemisorption of a gas on various metal surfaces may be approximated from its electronegativity (or a value) and $-\Delta H_0^0$. Such a general rule strongly suggests that the binding between adsorbate and adsorbent is chemical in nature, which is in accord with the general behaviour of chemical bonds.

The heat of adsorption of dinitrogen on a singly promoted iron catalyst was estimated as a function of coverage by Scholten *et al.*[52] on the basis of the Clapeyron–Clausius equation:

$$(\mathrm{d}\ln P/\mathrm{d}T)_\theta = \Delta H/RT^2,$$

where ΔH (equal to $-Q_{ads}$) is the increase in enthalpy due to adsorption when the surface coverage is θ and P is the pressure.

As is shown in Fig. 12, the heat of adsorption, and the activation energies for the adsorption and desorption change approximately linearly with the coverage. If we assume that the heat of adsorption decreases linearly with coverage, we obtain an adsorption isotherm:

$$\frac{\theta}{1-\theta} = b_0 \exp{(Q_{ad} - \alpha\theta)/RT)P}$$

or

$$A[\ln(1-\theta)/\theta + \ln BP] = \theta,$$

where A and B are constant at constant temperature and Q_{ad} is the differential heat of adsorption when $\theta = 0$. If we neglect $\ln[(1-\theta)/\theta]$, since at moderate

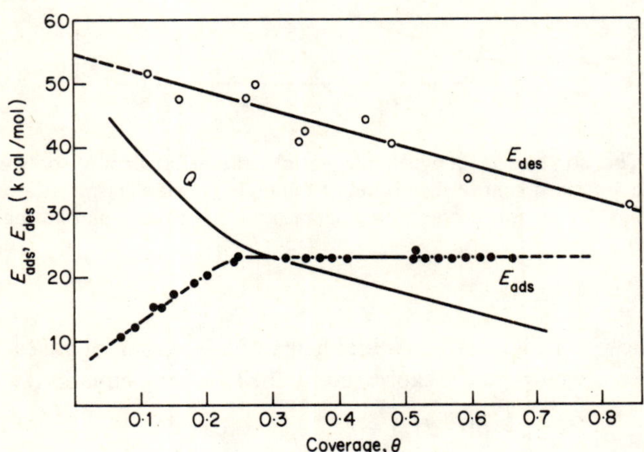

Fig. 12. Activation energies of adsorption and desorption, and heat of chemisorption, as a function of coverage, for nitrogen on a singly promoted intensively reduced catalyst. E_{ads} = activation energy of adsorption; E_{des} = activation energy of desorption; Q = heat of chemisorption = $E_{des} - E_{ads}$.[52]

surface coverages its value is small and relatively independent of θ, the following form of adsorption isotherm is obtained:

$$\theta = A \ln BP.$$

This is often called the Frumkin–Temkin isotherm.

IV. Metal Amides, Imides and Nitrides

Although rather few metals chemisorb dinitrogen gas below room temperature, most elements form nitrides. The properties of these nitrides are listed in Table 4.[53] The nitrides are classified as ionic, covalent or metallic (interstitial). The electropositive elements of Groups I and II form ionic nitrides which are apparently derived from trinegative nitrogen (N^{3-}) and are characterized by the usual valency formulae. They are hydrolysed by water to hydroxides and ammonia.

Covalent metal nitrides are also formed with the elements of Group III. Boron nitride has the graphite structure, whereas the nitrides of other Group IIIb elements have the wurtzite structure which is very much like that of diamond.

The transition metals of Groups III, IV and V form nitrides of the type MN (M = Sc, Ce, La, Pr, Ti, Zr, Hf, V, Nb and Ta) and other nitrides (Mo_2N, W_2N, Mn_4N, Fe_4N) result from closely related metals. These compounds generally crystallize in the cubic system with NaCl structures and are often (especially those of Groups IV and V metals) characterized by extremely high melting points, extreme hardness, metallic conductivity and metallic luster as well as by chemical inertness. They are true interstitial compounds with nitrogen atoms occupying the interstices in the metal structures, something like interstitial carbides and borides. They are prepared by heating the powdered metal in N_2 or NH_3 at 1100–1300°C.

The affinity towards nitrogen is large for Group IVb, but decreases towards Group VIII, where only Fe, Co and Ni form nitrides. Both Fe_2N and Fe_4N are prepared by reaction with NH_3 at 400–500°C. In these interstitial nitrides, nitrogen at the interstices expands the metal lattices. Accordingly, some metals form their nitrides by reacting with N_2, but some through indirect methods only. Those elements which are active for NH_3 synthesis are located in Groups IVb and VIII, for which the interstitial nitrides are formed.

Amides, imides and nitrides are related to ammonia via the successive replacement of hydrogen atoms by a metal atom. The alkali and alkaline earth metals and a few others (e.g. Ag and Zn) form amides. These compounds are comparatively high-melting crystalline solids which are ionic in character. Metal imides are less well characterized than the amides or the nitrides. Typical imides are those of Li and Pb.

Table 4. Standard heats of formation (kJ per N-atom 25°C and properties of nitrides.

1A	2A	3B	4B	5B	6B	7B	8			1B	2B	3A	4A	5A	6A	7A
Li_3N −197 ◎ (S)	$Be_3N_2^-$ −285 O >220											BN −134 O >3000	$(Cn)_2$ +155 O	N	ON +92 O	F_3N −109 △ S
Na_3N −151 △ 150	Mg_3N_2 −230 O 700											AlN −243 O 2000	Si_3N_4 −188 O H	PN −84 O 750	S_4N_4 +134 △ 178 Ex	Cl_3N +230 △ Ex
K_3N +84 △ L	Ca_3N_2 −218 ◎ H	ScN −285 O H	TiN −305 O H	VN −172 O >2300	CrN −121 O H	Mn_5N_2 −117 O >1200	Fe_4N −12 △ 440	Co_3N	Ni_3N +0 △	Cu_3N +75 △ 450	Zn_3N_2 −12 △ H	GaN −105 △ H	Ge_3N_4 −17 △ 450	As X	Se_4N_4 +176 △ U	Br_3N +335 △ U
Rb_3N +180 △ L	Sr_3N_2 −197 ◎ H	YN −301 O H	ZrN −343 O >3000	NbN −247 O >2300	Mo_2N −71 O H	TcN	Ru X	Rh X	Pd X	Ag_3N +285 △ Ex	Cd_3N_2 +79 △	InN −21 △ H	Sn_3N_4 △ <360	Sb X	Te_3N_4 △	I_3N +272 △ Ex
Cs_3N +314 △ L	Ba_3N_2 −184 ◎ H	LaN −301 O H	HfN −326 O H	TaN −243 O >3000	W_2N −71 O H	Re_2N	Os X	Ir X	Pt X	Au_3N △ Ex	Hg_3N_2 +8 △	Tl_3N +84 △	Pb X	BiN △		
					UN −335 O H											

◎ Reactive with N_2 directly <300°C; O Reactive with N_2 directly >300°C; △ Made from nitrogen compound; × Nitride unknown. Decomp. temp. (L: low temp, H: high temp. S: stable, U: unstable, Ex: explosive).

The heats of formation of nitrides are correlated with the heats of adsorption of dinitrogen on the corresponding metals.[54] However, it is to be noted here that the initial heat of chemisorption of nitrogen on iron is 293 kJ mol^{-1} whereas the heat of formation of Fe_4N is only 12·5 kJ mol^{-1}, which indicates that the chemisorption involves large surface energies at metal surfaces and, in some cases, N_2 chemisorption can take place over metals which do not form nitrides with N_2.

V. Chemisorption of Dinitrogen and Dihydrogen on Metals

The state of chemisorbed nitrogen on metals has been studied by various methods,[55] such as temperature programmed desorption techniques. After nitrogen had been adsorbed on polycrystalline tungsten wires, its desorption spectra were obtained by raising the temperature in vacuo, which designated α, β, and γ for the desorption at temperatures around 400 K, 1300 K and 150 K, respectively. Thus the β-state is responsible for the strong adsorption, which is resolved into two pressure peaks (β_1 and β_2) according to Oguri[55f] and others (see Fig. 13[55g]). Field emission or field ion microscopy shows that this type of adsorption occurs on all crystal planes except (110). The β_1-state desorbs according to first-order kinetics, while the β_2 state follows higher-order kinetics, which suggests that two or more unresolved sub-states might be involved.

Since the heats of adsorption of the β_1 and β_2 states are roughly equal at approximately 85 kcal mol^{-1}, it is rather improbable that the β_1 state is molecular adsorption, while the β_2 state is dissociative, as some workers have proposed on the basis of their different desorption kinetics. The adsorption states β_1 and β_2 may not be associated with different crystal planes, because Oguri[55f] and Delcher and Ehrlich[55m] observed both types of desorption from the (111) plane. Since isotope exchange experiments demonstrated a complete isotope mixing for the β state, at least, dissociative adsorption of dinitrogen occurs on the tungsten surface, although the state of most of the adsorbed species is not necessarily atomic.

α-State adsorption, on the other hand, gave no isotopic mixing, which indicates that it is a molecular chemisorption. β-State adsorption on a molybdenum surface was also reported by Pasternak and Wiesendanger[55a] and by Oguri.[55f] But in this case the α-state was not observed. The concentration of the α-state increases in the initial stage of adsorption, reaches a maximum and then decreases until finally it becomes imperceptible at saturation. This suggests that the α-state adsorption takes place on the sites which are subsequently used for the β-state adsorption.

The adsorption of dinitrogen on iron surfaces has been studied extensively, and gave different results from those obtained over the group A metals. At

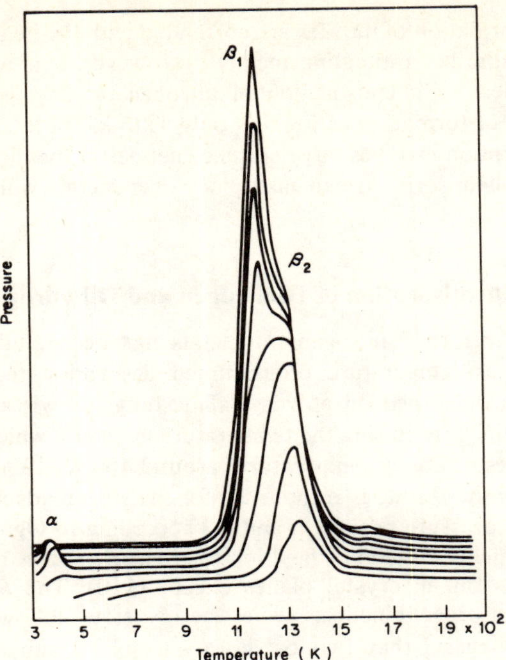

Fig. 13. Desorption spectra taken after dinitrogen was adsorbed on a polycrystalline tungsten wire at room temperature for successively greater periods of time. The top curve corresponds to saturation at a pressure of 10^{-6} Torr.[55g]

lower temperatures, such as 90 K, mobile molecular adsorption (the initial heat of adsorption is 10 kcal mol^{-1}), which lowers the work function, was observed.[8] However, at room temperature, stronger adsorption involving an initial heat of adsorption of 70 kcal mol^{-1} was observed, as shown in Fig. 14,[55u] where it is demonstrated that the heat drops steeply with increasing coverage to 16 kcal mol^{-1}, suggesting an effect of surface heterogeneity.

The chemisorption of dinitrogen has recently been studied using modern physical techniques such as Auger electron spectroscopy (AES), low-energy electron diffraction (LEED), photoelectron spectroscopy such as XPS and UPS, thermal-desorption spectroscopy, work-function measurements and also high-resolution Auger electron spectroscopy (HRAES). Ertl and his co-workers[56] studied dinitrogen adsorption on the (100), (111), and (110) surfaces of iron. Brill[57] stated that nitrogen is not adsorbed on Fe(100) and Fe(110) surface, being adsorbed predominantly on Fe(111). Ertl *et al.* concluded that the dissociative adsorption, which is identified in UPS by the appearance of a chemisorption level derived from the N2p-state at about 5 eV below the Fermi level, takes place above room temperature with an initial sticking coefficient of

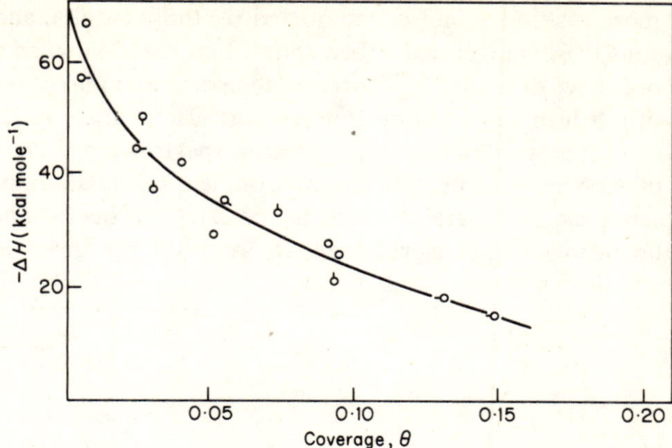

Fig. 14. The heat of adsorption of dinitrogen on an iron film at 295 K as a function of coverage θ. θ is defined as the ratio of the number of nitrogen molecules adsorbed to the number of hydrogen molecules adsorbed on similar films at a pressure of 0·1 Torr.[55u]

the order of 10^{-7} and with initial activation energies of about 5, 0 and 7 kcal mol^{-1} for (100), (111), and (110) surfaces, respectively, increasing with coverage. The activation energy for the desorption is estimated to be in the range of 50–60 kcal mol^{-1} for the three surfaces. (The LEED pattern showed the formation of a $c2 \times 2$ structure on an Fe(100) and surface reconstruction seems to take place on Fe(111) and (110) surfaces.)

Kishi and Roberts[22] observed two states of nitrogen adsorbed from dinitrogen at temperatures lower than room temperature. The two states are assigned to a molecularly adsorbed

$$\underline{\mathrm{N{\equiv}N}}$$

and linear

$$\begin{array}{c}\mathrm{N}\\ \|\!|\\ \mathrm{N}\\ \perp\end{array}$$

species, characterized by an N(1s) value of 400·2 eV for the former and 405·3 eV for the latter. At higher temperatures nitrogen is adsorbed with a very low sticking probability ($\leqslant 10^{-6}$) which gives an N(1s) value of 397·2 eV. This is a dissociatively chemisorbed

$$\begin{array}{c}\mathrm{N}\\ \underline{\|\!|}\end{array}$$

species. Gay and others[23] also reported similar results on an Fe(111) plane. It was also noted that the suggestion of Brill *et al.*[57] that the (111) surface is appreciably more reactive is not substantiated by those results, and also that the nitrogen thus dissociatively adsorbed showed no reaction when exposed to hydrogen at pressures of up to 10^{-5} Torr and temperature up to 650 K.

The state of nitrogen adsorbed on iron foil was also studied by Kunimori *et al.*[58] by means of high-resolution Auger electron spectroscopy. The shape and the position of AES peaks reflect on the environment of the atom, particularly when the valence electrons are involved in the Auger process. The HRAES spectra of the nitrogen chemisorbed on an iron foil are given in Fig. 15.

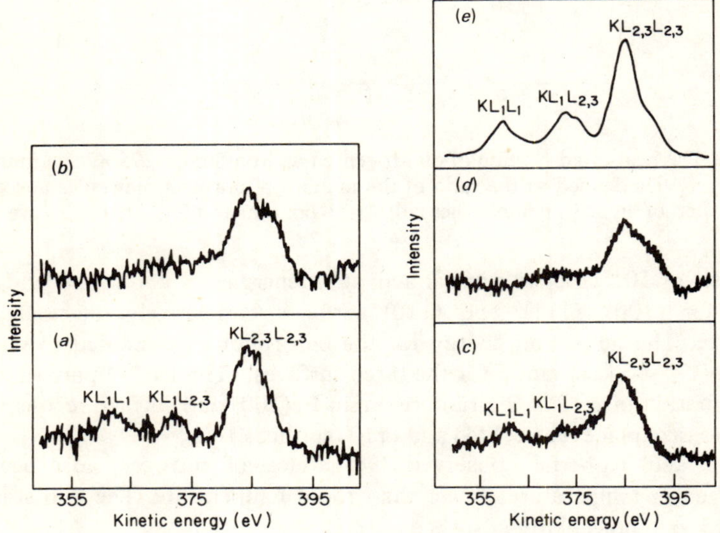

Fig. 15. Nitrogen KLL Auger spectra of nitrogen adsorbed on (*a*) Fe surface at 400°C, (*b*) Fe surface at 100°C ($P_{N_2} = 8 \times 10^{-7}$ Torr), (*c*) Mo surface at 25°C, (*d*) Mo surface at −190°C ($P_{N_2} = 1 \times 10^{-6}$ Torr). (*e*), Spectrum of Mo nitride, 450°C.[20]

Nitrogen chemisorbed at lower temperatures (100°C) exhibits different spectra from that at higher temperatures, the former being similar to that of the dinitrogen molecule, whereas the latter is similar to that of nitride, or dinitrogen dissociatively chemisorbed. It was also claimed by Wedler, Borgmann and Geuss,[59] on the basis of isotopic equilibrium, that up to a temperature of 370 K dinitrogen is adsorbed on an iron film in the molecular form. The heat of chemisorption of nitrogen on an iron catalyst was estimated as 33·6 kcal mol^{-1} from its adsorption isotherm on both singly and doubly promoted catalysts. An analysis of these isotherms showed the heat of chemisorption to decrease linearly with coverage. The elaborate results obtained by Scholten *et al.*[52] are given in Fig. 12.

The states of chemisorbed nitrogen were studied from the effects of the preadsorbed nitrogen on the successive chemisorption of carbon monoxide. The adsorption of nitrogen at 450°C did not increase the amount of CO chemisorption possible at −183° or −195°C. In the temperature range from 130°C to 300°C the inhibition increased with decrease in temperature. In this manner Takezawa and Emmett[60] concluded that nitrogen is chemisorbed in two different states, a molecular or undissociative form near 200°C and a dissociative form above 400°C. Takezawa[61] also reported that the rate of the hydrogenation of adsorbed nitrogen strongly depended upon the temperature of the nitrogen adsorption when the hydrogenation was carried out at fixed temperature, amount of adsorbed nitrogen and partial pressure of dihydrogen. In a similar manner he deduced the existence of two different adsorbed states of nitrogen on the basis of the dependence of the reactivity of the adsorbed nitrogen on the adsorption temperature. That is, the nitrogen chemisorbed at around 210°C (L-type nitrogen) exhibited higher reactivity toward dihydrogen at 200°C than that chemisorbed at around 440°C (H-type nitrogen). The activation energy of 12–15 kcal mol^{-1} for the hydrogenation of H-type nitrogen agrees reasonably well with that for iron nitride, which also supports the view of dissociative-type adsorption for the H-type nitrogen.

The molecular adsorption of dinitrogen has also been observed by means of infrared spectroscopy. Eischens and Jacknow[62] and van Hardeveld and van Montfoort[63] observed an infrared absorption band at about 2200 cm^{-1} when dinitrogen was adsorbed on silica-supported nickel at 300 K and pressures in excess of 10^{-2} Torr. This band was assigned as an NN stretching frequency which indicates that the molecule is undissociatively adsorbed. Such a molecular adsorption was reported on other metals such as iron and ruthenium in a similar manner.[64]

Broden *et al.*[65] summarized the chemisorption of dinitrogen on various metal surfaces as shown in Table 5, which demonstrates the same adsorption pattern as carbon monoxide which is isoelectronic with dinitrogen and has similar molecular orbitals.[66] The sensitivity of dissociation to surface geometry is found only for the borderline element tungsten where the (110) surface does not absorb dinitrogen at room temperature.[55m] In the case of dinitrogen on Ni, the γ-states are weakly chemisorbed rather than physisorbed, but when the temperature is raised to room temperature, the γ-state desorbs, whereas on Mo and Ti it changes to more tightly bound β-states.[55y] It seems therefore that the activation barrier to dinitrogen dissociation decreases as one goes to the left in the Periodic Table and that the dissociation takes place through a molecularly bound precursor state or γ-state.[55z]

The adsorption of dihydrogen on tungsten has been extensively studied[67] and clear evidence was obtained for the existence of both atomic and molecular species adsorbed on the surface under appropriate conditions. The γ-state

Table 5. Section of the Periodic Table showing the room temperature adsorption behaviour of dinitrogen

Sc	Ti	V	Cr	Mn	Fe	Co	Ni	Cu
	(D)		(D)		D		M	
Y	Zr	Nb	Mo	Tc	Ru	Rh	Pd	Ag
			(D)					
La	Hf	Ta	W	Re	Os	Ir	Pt	Au
		(D)	D					

M denotes molecular adsorption and D dissociative adsorption on at least one surface, single crystal or polycrystalline. (D) indicates that the adsorption is likely to be dissociative. The thick line in the Table gives the borderline between molecular and dissociative adsorption at room temperature.

which is weakly adsorbed at temperatures below 190 K does not give rise to isotopic mixing[67b,c] and lowers the work-function of tungsten, whereas the β-state,[67d,f] which desorbs at higher temperatures does give rise to isotopic mixing. Over polycrystalline tungsten surfaces the desorption spectra on flashing from 300 K to 600 K may be resolved into two, and sometimes three, substrates, which indicates appreciable discrepancies between the results obtained by different investigators. In some cases an intermediate state which desorbs between 190 and 300 K was also reported. Accumulated data on the heat of β-state adsorption of dihydrogen on a tungsten surface is given in Fig. 16,[67b,k] for which various types of tungsten samples, such as evaporated films, filaments, field emission tip, and foil were employed.[67b]

In recent years Rye[68] reported the results of dihydrogen adsorption on different crystal planes of a tungsten single crystal, which clearly indicate a high sensitivity to surface structure, as given in Figs. 17 and 18 and Table 6. The UPS difference spectra for β_1- and β_2-states of adsorbed D_2 on W(100) at 300 K are shown in Figs. 19 and 20.[69] Recently the adsorbed state of dihydrogen on a W(100) surface was studied by electron-energy-loss techniques, which clearly suggested that hydrogen is adsorbed in on-top sites at lower coverage and shifts to bridge sites at higher coverage, as shown in Figs. 21 and 22.[70]

The adsorption of dihydrogen on other metals also indicated different states of adsorption.[71] The work-function of a platinum film increases initially on adsorption of dihydrogen at 78 K, and then decreases until eventually the adsorbed layer as a whole becomes electropositive. Similar results were reported on nickel films. The heats of adsorption of dihydrogen on various evaporated metals, studied by Beeck, are given in Fig. 23.[72] Similar results

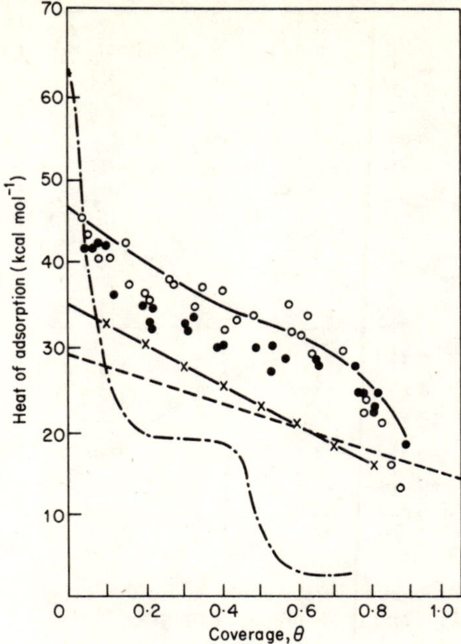

Fig. 16. Comparison of reported heats of adsorption of dihydrogen on tungsten.[50]

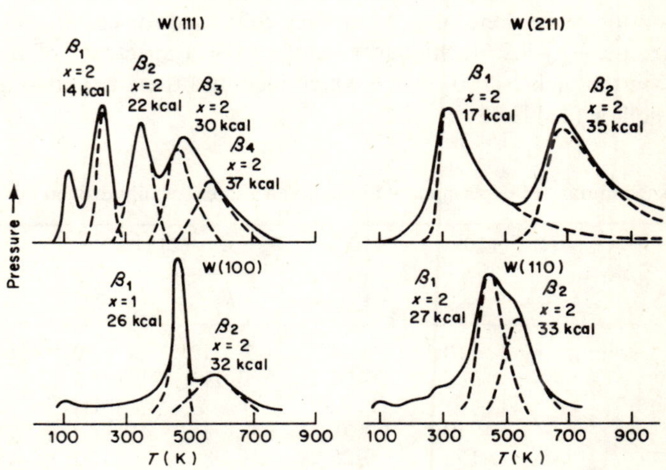

Fig. 17. Hydrogen flash desorption spectra from W(111), W(100), W(110), and W(211). The notation is the standard designation for peaks in spectra of this type. x is the order for the hydrogen desorption reaction, and the energy beside each peak is the activation energy for the desorption reaction represented by each peak.[68]

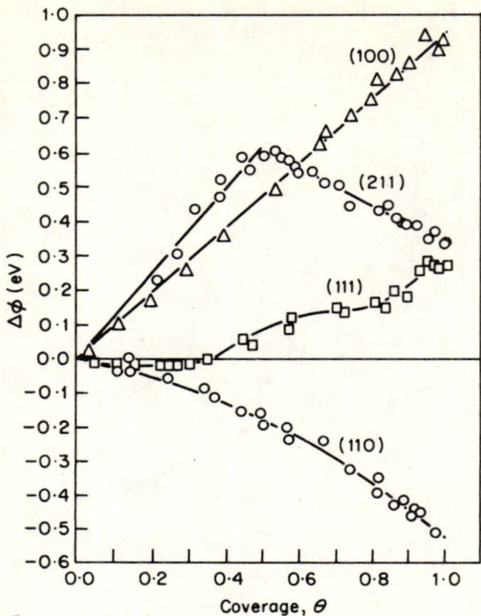

Fig. 18. Work function change versus coverage for dihydrogen adsorption at ~110 K on the (110), (100), (111) and (211) planes of tungsten.[68]

have been obtained by other workers. The heat of adsorption of dihydrogen on single-crystal planes has also been studied. Figure 24[73] gives results obtained by Ertl and his co-workers on a Ni surface examined for cleanliness by AES. The electron-energy-loss technique recently gave a spectrum of dihydrogen adsorbed on the nickel (100) plane, which identifies the adsorption site as the four-fold hollow (Table 7).[74]

Table 6. Summary of experimental and derived values for dihydrogen adsorption

	(110)	(100)	(211)		(111)
			β_1	β_2	
$n°$ (molecules/cm^2)	$9{\cdot}1 \times 10^{14}$	1×10^{15}	$4{\cdot}5 \times 10^{14}$	4×10^{14}	$9{\cdot}4 \times 10^{14}$
α	0·22	0·56	0·05	0·57	~1
$f(\theta)$	$(1-\theta)^2$	$(1-\theta)$	$(1-\theta)$	1	~1
$n°$ (calcd)	$7{\cdot}1 \times 10^{14}$	1×10^{15}	$8{\cdot}2 \times 10^{14}$		$8{\cdot}7 \times 10^{14}$
μ, D	0·15	0·25	0·08	0·20	complex

Row 1, monolayer coverages in molecules/cm^2; row 2, sticking probabilities for adsorption; row 3, coverage dependence for adsorption kinetics; row 4, calculated saturation coverages assuming one hydrogen atom adsorbed for each two missing nearest-neighbour tungsten atoms; row 5, surface dipole moments per hydrogen atom.

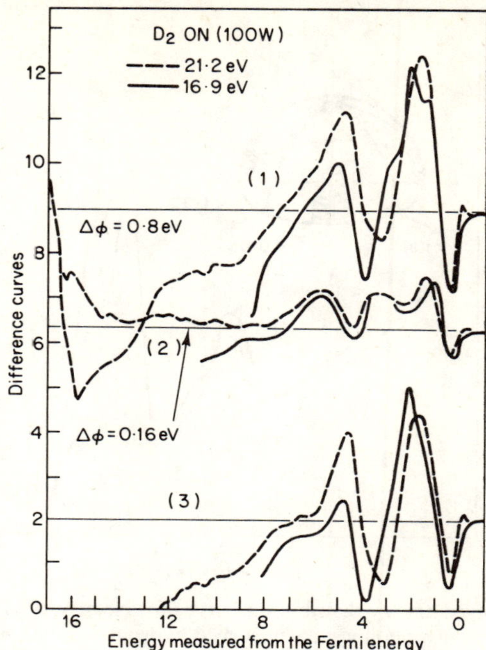

Fig. 19. UPS difference spectra for β_1 and β_2 states of adsorbed D_2 on W(100) at 300 K. Curve 1 is for saturation coverage where both the β_1 and β_2 states will desorb. Curve 2 is for low coverage where only the β_2 state will desorb. Curve 3 is the difference between curves 1 and 2, which should give the contribution from the β_1 state. ---, 21·2 eV; ——, 16·9 eV.[69a]

The adsorption of dihydrogen on Fe(100), (100) and (111) single-crystal planes has been studied by Ertl and co-workers[75] by means of LEED, thermal desorption spectroscopy, work-function measurements and UPS. Even at 140 K the isotope exchange reaction takes place, which revealed the atomic nature of the adsorbed species. The work-function increases on adsorption by 75 mV and 310 mV over Fe(100) and (111) surfaces, respectively, whereas it decreases by 95 mV over Fe(110). The chemisorption bond is characterized by a bonding level with an ionization energy 5·6 eV below the Fermi energy as identified by UPS, which is derived from coupling the H 1s state to the valence states of the metal. The initial adsorption energies were 26, 24 and 21 kcal mol^{-1} for the (110), (100) and (111) planes, respectively. The initial sticking coefficient on Fe(110) is $S_0 = 0{\cdot}16$ and it was found that with this plane a simple Langmuir-type law, $\theta = s_0(1 - \theta)^2$, is applicable to the coverage between $\theta = 0{\cdot}1$ and 1.

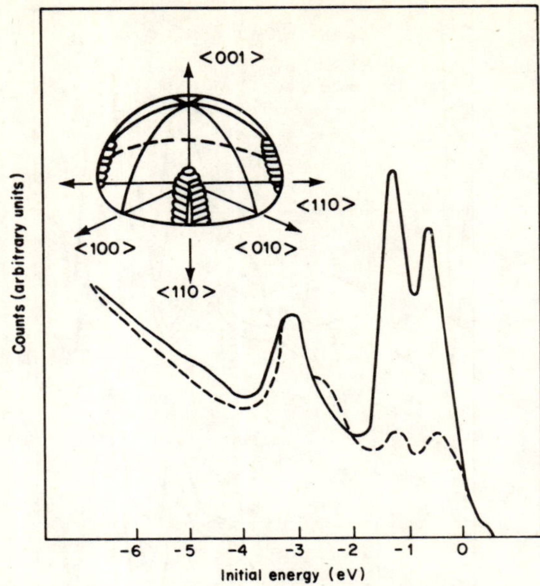

Fig. 20. Angle-resolved energy distribution curves for clean W(001) (dashed line) and for W(100) plus quarter monolayer chemisorbed hydrogen (solid line). The insert is an emission hemisphere which shows (shading) the angular properties of the hydrogen-induced structure: $h\nu = 16{\cdot}5$ eV, $\theta = 60°$, $\phi = \langle 110 \rangle$.[69b]

The adsorption of dihydrogen on a doubly promoted iron catalyst has also been studied by Emmett and Harkness,[76] where it was demonstrated by the desorption isobar that the adsorption is very complex in nature, suggesting the existence of three different types of adsorption, denoted A, B and C in Fig. 25. These three types of adsorption are all considered as chemisorption. However, Scholten and others observed only one H_2 chemisorption on highly reduced, singly promoted iron in the same temperature range.

Table 7. Calculated adsorption energies, E and vibrational energies,† $\hbar\omega$, for a hydrogen atom adsorbed on Ni(100)

Adsorption site	E (eV)	$\hbar\omega$ (meV)	Experimental $\hbar\omega$ (meV)
Top (A)	2·6	169	
Bridge (B)	2·5	118	
Centre (C)	2·3	77	74

† The vibrational energies were calculated from numerical fits to the binding energy curves.

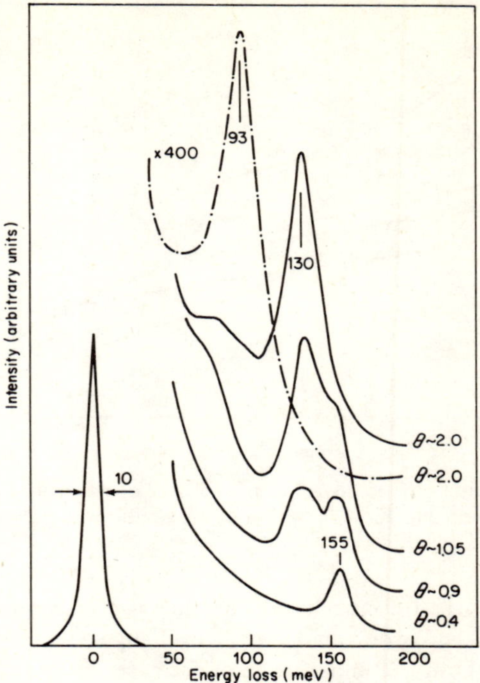

Fig. 21. Electron-energy-loss spectra of H (solid curve) and D (chain curve) on W(100). The two losses at 155 and 130 meV correspond to atomic hydrogen adsorbed in on-top and bridge sites, respectively. For deuterium the losses are shifted to lower energies by a factor $1/\sqrt{2}$.[70]

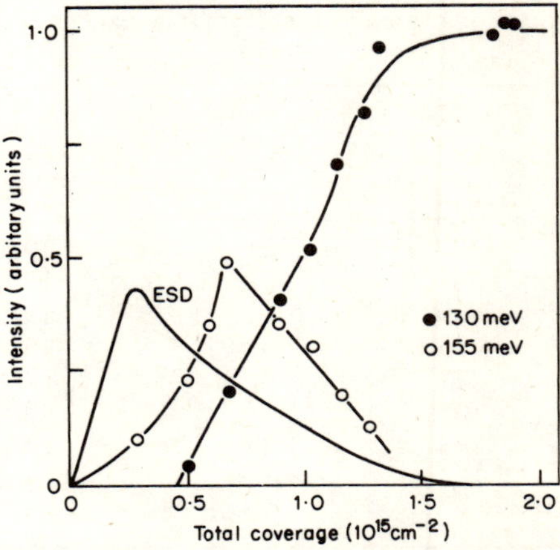

Fig. 22. Measured intensity of the two energy losses (□, 130 meV; ○, 155 meV) versus coverage for H over W(100). The intensity of electron-stimulated desorption (ESD) is shown for comparison.[70]

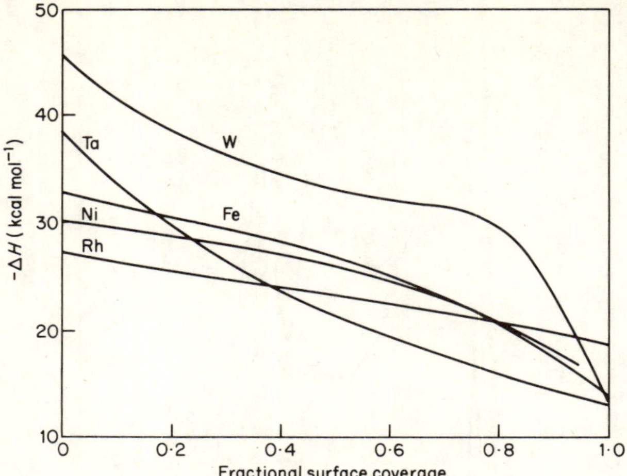

Fig. 23. Variation with surface coverage of the heat of adsorption of dihydrogen on different metal films at 296 K. Adsorption at an equilibrium pressure of 0·1 Torr is taken to represent full coverage.[72]

When D_2 and NH_3 are introduced onto metal surfaces such as platinum, rhodium, tungsten and iron, an isotopic exchange reaction takes place to produce only NH_2D in the initial stage of the reaction.[67j,77] Table 8 gives the kinetic parameters for ammonia exchange over metal films. The exchange reactions exhibits a kinetic dependence $P_{NH_3}^{0\cdot 6}$ $P_{D_2}^{0\cdot 5}$ over a random polycrystalline film, while over sintered nickel the dependence is $P_{NH_3}^{0\cdot 0}P_{D_2}^{0\cdot 5}$.[77]

Table 8. Kinetic parameters for ammonia exchange over metal films

	T_r (°C)†	E (kcal mol^{-1})	$\log_{10} A$ (A in molecules cm^{-2} sec^{-1})
Platinum	−109	5·2	19·2
Rhodium	−67	6·7	19·4
Palladium	−30	8·5	19·9
Nickel	25	9·3	19·1
Tungsten	72	9·2	18·1
Iron	105	12·5	19·5
Copper	143	13·4	19·3
Silver	172	14·1	19·2
Zinc	-------------	inactive	-------------

† T_r, temperature at which rate of adsorption/desorption equals 2×10^{12} molecules cm^{-2} sec^{-1}.

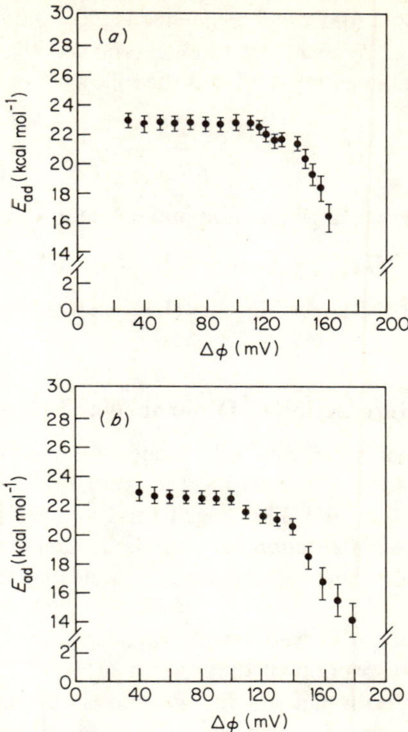

Fig. 24. Isosteric heats E_{ads} of H_2 adsorption (*a*) on Ni(100) and (*b*) on Ni(111) as a function of work-function change, $\Delta\phi$.[73]

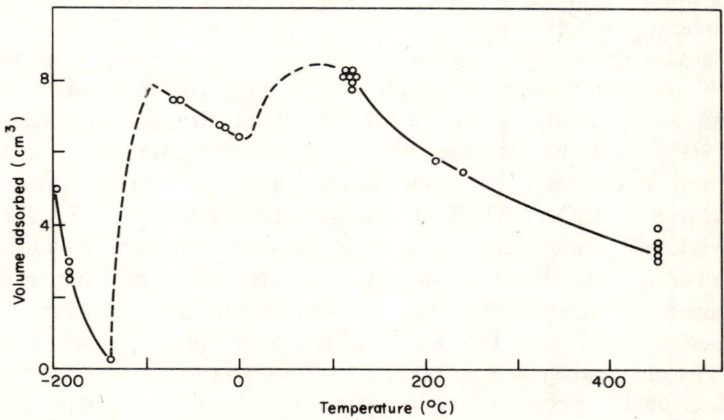

Fig. 25. The approximate position of the 760 mm isobars for the three types of dihydrogen adsorption.[76]

Over an iron evaporated film the dependence is $P^{0}_{NH_3} P^{0.5}_{D_2}$ but over a sintered iron film it is $P_{NH_3} P^{0.5}_{D_2}$. Weber and Laidler reported that exchange reaction over a synthetic ammonia catalyst follows the following rate equation:

$$v = \frac{P_{NH_3} P^{1/2}_{D_2}}{(1 + aP_{NH_3})^2}.$$

The mechanism of the exchange reaction may be interpreted as follows:

$$NH_3(g) + D(a) \rightarrow NH_2(a) + HD,$$

$$NH_2(a) + D_2 \rightarrow NH_2D + D(a).$$

VI. The Interaction of Dinitrogen and Dihydrogen

The adsorption of dinitrogen and dihydrogen on metal surfaces has been examined separately so far. But during the course of ammonia synthesis or decomposition, dinitrogen and dihydrogen (and ammonia) are present in the reacting system, and the interaction between dinitrogen and dihydrogen should also be examined, since some reports actually indicated the interaction not to be negligible.[78]

Brunauer and Emmett observed that the preadsorption of nitrogen at 390°C increases type-B chemisorption of dihydrogen at 100°C on singly promoted iron, but decreases on both unpromoted and doubly promoted iron. The effect of the preadsorbed dihydrogen on the rate of isotopic exchange of dinitrogen,

$$^{30}N_2 + {}^{28}N_2 = 2\,{}^{29}N_2,$$

was also studied, and indicated an enhancement over a promoted iron catalyst, but not over unpromoted iron.

Over doubly promoted iron catalysts the rate of dinitrogen chemisorption at 250°C increased markedly in the presence of dihydrogen, as shown in Fig. 26,[80] and was proportional to the amount of dihydrogen adsorbed on the catalyst (Fig. 27). No isotope effect was observed for dihydrogen and dideuterium in this case. A similar promotion effect with dihydrogen was observed over FeK_2O and FeK_2O—Al_2O_3, but not over Fe—Al_2O_3. Ertl and his co-workers[81] confirmed, using a clean Fe(100) surface, that dinitrogen is chemisorbed considerably more rapidly in the presence of dihydrogen than in its absence, although the main part of the chemisorbed dinitrogen is in its dissociated form. The adsorption equilibrium of nitrogen over a doubly promoted iron catalyst was also reported[80] as given in Fig. 28, where the amount of chemisorbed dinitrogen increases with co-adsorbed dihydrogen. Similar results were reported at elevated pressures and temperatures up to 200°C.

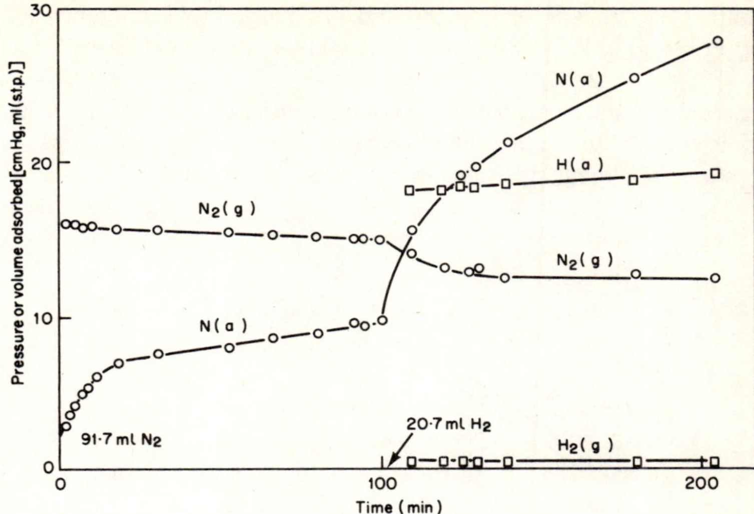

Fig. 26. Effect of dihydrogen on dinitrogen chemisorption at 250°C.[80]

In the case of an evaporated iron film dihydrogen displaces nitrogen preadsorbed on the surface,[89] whereas dinitrogen does not displace preadsorbed dihydrogen. No evidence for the formation of N–H adducts was detected in the thermal desorption experiments. The formation of N–H adducts has, however, been reported using IR spectroscopic techniques by Nakata and Matsushita, Brill, Kiru and Schulz, and Okawa *et al.*,[84] who observed N–H

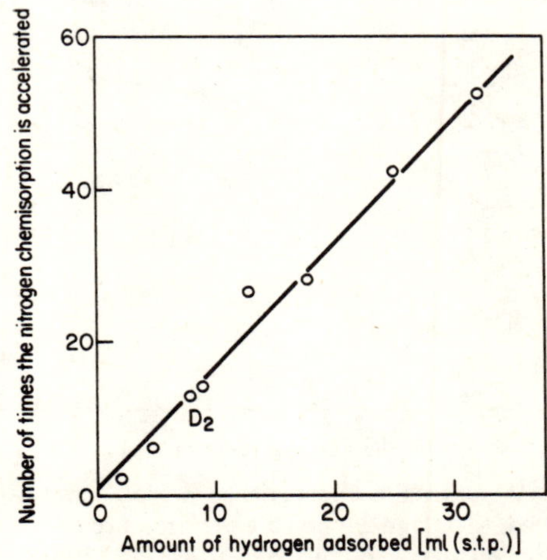

Fig. 27. Acceleration of dinitrogen chemisorption by dihydrogen at 250°C.[80]

stretching, NH_2 bending or wagging and symmetric rocking frequencies, and also by Propst *et al.*[85] using electron energy loss spectroscopy over W (see Fig. 29).

Palladium is a poor catalyst for ammonia synthesis. The reason for its poor activity is ascribed to the fact that dinitrogen is not chemisorbed on palladium. However, by employing a strong electron beam in the presence of dinitrogen it

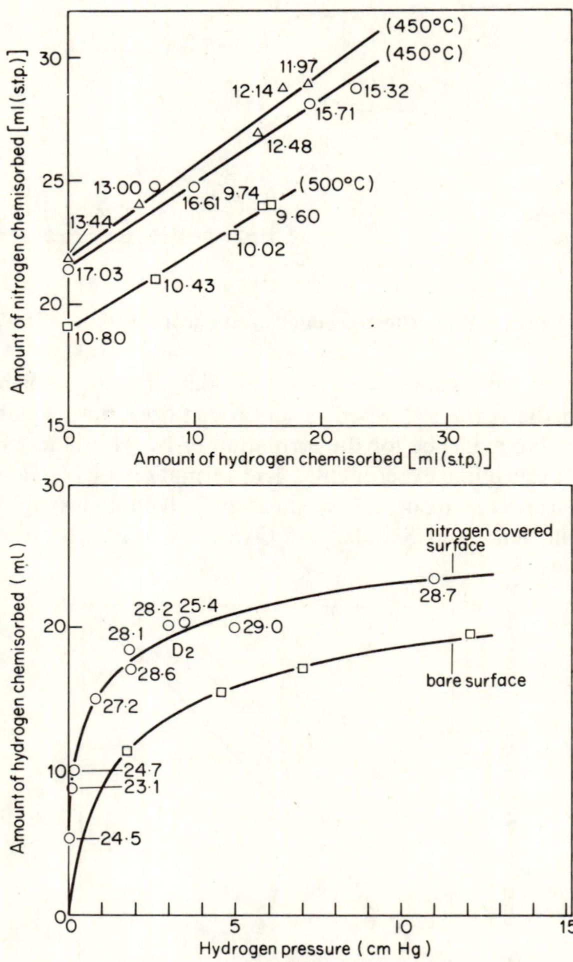

Fig. 28. (*a*) Amount of dinitrogen chemisorbed at 450° and 500°C plotted against dihydrogen co-adsorbed. (The figures indicate the pressure of dinitrogen in cm Hg.) (*b*) Dihydrogen adsorption on nitrogen-covered surface at 450°C. (The figures indicate the amount of dinitrogen chemisorbed in ml (s.t.p.).)[80]

was found that dinitrogen is dissociatively chemisorbed on palladium, as shown in Fig. 30(*d*)[86] by high-resolution Auger electron spectroscopy. The dinitrogen thus chemisorbed can readily be hydrogenated to form NH_x(a) by introducing dihydrogen (Fig. 30(*c*)) and is also obtained from ammonia adsorption (Fig. 30) followed by heat treatment (Figs. 30(*a*), (*b*)). It is thus seen that high-resolution Auger electron spectroscopy can follow the behaviour of

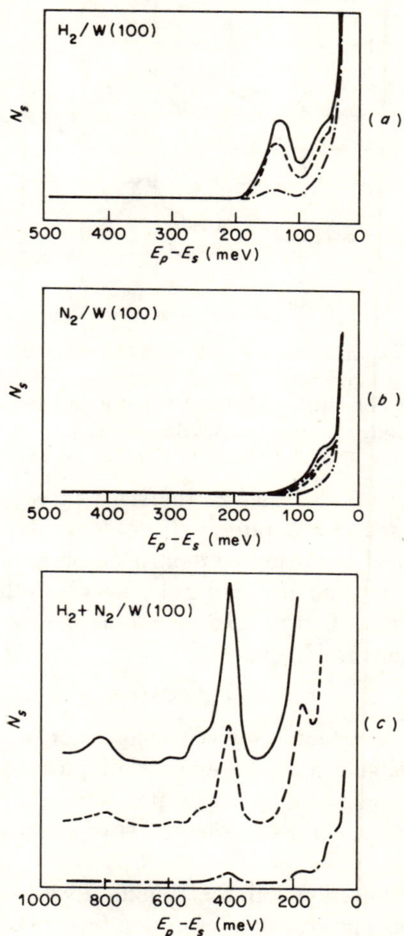

Fig. 29. Energy-loss spectra over W(100) surface. $E_p = 4{\cdot}5$ eV. (*a*) H_2: ——, $4{\cdot}0 \times 10^{-6}$ Torr s; ----, $2{\cdot}3 \times 10^{-6}$ Torr s; —·— $0{\cdot}5 \times 10^{-6}$ Torr s. (*b*) N_2: ——, 31×10^{-6} Torr s; —·— 17×10^{-6} Torr s; --- 4×10^{-6} Torr s; –··— $0{\cdot}5 \times 10^{-6}$ Torr s. (*c*) $H_2 + N_2$: ——, ×20; --- × 10; —·— × 1.[85]

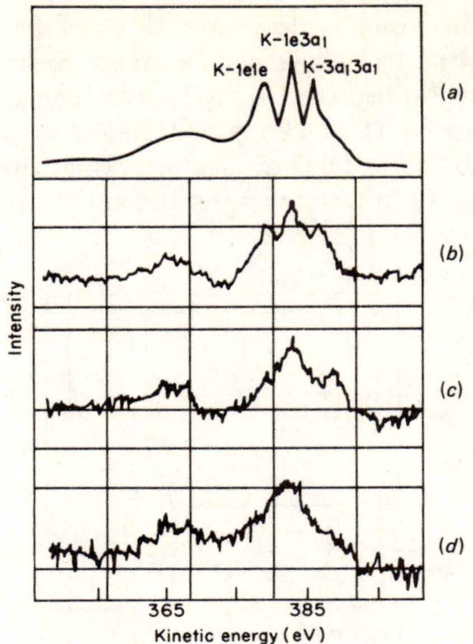

Fig. 30. Nitrogen species adsorbed on a Pd surface observed HRAES. The nitrogen KLL Auger spectra of (*a*) NH_3 gas; (*b*) NH_3 adsorbed on Pd at 25°C; (*c*) $NH_3 + O_2$ adsorbed on Pd at 25°C or N(a) + H_2 flow; (*d*) atomic nitrogen on Pd dissociatively adsorbed from N_2 by a long electron bombardment.[86]

chemisorbed species, and also that all the steps except the chemisorption of dinitrogen proceed at reasonable rates.

The hydrogenation of chemisorbed dinitrogen on an iron catalyst was also studied quantitatively by Tamaru[9] and Takezawa.[61] Both have confirmed that the rate (r_h) of hydrogenation to form ammonia is expressed as follows at a constant partial pressure of dihydrogen:

$$r_h = k_h P^n_{H_2} \exp(mx),$$

where m and k_h are constants at given temperatures of hydrogenation and dinitrogen adsorption, and x is the amount of preadsorbed dinitrogen. For dependence of the rate upon dihydrogen pressure (n), Tamaru reported first order kinetics, and Takezawa 0·95 order for H-type nitrogen[61] and 1·35 order for L-type nitrogen.

As mentioned in connection with the Temkin–Pyzhev mechanism, the rates of dinitrogen chemisorption (r_a) and desorption (r_d) are expressed as

$$r_a = k_a P_{N_2} \exp(-gx), \qquad r_d = \exp(hx),$$

respectively, whereas the rate of hydrogenation of chemisorbed dinitrogen is

$$r_h = k_h P^n_{H_2} \exp(mx),$$

where n is roughly unity. If P_{H_2} is extremely low, r_a is much larger than r_h, which means the hydrogenation step is rate-determining. In ammonia synthesis if the flow rate is high enough, the backward reaction of hydrogenation of chemisorbed dinitrogen may be neglected, since ammonia pressure will be negligible. Consequently, in the steady state supply and consumption of chemisorbed dinitrogen should be balanced according to the following equation:

$$k_2 P_{N_2} \exp(-gx) = k_d \exp(hx) + k_h P_{H_2} \exp(mx).$$

If x is too high the left-hand side of the equation, the supply of the chemisorbed nitrogen, would be slower than its consumption, the right-hand side of the equation, and vice versa if x is too low.

Table 9. Comparison of N_2 surface coverage during synthesis with N_2 coverage expected at equilibrium

Run No.	T (°C)	θ_s	θ_{isoth}
1	64·7	0·16	0·15
2	165·4	0·27	0·27
3	210·0	0·44	0·40
4	248·8	0·45	0·47
5	275·5	0·51	0·55
6	294·9	0·54	0·57
7	294·9	0·60	0·60

If we take the ratio r_d/r_h as a criterion of the rate-determining step,[87] the activation energy for the desorption of chemisorbed dinitrogen being considerably higher than that for the hydrogenation, as the former includes the heat of adsorption of dinitrogen, r_d/r_h would become much more than unity at higher temperatures and at lower dihydrogen pressures, which means that the hydrogenation step is rate-determining. At lower temperatures and higher pressures of dihydrogen, accordingly, nitrogen adsorption would be rate-determining (see Tamaru[9] and Takezawa and Toyoshima[88]).

For iron catalysts, however, Scholten *et al.*[52] measured dinitrogen adsorption during the course of reaction and confirmed that the adsorption of dinitrogen is in equilibrium with H_2 and NH_3 if it is assumed that nitrogen is chemisorbed as dinitrogen in the absence of dihydrogen (Table 9). In this case the pressure of ammonia (and dihydrogen) was taken at the exit of the catalyst bed, assumed to be the same as for the overall catalyst. Since the pressure of dinitrogen in equilibrium with ammonia and dihydrogen is proportional to $P^2_{NH_3}/P^3_{H_2}$, the ammonia pressure for the overall catalyst should actually be lower and dinitrogen chemisorbed to greater extent than that for the equilibrium with ammonia and hydrogen. Moreover, during the course of the reaction, nitrogen on the catalyst surface may be adsorbed not as dinitrogen but as $NH_x(a)$, as has been observed by infrared techniques. However, the pioneer work of Emmett and co-workers suggested that dinitrogen chemisorption is the rate-determining step on the basis of similar rates for dinitrogen chemisorption and ammonia synthesis.[89] The fact that nitrogen isotope exchange takes place only at the temperature of ammonia synthesis also favours this conclusion,[5] although the treatments were not necessarily quantitatively exact.

The determination of the stoichiometric number of the rate-determining step of the overall reaction excludes the possibility of steps of stoichiometric number other than unity being rate-determining, and both the hydrogenation of molecularly adsorbed dinitrogen, and dinitrogen adsorption itself can be rate-determining.

With some metals, such as tungsten, dinitrogen is easily chemisorbed at room temperature, but its hydrogenation proceeds with difficulty which would make the hydrogenation step rate-determining. But in the case of metals such as platinum and nickel, for which dinitrogen chemisorption is not easy, it is natural that dinitrogen chemisorption is rate-determining, as has actually been demonstrated in the case of ammonia decomposition on a nickel surface. Dinitrogen chemisorption during the decomposition is dependent upon $P^2_{NH_3}/P^3_{H_2}$, whereas the rate of the decomposition depends upon $P_{NH_3}/P^{1/2}_{H_2}$. This suggests that chemisorption on the nickel surface takes place in the form of $NH_2(a)$, while the rate-determining step of the overall decomposition is the desorption of nitrogen.[40]

The kinetic isotope effect in $N_2 + H_2$ and $N_2 + D_2$ has also been examined. If N_2 chemisorption is rate-determining in ammonia synthesis, no kinetic isotope effect is expected in the reaction. Ozaki, Taylor and Boudart[90] tested this over two different doubly promoted iron catalysts at 218°C and 302°C. Surprisingly the synthesis rate with D_2 was faster than with H_2. To explain this kinetic isotope effect, they assumed on the basis of the rate equation that NH is the main species adsorbed during the synthesis and the coverage of the imine radical was calculated to be higher for NH than for ND, dinitrogen chemisorption taking place more rapidly for $N_2 + D_2$ than for $N_2 + H_2$. In this manner they interpreted the apparent kinetic isotope effect as being not kinetic but thermodynamic, in accordance with N_2 chemisorption being rate-determining. However, Tamaru observed no such isotope effect in the adsorption of dinitrogen when $N_2 + H_2$ and $N_2 + D_2$ were introduced onto a doubly promoted iron catalyst.[80]

Shapatina *et al.*[91] similarly observed the inverse isotope effect at 400–475°C over a doubly promoted iron catalyst. Their kinetic equation, however, suggests that the chemisorbed species during the reaction is N rather than NH.

Takezawa[61] studied the hydrogenation of chemisorbed nitrogen with high H_2 (or D_2) space velocity and concluded that the formation of deutero-ammonium from L-type nitrogen (chemisorbed at lower temperatures, supposedly in undissociative form) reacting more rapidly than H-type nitrogen (chemisorbed at higher temperatures) is faster than that of normal ammonia. However, the effect is reversed in the deuteration of H-type nitrogen.

Transient species on the catalyst surface during ammonia decomposition and synthesis were studied by argon-ion bombardment which removes these species from the surface. In the decomposition reaction NH seemed to be an intermediate and the FeN_2^+ ion was also identified, indicating non-dissociative chemisorption of N_2. The intensities of the FeN_2^+ ion reach a maximum at 400°C, which suggests that ammonia is formed from FeN_2.

The field-emission technique was also applied to analysis of surface species.[92] Schmidt[93] observed ions such as H_2^+, N_2^+· N_2H^+ and $N_2H_3^+$, but no N^+ ion even at 200°C. These results were interpreted as suggesting that ammonia synthesis would proceed through a process such as,

$$N_2 \rightarrow Fe{-}N_2 \rightarrow Fe\langle{}^{NH}_{NH} \rightarrow NH_3.$$

Such a mechanism for hydrogenation of undissociatively chemisorbed nitrogen was also proposed by Brill and others.[57,94]

VII. Conventional Iron Catalysts

A "doubly promoted iron catalyst" for ammonia synthesis generally contains 0·6–2·0% Al_2O_3 and 0·3–1·5% K_2O.[29a] The catalysts which are used in industry sometimes contain CaO, MgO or SiO_2 in addition to Al_2O_3 and K_2O. To prepare these catalysts pure iron is oxidized in an atmosphere of oxygen, to which potassium nitrate and alumina are added and melted at temperatures above 1600°C in an electric furnace.[95] It is generally accepted on the basis of X-ray studies that a solid solution is formed between Fe_3O_4 and Al_2O_3.[96] The reduction of such a mixed oxide can be carried out above 350°C by using H_2 or synthesis gas. The external volume of the catalyst particles is almost unchanged during the reduction. The density of the reduced catalyst particles ranges from 2·7 g cm^{-3} to 3·78 g cm^{-3}, which is less than half of that of pure iron (7·68 g cm^{-3}).[29a]

The surface area of the catalyst per unit weight increases approximately 20-fold on addition of Al_2O_3, whereas K_2O gives no increase.[97] The specific surface area of Fe–Al_2O_3 catalysts is given in Fig. 31[98] as a function of Al_2O_3 content. The change in surface area and CO chemisorption (bare iron surface) in the course of reduction at 450°C is illustrated in Fig. 32.[99]

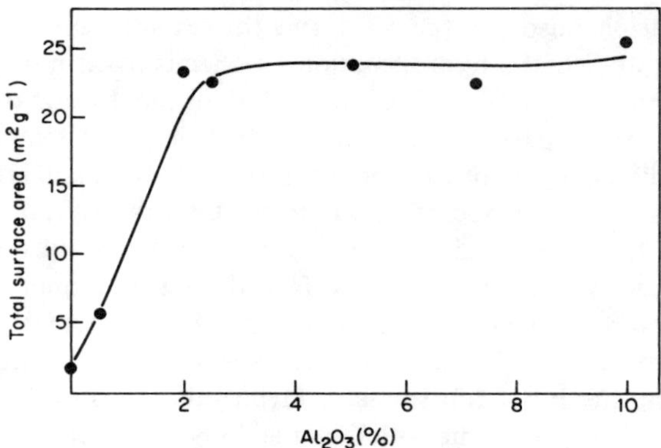

Fig. 31. The change in total surface area with the addition of alumina.[98]

The distribution of the promoters in the surface was studied using chemisorption techniques by Emmett and Brunauer.[100] In the case of singly promoted iron catalysts (Fe—Al_2O_3), the amount of adsorbed CO_2 at $-78\,°C$ was equal to that of N_2 at $-183\,°C$, whereas iron containing K_2O adsorbed CO_2 to a considerably larger extent than dinitrogen. On this basis it was suggested that 1% Al_2O_3 and 1% K_2O cover 35% and 90%, respectively of the catalyst surface.

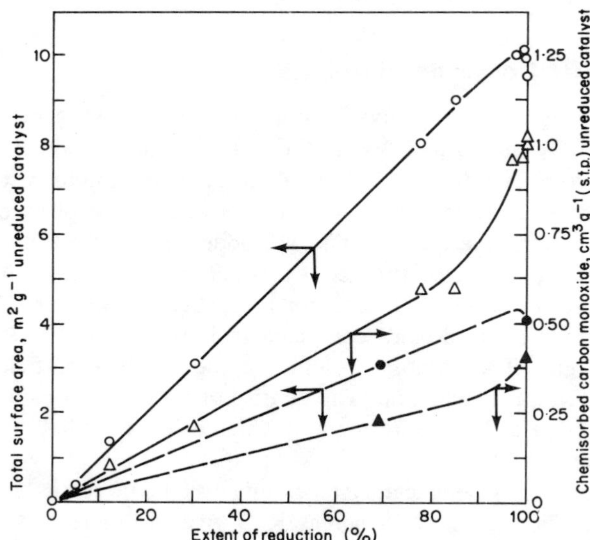

Fig. 32. Variation of surface area and of carbon monoxide chemisorption with extent of reduction. ○, Surface areas of samples reduced at 450 °C; △, their carbon monoxide chemisorptions; ●, samples reduced at 550 °C.[99]

The concentration of promoters in the surface was also examined by an exchange method. $H_2{}^{18}O$ in H_2 was introduced to the catalyst surface at 450°C and from the equilibrium point of oxygen isotope exchange the extent of oxide surface in the catalyst surface was estimated.[101] Surface coverage by promoters was estimated as about 60% for a doubly promoted catalyst containing 1·06% Al_2O_3 and 0·52% K_2O, which agrees reasonably with that estimated from chemisorption of CO at −195°C.

The concentration of promoters in the doubly promoted iron catalyst was also studied by AES[102] and XPS[103], which is the most effective method for estimating the concentration of solid surfaces. The results are given in Fig. 33, where one can easily recognize that the surface concentration of the promoters, K and Al, is much higher than in the bulk.

Secondary emission micrography of the catalyst surface demonstrated that some parts of the catalyst surface are iron sulphide containing neither Al, K

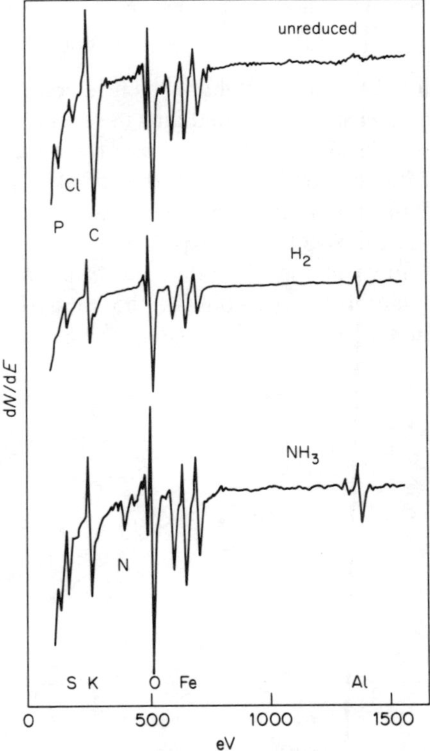

Fig. 33. The change in surface composition of a doubly promoted iron catalyst on reduction and NH_3 treatment.[102]

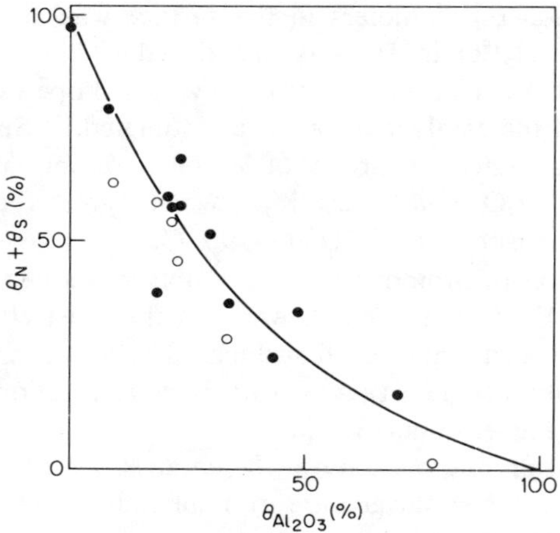

Fig. 34. The dependence of the coverage of nitrogen and sulphur upon that of alumina on a promoted iron catalyst studied by means of scanning Auger electron spectroscopy.[102]

nor oxygen. It is well known that sulphur which is contained in metals to an extremely small extent is easily segregated into the surface, and this is also the case in the ammonia synthesis catalyst. If we plot the concentration of sulphur plus nitrogen in the catalyst surface against that of alumina, it is clearly seen that alumina and sulphur inhibit dinitrogen chemisorption (Fig. 34).[102]

The surface area of Fe–Al_2O_3 catalysts increases with Al_2O_3 content to reach saturation, as shown in Fig. 31, while the fraction of free iron surface should decrease with alumina. Accordingly, the catalytic activity exhibits a maximum when plotted against Al_2O_3 content, since it would primarily be

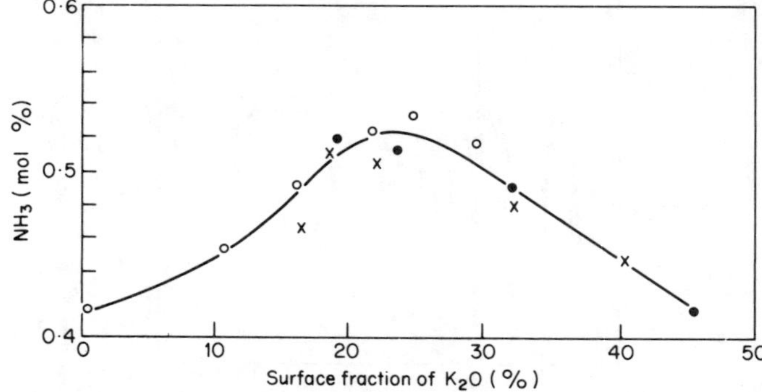

Fig. 35. Variation of ammonia synthesis activity with per cent free K_2O surface over total surface of Fe–Al_2O_3–K_2O–CaO(MgO)–SiO_2 catalyst. ●, 0% SiO_2; ○, 2·2% SiO_2; ×, 4·2% SiO_2.[104]

dependent upon the free iron surface area. When K_2O is added to an Fe—Al_2O_3 catalyst, its catalytic activity increases markedly (see Fig. 35,[104] where K_2O is added to Fe—$Al_2O_3 \cdot$ CaO—SiO_2 catalysts). Total surface area as well as the fraction of free iron surface decreases on addition of K_2O. It is accordingly said that alumina is a structural promoter which prevents sintering, while K_2O is an electronic promoter which improves the specific catalytic activity by modifying the electronic properties of the surface.

The role of K_2O and the reason why its addition increases catalytic activity has been variously speculated upon. Since dinitrogen chemisorption is accelerated in the presence of K_2O, it is assumed by Ozaki, Taylor and Boudart[90] that the main adsorbed species at lower temperatures is NH on Fe—K_2O—Al_2O_3 and N on Fe and Fe—Al_2O_3.[105] It is also proposed by Tamaru[106] that the sulphur impurity segregated or deposited onto the catalyst surface can be scavenged by potassium as follows:

$$K_2O + FeS \rightarrow K_2S + FeO,$$

$$2K + FeS \rightarrow Fe + K_2S,$$

both reactions being exothermic or spontaneous reactions on the surface.

Potassium oxide retained in alumina may also spill over the iron surface as potassium and lower the work-function enough to facilitate donation of electrons more easily to the dinitrogen molecule which chemisorbs in a negatively charged state. This possibility is also indirectly supported by the higher heat of adsorption of CO, a molecule isoelectronic with dinitrogen, in the presence of K_2O. It was also demonstrated by Ozaki and co-workers that the addition of alkali metals to iron and ruthenium catalysts increases the catalytic activity, for which a similar interpretation is proposed.[107,108] Recently Ertl, Weiss and Lee[109] added potassium to a Fe(100) surface and studied the rate of dissociative adsorption of nitrogen. It was demonstrated that a K concentration of $1{\cdot}5 \times 10^{14}$ atoms cm^{-2} increases the rate by a factor of about 300 at 430 K and removes the activation energy for adsorption. Such an effect was attributed to the increased adsorption energy of molecular N_2 (11·5 kcal mol^{-1}) which is related to the pronounced transfer of electronic charge from K to the Fe surface.

VIII. Summary and Conclusions

The role of promoters in the synthesis and decomposition of ammonia over iron catalysts has been studied in the ways discussed above. Since we have started to collect varied information on adsorption and the interactions and reactions of dihydrogen, dinitrogen and ammonia on better-defined metal surfaces, our understanding of the nature of the catalytic reactions has advanced markedly. As has been suggested by dinitrogen chemisorption on different crystal planes of iron,[56] ammonia synthesis on metal catalysts seems to be a structure-sensitive reaction, as Boudart and co-workers[110] demonstrated using iron particles of different sizes. Accordingly, the mechanism of ammonia synthesis will certainly be further elaborated on differently characterized sites, and in the presence of various kinds and concentrations of promoters, helping us to elucidate not only how the reaction proceeds, but also the reasons why catalysis takes place and why some metals such as iron are good catalysts. These are among the most fundamental problems in the field of heterogeneous catalysis in general.

References

1. E. K. Rideal and N. S. Taylor, "Catalysis in Theory and Practice" 2nd edition, p. 305. MacMillan, London (1926).
2. F. Haber and G. van Oordt, *Z. Anorg. Chem.* (1905) **43**, 111; (1905) **44**, 341; F. Haber, *Z. Elektrochem.* (1914) **20**, 597; F. Haber, S. Tamaru, and C. Ponnaz, *Z. Elektrochem.* (1915) **21**, 89, 128, 241; F. Haber and S. Tamaru, *Z. Elektrochem.* (1915) **21**, 191, 206, 228.
3. S. Brunauer and P. H. Emmett, *J. Amer. Chem. Soc.* (1940) **62**, 1732; P. H. Emmett and R. W. Harkness, *J. Amer. Chem. Soc.* (1935) **57**, 1631; and others.
4. M. I. Temkin and V. M. Pyzhev, *Acta Physicochim* (1940) **12**, 327; and others.
5. J. T. Kummer and P. H. Emmett, *J. Chem. Phys.* (1951) **19**, 289; and others.
6. R. Eischens and J. Jacknow, "Proc. 3rd Internat. Congress on Catalysis" (Ed. W. M. H. Sachtler, G. C. A. Schuit and P. Zwietering), p. 627. North–Holland, Amsterdam (1965).
7. F. M. Propst and T. C. Piper, *J. Vac. Sci. Tech.* (1967) **4**, 53; and others.
8. V. Ponec and Z. Knor, *J. Catalysis* (1968) **10**, 73; M. Grunze, F. Bozso, G. Ertl and M. Weiss, *Appl. Surf. Sci.* (1978) **1**, 241; and others.
9. K. Tamaru, "Proc. 3rd Internat. Congress on Catalysis, 1965" (Ed. W. M. H. Sachtler, G. C. A. Schuit and P. Zwietering), p. 664. North–Holland, Amsterdam (1965).
10. C. N. Hinshelwood and R. E. Burk, *J. Chem. Soc.* (1925) **127**, 1105.
11. W. Frankenburger and A. Holder, *Trans. Faraday Soc.* (1932) **28**, 229.
12. J. C. Jungers and H. S. Taylor, *J. Amer. Chem. Soc.* (1935) **57**, 679.
13. R. M. Barrer, *Trans. Faraday Soc.* (1936) **32**, 490.
14. K. Tamaru, *Trans. Faraday Soc.* (1961) **57**, 1410.
15. K. Matsushita and R. S. Hansen, *J. Chem. Phys.* (1970) **52**, 4877.
16. J. McAllister and R. S. Hansen, *J. Chem. Phys.* (1973) **59**, 414.
17. H. Shindo, C. Egawa, T. Onishi and K. Tamaru, to be published.
18. H. Shindo, C. Egawa, T. Onishi and K. Tamaru, *Z. Naturforsch.* to be published.
19. C. Egawa, H. Shindo, T. Onishi and K. Tamaru, to be published.
20. T. Kawai, K. Kunimori, T. Kondow, T. Onishi and K. Tamaru, Proc. 2nd Internat. Conf. on Solid Surfaces Japan, 1974, *Appl. Phys.* (1974) **2**, 513; *Phys. Rev. Lett.* (1974) **83**, 533.
21. (a) S. R. Logan, R. L. Moss and C. Kemball, *Trans. Faraday Soc.* (1958) **54**, 922.

(b) H. J. Grabke, *Ber. Bunsenges. Phys. Chem.* (1968) **72**, 533.
(c) V. I. Shvachko and Y. M. Fogel, *Kinet. Katal.* (1960) **7**, 635 (English edition).
(d) N. Takezawa and R. Mezaki, *Can. J. Chem. Eng.* (1970) **48**, 428.
(e) D. G. Loffler and L. D. Schmidt, *J. Catalysis* (1976) **44**, 244.
(f) P. T. Dawson and R. S. Hansen, *J. Chem. Phys.* (1966) **45**, 3148.
(g) P. G. Estrup and A. Anderson, *J. Chem. Phys.* (1968) **49**, 523.
(h) J. W. May, R. J. Szostak and L. H. Germer, *Surf. Sci.* (1969) **15**, 37.
(i) Y. K. Peng and P. T. Dawson, *J. Chem. Phys.* (1971) **54**, 950.

22. K. Kishi and M. W. Roberts, *Surf. Sci.* (1977) **62**, 252.
23. I. D. Gay, M. Textor, R. Mason and Y. Iwasawa, *Proc. Roy. Soc.* (1977) **A356**, 25.
24. (a) P. M. Gundry, J. Haber and F. C. Tompkins, *J. Catalysis* (1962) **1**, 363.
 (b) T. Nakata and S. Matsushita, *J. Phys. Chem.* (1968) **72**, 458.
 (c) W. A. Schmidt, *Angew. Chem.* (Internat. Edn.) (1968) **7**, 139.
 (d) R. Brill, P. Jiru and G. Schulz, *Z. Physik. Chem., N.F.* (1969) **64**, 215.
 (e) M. Grunze and G. Ertl, *Proc. 7th Internat. Vacuum Congress, Vienna* (1977), p. 1137.
25. M. Weiss, G. Ertl and F. Nitschke, *Appl. Surf. Sci.* (1979) **21**, 615.
26. G. Ertl, M. Weiss, M. Drechsler, H. Hoinkes, H. Kaarmans and H. Wilsch, to be published.
27. M. Grunze, F. Bozso, G. Ertl, and M. Weiss, *Appl. Surf. Sci.* (1978) **1**, 241.
28. K. Tamaru, *Trans. Faraday Soc.* (1963) **59**, 979.
29. (a) A. Nielsen, "An Investigation of Promoted Iron Catalysts for the Synthesis of Ammonia." J. Gjellerips, Copenhagen (1959).
 (b) M. I. Temkin and V. M. Pyzkev, *Acta Physicochim.* (1940) **12**, 327.
 (c) R. Brill, *J. Chem. Phys.* (1951) **19**, 1047.
 (d) C. Bokhoven, C. van Heerden, R. Westrik and P. Zwietering, "Catalysis" (Ed. P. H. Emmett), Vol. 3, 265. Reinhold, New York (1955).
 (e) R. Krabetz and C. Peters, *Ber. Bunsenges. Phys. Chem.* (1963) **67**, 381.
 (f) C. Peters and R. Krabetz, *Z. Elektrochem.* (1956) **60**, 859.
 (g) P. H. Emmett and J. T. Kummer, *Ind. Eng. Chem.* (1943) **35**, 677.
 (h) A. Nielsen, J. Kjaer and B. Hansen, *J. Catalysis* (1964) **3**, 68.
 (i) I. P. Sidrov and V. D. Livshits, *Zhur. Fiz. Khim.* (1947) **21**, 1177.
30. S. Kiperman, *Zhur. Fiz. Khim.* (1947) **21**, 1435.
31. S. Kiperman and M. I. Temkin, *Acta Physicochim.* (1946) **21**, 267.
32. S. Kiperman and M. I. Temkin, *Zhur. Fiz. Khim.* (1946) **20**, 623.
33. S. Kiperman and V. Granovskaya, *Zhur. Fiz. Khim.* (1951) **25**, 557.
34. (a) S. R. Logan and C. Kemball, *Trans. Faraday Soc.* (1960) **56**, 144.
 (b) G. M. Schwab and F. H. Schmidt, *Z. Elektrochem.* (1929) **35**, 605.
 (c) L. O. Apelbaum and M. I. Temkin, *Zhur. Fiz. Khim.* (1959) **33**, 2697.
35. (a) C. N. Hinshelwood and R. E. Burk, *J. Chem. Soc.* (1925) **127**, 1105.
 (b) J. K. Dixon, *J. Amer. Chem. Soc.* (1931) **53**, 2071.
36. (a) J. P. McGeer and H. S. Taylor, *J. Amer. Chem. Soc.* (1951) **73**, 2743.
37. J. K. Dixon, *J. Amer. Chem. Soc.* (1931) **53**, 1763.
38. (a) A. Amano and H. Taylor, *J. Amer. Chem. Soc.* (1954) **76**, 4210.
 (b) S. Kiperman, *Zhur. Fiz. Khim.* (1947) **21**, 1435.
39. M. I. Rusov and T. V. Pevsner, *Zhur. Fiz. Khim.* (1954) **28**, 1765.
40. K. Tamaru, K. Tanaka, S. Fukasaku and S. Ishida, *Trans. Faraday Soc.* (1965) **61**, 765.

41. (a) S. R. Logan, R. L. Moss and C. Kemball, *Trans. Faraday Soc.* (1958) **54**, 922.
(b) G. M. Schweb and R. Krabetz, *Z. Elektrochem.* (1956) **60**, 855.
(c) E. Winter, *Z. Physik. Chem.* (1931) **B13**, 401.
(d) K. S. Love and P. H. Emmett, *J. Amer. Chem. Soc.* (1941) **63**, 3297.
(e) N. Takezawa and I. Toyoshima, *J. Phys. Chem.* (1966) **70**, 594.
42. Y. Morikawa, and A. Ozaki, *J. Catalysis* (1971) **23**, 97.
43. (a) C. H. Kunsman, *J. Amer. Chem. Soc.* (1928) **50**, 2100.
(b) K. Tamaru, *Trans. Faraday Soc.* (1961) **57**, 1410.
(c) J. C. Jungers and H. S. Taylor, *J. Amer. Chem. Soc.* (1935) **57**, 679.
(d) R. M. Barrer, *Trans. Faraday Soc.* (1936) **32**, 490.
(e) C. H. Kunsman, *J. Am. Chem. Soc.* (1928) **50**, 2100.
(f) H. R. Hailes, *Trans. Faraday Soc.* (1931) **27**, 601.
(g) J. McAllister and R. S. Hansen, *J. Chem. Phys.* (1973) **59**, 414.
(h) H. Shindo, C. Egawa, T. Onishi and K. Tamaru, *J.C.S.* (*Faraday*) (1980) **76**, 280.
44. K. Tanaka and K. Tamaru, *Kinet. Katal.* (1966) **7**, 242.
45. M. G. Evans and M. Polanyi, *Trans. Faraday Soc.* (1938) **34**, 11.
46. J. Horiuti, *Proc. Japan Academy* (1953) **29**, 160, 164; *Adv. in Catalysis*, (1957) **IX**, 339.
47. K. Tanaka, O. Yamamoto and A. Matsuyama, "Proc. 3rd Internat. Congress on Catalysis" (Ed. W. M. H. Sachtler, C. C. A. Schuit and P. Zwietering), p. 676. North–Holland Amsterdam (1965). K. Tanaka and A. Matsuyama, *J. Res. Inst. Catalysis, Hokkaido Univ.* (1971) **19**, 63.
48. B. M. W. Trapnell, *Proc. Roy. Soc.* (1953) **A218**, 566.
49. G. C. Bond, "Catalysis by Metals". Academic Press, New York and London (1962).
50. D. O. Hayward, "Chemisorption and Reactions on Metallic Films" (Ed. J. R. Anderson), p. 255. Academic Press, New York and London (1971).
51. K. Tanaka and K. Tamaru, *J. Catalysis* (1963) **2**, 366.
52. J. J. F. Scholten, J. A. Konvalinka and P. Zwietering, *Trans. Faraday Soc.* (1960) **56**, 262.
53. A. Ozaki and K. Aika, "Dinitrogen Fixation" (Ed. R. W. F. Hardy). Wiley–Interscience, New York (1979).
54. W. M. H. Sachtler and L. L. van Rijen, *J. Res. Inst. Catalysis Hokkaido Univ.* (1962) **10**, 87.
55. (a) O. Beeck, W. A. Cole and A. Wheeler, *Disc. Faraday Soc.* (1950) **8**, 314.
(b) G. Ehrlich and F. G. Hudda, *J. Chem. Phys.* (1961) **34**, 29; (1961) **35**, 1421; (1962) **36**, 3233.
(c) E. Greenhalgh, N. Slack and B. M. W. Trapnell, *Trans. Faraday Soc.* (1956) **52**, 865.
(d) T. W. Hickmott and G. Ehrlich, *J. Phys. Chem. Solids* (1958) 47.
(e) G. Ehrlich, *J. Chem. Phys.* (1961) **34**, 29; (1962) **36**, 1171.
(f) T. Oguri, *J. Phys. Soc. Japan* (1963) **18**, 1280; (1964) **19**, 77, 83.
(g) L. J. Righby, *Can. J. Phys.* (1965) **43**, 532, 1020.
(h) J. T. Yates and T. E. Madey, *J. Chem. Phys.* (1965) **43**, 1055.
(i) P. Kisliuk, *J. Chem. Phys.* (1959) **31**, 174, 1605.
(j) P. J. Estrup and J. Anderson, *J. Chem. Phys.* (1967) **46**, 567.
(k) T. M. Madey and J. T. Yates, *J. Chem. Phys.* (1965) **44**, 1675.
(l) J. L. Robins, W. K. Warburton and T. N. Rhodin, *J. Chem. Phys.* (1967) **46**, 665.

(m) T. A. Delcher and G. Ehrlich, *J. Chem. Phys.* (1965) **42**, 2868.
(n) D. O. Hayward, D. A. King and F. C. Tompkins, *Proc. Roy. Soc.* (1967) **A297**, 305, 321.
(o) R. A. Pasternak and H. U. D. Wiesendanger, *J. Chem. Phys.* (1961) **34**, 2062.
(p) M. W. Roberts, *Trans. Faraday Soc.* (1963) **59**, 698.
(q) R. A. Pasternak, N. Endow and B. Bergsnov-Hansen, *J. Phys. Chem.* (1966) **70**, 1304.
(r) D. A. King and F. C. Tompkins, *Trans. Faraday Soc.* (1968) **64**, 496.
(s) O. Beeck, *Adv. in Catalysis* (1950) **2**, 151.
(t) P. M. Gundry, J. Haber and F. C. Tompkins, *J. Catalysis* (1962) **1**, 363.
(u) J. Bagg and F. C. Tompkins, *Trans. Faraday Soc.* (1955) **51**, 1071.
(v) L. W. Anders, R. S. Hansen and L. S. Bartell, *J. Chem. Phys.* (1975) **62**, 1641.
(w) B. E. Nieuwenhuys, O. G. van Aardenne and W M. H. Sachtler, *Thin Surf. Films* (1973) **17**, S7.
(x) M. Housley and D. A. King, *Surf. Sci.* (1977) **62**, 93.
(y) D. A. King, *Surf. Sci.* (1968) **9**, 375.
(z) D. O. Hayward and B. M. W. Trapnell, "Chemisorption", pp. 90, 115. Butterworths, London (1964). H. P. Bonzel and R. Ku, *Surf. Sci.* (1973) **40**, 85.

56. F. Bozso, G. Ertl, M. Grunze and M. Weiss, *J. Catalysis* (1977) **49**, 18; F. Bozso, G. Ertl and M. Weiss, *J. Catalysis* (1977) **50**, 519.
57. R. Brill, P. Jiru and G. Schultz, *Z. Phys. Chem., N. F.* (1969) **64**, 215: R. Brill, E. L. Richter and E. Ruch, *Angew. Chem.* (1967) **79**, 905; R. Brill, *J. Catalysis* (1970) **16**, 16.
58. K. Kunimori, T. Kawai, T. Kondow, T. Onishi and K. Tamaru, *Surf. Sci.* (1976) **54**, 525.
59. G. Wedler, D. Borgmann and K. P. Geuss, *Surf. Sci.* (1975) **47**, 592.
60. N. Takezawa and P. H. Emmett, *J. Catalysis* (1968) **11**, 131.
61. N. Takezawa, *J. Catalysis* (1972) **24**, 417.
62. R. van Hardeveld and A. van Montfort, *Surf. Sci.* (1966) **4**, 396.
63. R. Eischens and J. Jacknow, "Proc. 3rd Internat. Congress Catalysis" (ed. W. M. H. Sachtler, G. C. A. Schuit and P. Zwietering), p. 627. North–Holland, Amsterdam (1965).
64. D. A. King, *Surf. Sci.* (1968) **9**, 375; T. Okawa, T. Onishi and K. Tamaru, *Z. Phys. Chem. N.F.* (1977) **107**, 239; M. Ohkita, K. Aika, K. Urabe and A Ozaki, *J.C.S. Chem. Commun.* (1975) **147**, *J. Catalysis* (1976) **44**, 460.
65. Broden, T. N., Rhodin, C. Brucker, R. Benlow and Z. Hurych, *Surf. Sci.* (1976) **59**, 593.
66. W. L. Jorgensen and L. Salem, "The Organic Chemist's Book of Orbitals", pp. 78, 80. Academic Press, New York and London (1973).
67. (a) T. W. Hickmott, *J. Chem. Phys.* (1960) **32**, 810.
(b) V. J. Mimeault and R. S. Hansen, *J. Chem. Phys.* (1966) **45**, 2240.
(c) F. Ricca, R. Medana and G. Saini, *Trans. Faraday Soc.* (1965) **61**, 1492; (1966) **62**, 2273.
(d) J. C. P. Mignolet, *Recl. Trav. chim.* (1955) **74**, 685, 701.
(e) R. Gomer and J. K. Hulm, *J. Chem. Phys.* (1957) **27**, 1363.
(f) W. J. M. Rootsaert, L. L. van Reijen and W. M. H. Sachtler, *J. Catalysis* (1962) **1**, 416.
(g) L. J. Rigby, *Can. J. Phys.* (1965) **43**, 1020.

(h) P. A. Redhead, *Proc. Symp. Electronic and Vacuum Physics, Balalonfoldvar, Hungary*, p. 89.
(i) O. Beeck, *Adv. in Catalysis* (1950) **2**, 151.
(j) M. Wahba and C. Kemball, *Trans. Faraday Soc.* (1953) **49**, 1351.
(k) D. Brennan and F. H. Hayes, *Trans. Faraday Soc.* (1964) **60**, 589.
(l) P. J. Estrup and J. Anderson, *J. Chem. Phys.* (1966) **45**, 2254; (1967) **46**, 567.
(m) J. C. Cavalier and E. Chornet, *Surf. Sci.* (1976) **60**, 125.
68. R. R. Rye, *Accnts. Chem. Res.* (1975) **8**, 347.
69. (a) E. W. Plummer, *In* "Topics in Applied Physics," (Ed. R. Gomer), Vol. 4. Springer–Verlag, Berlin (1975).
(b) J. Anderson and G. J. Lapeyre, *Phys. Rev. Lett.* (1976) **36**, 376.
70. H. Froitzheim, H. Ibach and S. Lehwald, *Phys. Rev. Lett.* (1976) **36**, 1549.
71. (a) D. O. Hayward, N. Taylor and F. C. Tompkins, *Disc. Faraday Soc.* (1966) **41**, 75.
(b) J. C. P. Mignolet, *Disc. Faraday Soc.* (1950) **8**, 105; *J. Chim. Phys.* (1957) **54**, 19.
(c) T. A. Delchar and F. C. Tompkins, *Trans. Faraday Soc.* (1968) **64**, 1915.
(d) D. D. Eley, D. M. Moran and C. H. Rochester, *Trans. Faraday Soc.* (1968) **64**, 2168.
(e) V. J. Mimeault and R. S. Hansen, *J. Chem. Phys.* (1966) **45**, 2240.
(f) F. Ricca, R. Medana and G. Saini, *Trans. Faraday Soc.* (1966) **62**, 2273.
(g) J. Bagg and F. C. Tompkins, *Trans. Faraday Soc.* (1955) **51**, 1071.
(h) M. Wahba and C. Kemball, *Trans. Faraday Soc.* (1953) **49**, 1351.
(i) D. Brennan and F. H. Hayes, *Trans. Faraday Soc.* (1964) **60**, 589.
(j) D. F. Klemperer and F. S. Stone, *Proc. Roy. Soc.* (1957) **A243**, 375.
(k) S. Černý, V. Ponec and L. Hládek, *J. Catalysis* (1966) **5**, 27.
(l) E. K. Rideal and F. Sweett, *Proc. Roy. Soc.* (1960) **A257**, 291.
(m) D. D. Eley and P. R. Norton, *Proc. Roy. Soc.* (1970) **A314**, 315.
(n) R. P. Gasser, K. Roberts and A. J. Stevens, *Trans. Faraday Soc.* (1969) **65**, 3105; *Surf. Sci.* (1970) **20**, 123.
(o) W. P. Gilbreath and D. E. Wilson, *J. Vac. Sci. Tech.* (1971) **8**, 45.
(p) J. Volter and M. Procop, *Z. Phys. Chem.* (Leipzig) (1972) **249**, 344: (1973) **253**, 33.
(q) G. Wedler, H. Reichenberger and H. Wenzel, *Z. Naturforsch.* (1971) **26a**, 1452.
(r) G. Wedler, G. Fisch and H. Papp, *Ber. Bunsenges.* (1970) **74**, 186; (1971) **75**, 1026, 1160.
(s) G. Wedler, F. J. Bröcker, G. Fisch and G. Schroll, *Z. Phys. Chem., N.F.* (1971) **76**, 212.
(t) J. E. Demuth and T. N. Rhodin, *Surf. Sci.* (1974) **42**, 261; (1974) **45**, 249.
(u) S. L. Bernasek, W. J. Siekhaus and G. A. Somorjai, *Phys. Rev. Lett.* (1973) **30**, 1202.
(v) A. E. Morgan and G. A. Somorjai, *Surf. Sci.* (1968) **12**, 405.
(w) B. Lang, R. W. Joyner and G. A. Somorjai, *Surf. Sci.* (1972) **30**, 454.
(x) K. E. Lu and R. R. Rye, *J. Vac. Sci. Tech.* (1975) **12**, 334.
(y) G. Doyen and G. Ertl, *Surf. Sci.* (1977) **69**, 157.
(z) A. S. Porter and F. C. Tompkins, *Proc. Roy. Soc.* (1953) **217A**, 529.
72. O. Beeck, *Disc. Faraday Soc.* (1950) **8**, 118.

73. K. Christmann, O. Schober, G. Ertl and M. Neumann, *J. Chem. Phys.* (1974) **60**, 4528.
74. S. Andersson, *Chem. Phys. Lett.* (1978) **55**, 185.
75. F. Bozso, G. Ertl, M. Grunze and M. Weiss, *Appl. Surf. Sci.* (1977) **1**, 103.
76. P. H. Emmett and R. W. Harkness, *J. Amer. Chem. Soc.* (1935) **57**, 1631.
77. J. Singleton, E. Roberts and E. Winter, *Trans. Faraday Soc.* (1951) **47**, 1318; A. Farkas, *Trans. Faraday Soc.* (1936) **32**, 416; C. Kemball, *Proc. Roy. Soc.* (1952) **A214**, 413; *Trans. Faraday Soc.* (1952) **48**, 254; J. Weber and K. J. Laidler, *J. Chem. Phys.* (1951) **19**, 1089.
78. (a) A. S. Porter and F. C. Tompkins, *Proc. Roy. Soc.* (1953) **A217**, 529.
    (b) J. Bragg and F. C. Tompkins, *Trans. Faraday Soc.* (1955) **51**, 1071.
    (c) P. M. Gundry, J. Haber and F. C. Tompkins, *J. Catalysis* (1962) **1**, 363.
    (d) V. Ponec and Z. Knor, *J. Catalysis* (1968) **10**, 73.
79. S. Brunauer and P. H. Emmett, *J. Amer. Chem. Soc.* (1940) **62**, 1732.
80. K. Tamaru, *Trans. Faraday Soc.* (1963) **59**, 979.
81. M. Grunze, F. Bozso, G. Ertl and M. Weiss, *Appl. Surf. Sci.* (1978) **1**, 241.
82. M. V. C. Sastri and H. Srikant, *J. Sci. Ind. Res.* (1961) **20D**, 321.
82. G. Wedler and D. Borgmann, *J. Catalysis* (1976) **44**, 139.
84. (a) T. Nakata and S. Matsushita, *J. Phys. Chem.* (1968) **72**, 458.
    (b) R. Brill, P. Jirů and G. Schulz, *Z. Phys. Chem., N.F.* (1969) **64**, 215.
    (c) T. Okawa, T. Onishi and K. Tamaru, *Z. Phys. Chem., N.F.* (1977) **107**, 239; *Chem. Lett.* (1977) 1077.
85. F. M. Propst and T. C. Piper, *J. Vac. Sci. Tech.* (1967) **4**, 53.
86. K. Kunimori, T. Kawai, T. Kondow, T. Onishi and K. Tamaru *Surf. Sci.* (1976) **59**, 302.
87. K. Tamaru, "Dynamic Heterogeneous Catalysis". Academic Press, London and New York. (1978).
88. N. Takezawa and I. Toyoshima, *J. Phys. Chem.* (1966) **70**, 594; *J. Catalysis* (1966) **6**, 145.
89. P. H. Emmett, *In* "The Physical Basis for Heterogeneous Catalysis" (Ed. E. Drauglis and R. I. Jaffee), p. 3. Plenum Press, New York (1975); P. H. Emmett and S. Brunauer, *J. Am. Chem. Soc.* (1934) **56**, 35.
90. A. Ozaki, H. S. Taylor and M. Boudart, *Proc. Roy. Soc.* (1960) **258A**, 47; G. Schulz-Ekloff, *Ber. Bunsen Gesel.* (1976) **80**, 352; S. R. Logan and J. Philp, *J. Catalysis* (1968 **11**, 1.
91. E. N. Shapatina, B. L. Kutaev and M. I. Temkin, *Kinet Katal.* (1971) **12**, 1476.
92. R. Brill, E. L. Richter and E. Ruch, *Angew. Chem.* (1967) **79**, 915.
93. W. A. Schmidt, *Angew. Chem.* (1968) **80**, 151.
94. S. Carrà and R. Ugo, *J. Catalysis* (1969) **15**, 435.
95. A. Mittasch, *Adv. in Catalysis* (1950) **2**, 81; A. T. Larson and C. N. Richardson, *Ind. Eng. Chem.* (1925) **17**, 971; G. Shima and H. Uchida, *Rept. Govt. Chem. Ind. Res. Inst. Tokyo* (1950) **45**, 369.
96. H. Ludwiczek, A. Preislinger, A. Fischer, R. Hosemann, A. Schonfeld and W. Vogel, *J. Catalysis* (1978) **51**, 326.
97. P. H. Emmett and S. Brunauer, *J. Amer. Chem. Soc.* (1937) **59**, 1553.
98. C. Peters, K. Schaefer and R. Krabetz, *Z. Elektrochem.* (1960) **64**, 1194.
99. W. K. Hall, W. H. Tarn and R. B. Anderson, *J. Amer. Chem. Soc.* (1950) **72**, 5436.
100. P. H. Emmett and S. Brunauer, *J. Amer. Chem. Soc.* (1937) **59**, 310; S. Brunauer and P. H. Emmett, *J. Amer. Chem. Soc.* (1940) **62**, 1732.

101. V. Solbakken, A. Solbakken and P. H. Emmett, *J. Catalysis* (1969) **15**, 90.
102. K. Hanji, H. Shimizu, H. Shindo, T. Onishi and K. Tamaru, to be published.
103. G. Ertl and N. Thiele, *Appl. Suf. Sci.* (1979) **3**, 99.
104. R. Krabetz and C. Peters, *Angew. Chem.* (1965) **77**, 333.
105. K. Aika and A. Ozaki, *J. Catalysis* (1969) **13** 232; (1970) **19**, 350.
106. K. Tamaru, private communication.
107. A. Ozaki, A. Aika and Y. Morikawa, *Proc. 5th Internat. Congress on Catalysis, 1973*, p. 1251. North Holland, Amsterdam; Elsevier, New York.
108. K. Aika, H. Hori and A. Ozaki, *J. Catalysis* (1972) **27**, 424.
109. G. Ertl, M. Weiss and S. B. Lee, to be published.
110. J. A. Dumesic, H. Topsøe, S. Khammouma and M. Boudart, *J. Catalysis* (1975) **37**, 503.

3

Fixation of Molecular Nitrogen in Aprotic Media

M. E. VOL'PIN and V. B. SHUR

Institute of Organoelement Compounds, USSR Academy of Sciences, USSR

FIXATION of molecular nitrogen under mild conditions is one of the intriguing problems of chemistry. For long a puzzling inconsistency existed between the chemical inertness of this molecule and the ability of some microorganisms to fix dinitrogen from the atmosphere with considerable efficacy. Repeated attempts to make dinitrogen react under conditions approaching those of enzymatic action proved vain. The first indications of its successful accomplishment appeared 16 years ago when it was shown that N_2 could be reduced in aprotic media in the presence of transition metal compounds.

When in 1962 we began our study of chemical nitrogen fixation, we started with the following considerations:

(1) Since biological fixation of dinitrogen takes place, the corresponding chemical reactions must exist.

(2) The important role of molybdenum and iron compounds in the enzymatic process suggests that these and other transition metals might possibly be exploited for the chemical activation of N_2.

(3) Similar features in the structure and some properties of N_2 and of the isoelectronic molecule CO gave rise to the idea of the possibility of complexing dinitrogen by transition metal compounds which could result in its activation.

(4) From a comparison of N_2 with CO and acetylenes one could have expected that in the complexing of N_2 with transition metals a special role would be played by back donation, so that in nitrogen fixation the lower-valent compounds of the transition metals should be the active ones.

On the basis of these considerations we were able in 1964 to find the first transition metal systems capable of reducing N_2 in solution.[1] As reducing agents use was made of organomagnesium, aluminium and lithium compounds, lithium–aluminium hydride and metallic magnesium.[1–6] It is remarkable that such generally simple systems, many of which had been widely used earlier as catalysts of polymerization and other processes, turned out to be exceptionally reactive towards dinitrogen, fixing it very rapidly at room temperature and atmospheric pressure.

The first systems employed reduced N_2 stoichiometrically to the level of ammonia. Later multifarious systems were developed, capable of transforming N_2 to hydrazine and amines as well as ammonia, and also of reducing N_2 catalytically. The next important step was to find systems that could function in protic media[7] (see Chapters 4–9). Finally a paper by Allen and Senoff[8] on the synthesis of the first dinitrogen complex, appearing in 1965, was the progenitor of a new direction: the chemistry of complexes of molecular nitrogen, which has shed and continues to shed considerable light on the nitrogen-fixing mechanism (Part 3).

In the present chapter we shall give an overview of the data on the reduction of N_2 in aprotic solvents, dwelling on those points which we consider of general importance to all nitrogen fixation processes, both chemical and enzymatic. Because N_2 complexes are treated in detail in subsequent chapters we shall here discuss only those studies in this area that are directly concerned with N_2 reduction in aprotic media.

I. Stoichiometric Reduction of Molecular Nitrogen by Means of Transition Metal Compounds

A considerable number of systems capable of reducing molecular nitrogen in aprotic solvents have now been found and extensively investigated. The pertinent reactions have received comprehensive coverage in a number of reviews[9–17] so that in this section we consider only the basic relationships.

A. General characteristics of the reaction

Table 1 lists the most typical nitrogen-fixing systems. They all consist of two components: a transition metal compound for complexing and activating N_2 and a sufficiently strong reducing agent acting as an electron source for the activated dinitrogen molecule. Both components are required for the system to manifest nitrogen-fixing capacity and no N_2 reduction is observed when one or the other is absent. The reactions proceed at room temperature in organic solvents such as ethers or hydrocarbons.

As a rule the final N_2 reduction products are nitride-like compounds, differing, however, from the simple nitrides formed in the high-temperature reactions of the metals with N_2. Hydrolysis of the products yields ammonia. In a number of cases hydrazine in varying amounts can be detected along with ammonia, and some organometallic compounds used as reducing agents can also cause amine formation (Section III).

The nitrogen-fixing activity of the systems varies widely depending on the reagents and reaction conditions. A crucial factor here is the nature of the metal in the initial salt or in the complex. The strongest N_2 reducing capacity is displayed by transition metals of groups IV, V and VI of the Periodic Table namely Ti, VCr, Mo and W.[15] Particularly active are titanium compounds. With the first row transition metals ammonia yield generally falls on moving from left to right, correlating with the diminishing strength of the metal–nitride bond.[18] Of interest, however, is that as a rule iron compounds are more efficient than manganese compounds. Zirconium compounds are usually less active than those of titanium. Systems based on Co, Ni, Pd, Pt and Cu are of low activity or totally inactive.

The ligand environment of the transition metal can differ within rather wide limits. Systems using halides, alkoxides, acetylacetonates (acac) and π-cyclopentadienyl (Cp) complexes have been reported.[9,14,15,17] Tertiary phosphine complexes have also been tried.[19]

Another important factor is the reducing agent. At present a wide variety of such agents has been tested, among them organometallic compounds (RMgX, RLi, R_3Al),[15,17,20,21] metal hydrides ($LiAlH_4$, LiH),[15] metals (alkali, alkaline earth, rare earth),[9,14,15,20,22–30] amalgams (Li/Hg, Na/Hg),[31] radical-anions and dianions of aromatic hydrocarbons,[11,14,21,32–39,91] and molecular hydrogen (Section II). It has also been reported that dinitrogen can be reduced electrochemically.[14,40] To achieve maximum reaction rates the reducing agent is often used in considerable excess with respect to the transition metal compound (up to 10–60 : 1). The stoichiometric ratios required for maximum possible ammonia production are much lower (e.g. 4 : 1 in the case of $[Cp_2TiCl_2]$ + Li/Hg(Na/Hg) in tetrahydrofuran (THF)[31] and 2·5 : 1 in the case of $[TiCl_3(THF)_3]$ + Mg and $[VCl_3(THF)_3]$ + Mg in THF)[25].

It is to be noted that if the transition metal compound is in a sufficiently reduced state, under certain conditions it could itself reduce dinitrogen, as has been observed for the complexes $C_6H_6 \cdot TiCl_2 \cdot 2AlCl_3$[41–43] and $[Cp_2TiH]_2$[35] and in the thermolysis of $[Cp_2TiR_2]$(R = Ph, Me)[15,44] and $[Cp_2TiR]$ (R = 2-MeC_6H_4, 2,6-$Me_2C_6H_3$)[91,92] under N_2.

The first nitrogen-fixing systems based on transition metal halides and Mg- and Al-organic compounds displayed relatively little activity. Ammonia yields did not exceed 0·2–0·3 moles per mole of initial salt ($P_{N_2} \sim 150$ atm.)[1,3,9,15]. Subsequently, by changing the ligand environment of the transition metal the activity of these systems was significantly enhanced. The system $[Cp_2TiCl_2]$ + EtMgBr in ether turned out to be especially efficient as it could reduce dinitrogen at room temperature and atmospheric pressure to give 0·7 mole of

Table 1. Nitrogen-fixing systems[a]

System	Molar ratio of reagents	Solvent	Pressure of N_2 (atm.)	Reaction time (h)	NH_3 yield, (mol/g-at. of transition metal)	Ref.
$TiCl_4$ + EtMgBr	1:9	ether	100	11	0·12	*1, 15*
$CrCl_3$ + MeMgI	1:9	ether	150	11	0·04	*3, 15*
$CrCl_3$ + EtMgBr	1:9	ether	150	11	0·17	*1, 15*
$CrCl_3$ + EtMgBr	1:9	ether	150	21	0·21	*3*
$CrCl_3$ + EtMgBr	1:9	ether	1	7	0·01	*1*
$CrCl_3$ + EtMgBr	1:9	THF	150	11	0·03	*1, 15*
$CrCl_3$ + n-PrMgBr	1:9	ether	150	11	0·30	*3, 15*
$CrCl_3$ + n-BuMgBr	1:9	ether	150	11	0·27	*3, 15*
$CrCl_3$ + i-PrMgBr	1:9	ether	150	11	0·26	*3, 15*
$CrCl_3$ + $PhCH_2MgBr$	1:9	ether	150	11	0·02	*3, 15*
$CrCl_3$ + PhMgBr	1:9	ether	150	11	0·04	*1, 15*
$CrCl_3$ + PhMgBr	1:9	THF	150	11	0·01	*1, 15*
$MoCl_5$ + EtMgBr	1:9	ether	150	11	0·08	*1, 15*
$MoCl_5$ + i-PrMgCl	1:10	not given	10	2	0·16[b]	*56*
WCl_6 + EtMgBr	1:9	ether	150	11	0·15	*1, 15*
$MnCl_2$ + EtMgBr	1:9	ether	150	11	0·02	*1, 15*
$FeCl_3$ + EtMgBr	1:9	ether	150	11	0·09	*1, 15*
$FeCl_3$ + i-PrMgCl	1:50	ether	>20	6	0·30[c]	*56*
$Ti(OEt)_4$ + EtMgBr	1:9	ether	100	10	0·08	*45*
$[Cp_2TiCl_2]$ + EtMgBr	1:9	ether	1	9	0·67	*2, 3, 15*
$[Cp_2TiCl_2]$ + EtMgBr	1:9	ether	150	31	0·93	*2, 3, 15*
$[Cp_2TiCl_2]$ + EtMgBr	1:9	THF	150	11	0·40	*2, 3, 15*
$[Cp_2TiCl_2]$ + n-PrMgBr	1:9	ether	150	11	0·70	*2*
$[Cp_2TiCl_2]$ + i-PrMgCl	1:5	1,2-DME	1	1	0·52	*20*
$[CpTiCl_3]$ + EtMgBr	1:9	ether	100	11	0·23	*15*
$[TiCl_4(Ph_3P)_2]$ + EtMgBr	1:9	ether	150	10	0·10	*19*

$[FeCl_3(Ph_3P)_2]$ + EtMgBr	1:9	ether	150	10	0·03	*19*
$[VO(acac)_2]$ + EtMgBr	1:9	ether	100	10	0·12	*45*
$[MoO_2(acac)_2]$ + EtMgBr	1:9	ether	100	10	0·14	*45*
$[Cr(acac)_3]$ + EtMgBr	1:9	ether	100	10	0·03	*45*
$[Fe(acac)_3]$ + EtMgBr	1:9	ether	100	10	0·02	*45*
$TiCl_4$ + i-Bu_3Al	1:3	n-heptane	150	11	0·25	*1*
$TiCl_4$ + i-Bu_3Al	1:6	toluene	100	11	0·32	*9, 15*
$Ti(OEt)_4$ + i-Bu_3Al	1:6	n-heptane	100	10	0·06	*45*
$Ti(OEt)_4$ + i-Bu_3Al	1:6	toluene	30	10	0·33	*15, 45*
$Tu(OBu)_4$ + i-Bu_3Al	1:3	n-heptane	50	12	0·08	*104*
$[CpTiCl_3]$ + i-Bu_3Al	1:6	toluene	90	11	0·15	*15*
$[CpTi(OEt)_3]$ + i-Bu_3Al	1:6	toluene	30	7	0·22	*15*
$[TiCl_4(Ph_3P)_2]$ + i-Bu_3Al	1:3	n-heptane	150	10	0·12	*19*
$[VO(acac)_2]$ + i-Bu_3Al	1:6	toluene	100	10	0·07	*45*
$[Cr(acac)_3]$ + i-Bu_3Al	1:6	toluene	100	10	0·14	*45*
$[MoO_2(acac)_2]$ + i-Bu_3Al	1:6	toluene	100	10	0·09	*45*
$[Mn(acac)_3]$ + i-Bu_3Al	1:6	toluene	100	10	0·09	*45*
$[Ni(acac)_3]$ + i-Bu_3Al	1:6	toluene	100	10	0·04	*45*
$TiCl_4$ + n-BuLi	1:9	n-heptane	100	11	0·29	*9, 15*
$FeCl_3$ + n-BuLi	1:10	Et_2O/PhMe	100	not given	0·14	*63*
$Ti(OEt)_4$ + n-BuLi	1:9	n-heptane	100	10	0·11	*45*
$[CpTiCl_3]$ + n-BuLi	1:9	n-heptane	100	11	0·07	*9, 15*
$[Cp_2TiCl_2]$ + n-BuLi	1:9	n-heptane	100	11	0·50	*9, 15*
$[VO(acac)_2]$ + n-BuLi	1:9	n-heptane	75	10	0·32	*45*
$[VO(acac)_2]$ + n-BuLi	1:9	toluene	83	10	0·35	*45*
$[MoO_2(acac)_2]$ + n-BuLi	1:9	n-heptane	75	10	0·07	*45*
$[MoO_2(acac)_2]$ + n-BuLi	1:9	toluene	83	10	0·07	*45*
$[Fe(acac)_3]$ + n-BuLi	1:9	n-heptane	100	10	0·05	*45*
$TiCl_4$ + PhLi	1:5	ether	100	24	0·53[d]	*121*
$FeCl_3$ + PhLi	1:20	ether	1	24	0·15[e]	*58*
$Ti(OBu)_4$ + PhLi	1:5	ether	100	24	0·60[d]	*121*
$[CpTiCl_3]$ + PhLi	1:5	ether	100	24	0·37[d]	*121*

Table 1 (cont.)

System	Molar ratio of reagents	Solvent	Pressure of N_2 (atm.)	Reaction time (h)	NH_3 yield, (mol/g-at. of transition metal)	Ref.
$[Cp_2TiCl_2]$ + PhLi	1:5	ether	100	24	0·65[d]	*15, 121*
$[Cp_2TiPh_2]$ + PhLi	1:3·3	ether	100	24	0·57[d]	*15, 121*
$[Cp_2TiCl_2]$ + n-MeC_6H_4Li	1:5	ether	100	24	0·25[f]	*15, 121*
$[Cp_2TiCl_2]$ + m-$CH_3C_6H_4Li$	1:5	ether	100	24	0·22[f]	*15, 121*
$[Cp_2TiCl_2]$ + o-MeC_6H_4Li	1:5	ether	100	24	0·41[f]	*15, 121*
$CrCl_3 + LiAlH_4$	1:9	ether	150	12	0·07	*1, 15*
$CrCl_3 + LiAlH_4$	1:9	ether	1	7	0·02	*1, 15*
$CrCl_3 + LiAlH_4$	1:9	THF	150	12	0·04	*1, 15*
$CrCl_3 + LiAlH_4$	1:9	1,2-DME	150	12	0·03	*1, 15*
$TiCl_4 + Mg + MgI_2$	1:14:5	Et_2O/C_6H_6	80–100	7	1·3	*5, 9, 15*
$CrCl_3 + Mg + MgI_2$	1:14:5	Et_2O/C_6H_6	80–100	7	0·35	*5, 9, 15*
$MoCl_5 + Mg + MgI_2$	1:14:5	Et_2O/C_6H_6	80–100	7	0·36	*5, 9, 15*
$WCl_6 + Mg + MgI_2$	1:14:5	Et_2O/C_6H_6	80–100	7	0·58	*5, 9, 15*
$[Cp_2TiCl_2] + Mg + MgI_2$	1:8:3	Et_2O/C_6H_6	80–100	7	0·8–1·0	*5, 15, 24*
$[Cp_2TiCl_2]$ + Li + LiI	1:10:10	ether	80–100	7	0·30	*15*
$[TiCl_3(THF)_3]$ + Mg	1:8·5	THF	1	2–3	ca. 1·0	*25*
$TiCl_4$ + Na	excess Na	THF	1	144	0·44	*22*
$[Cp_2TiCl_2]$ + Li	excess Li	THF	1	35	0·75	*22*
$[Cp_2TiPh_2]$ + Li	1:10	THF	100	72	1·1[d]	*123*
$[Cp_2TiCl_2]$ + Na	1:4	THF	1	18	0·56	*22*
$[Cp_2TiCl_2]$ + K	excess K	THF	1	50	0·09	*22*
$[Cp_2TiCl_2]$ + Rb	excess Rb	THF	1	18	0·30	*22*
$[Cp_2TiCl_2]$ + Cs	1:4	THF	1	18	0·44	*22*
$[Cp_2TiCl_2]$ + Mg	1:2	THF	1	120	0·46	*22*
$[Cp_2TiCl_2]$ + Ca	excess Ca	THF	1	96	0·35	*22*
$[Cp_2TiCl_2]$ + La	excess La	THF	1	72	0·46	*22*

$[Cp_2TiCl_2] + Ce$	excess Ce	THF	1	40	0·64	*22*
$[CpTiCl_3] + Na$	1:4	THF	1	5	0·47	*22*
$[CpTiCl_3] + Mg$	1:2	THF	1	24	0·46	*22*
$[CpTiCl_3] + Ce$	excess Ce	THF	1	24	0·35	*22*
$[Cp_2TiCl]_2 + Na$	1:6	THF	1	72	0·51	*22*
$[Cp_2TiBH_4] + Na$	excess Na	THF	1	24	0·46	*22*
$[Cp_2ZrCl_2] + Li$	excess Li	THF	1	14	0·12	*22*
$[Cp_2ZrCl_2] + Na$	1:4	THF	1	24	0·07	*22*
$[Cp_2TiCl_2] + Li/Hg$	1:10	THF	1	72	0·70	*31*
$[Cp_2TiCl_2] + Mg/Hg$	1:5	THF	150	72	0·74	*20*
$TiCl_4 + C_{10}H_8^{\div}Li^+$	1:15	THF	120	0·5	1·7	*11, 48*
$VCl_3 + C_{10}H_8^{\div}Li^+$	1:7	THF	120	0·5	2·0	*11, 48*
$VCl_3 + C_{10}H_8^{\div}Li^+$	1:10	THF	1	0·5	0·9	*11, 48*
$CrCl_3 + C_{10}H_8^{\div}Li^+$	1:10	THF	120	0·5	1·2	*11, 48*
$CrCl_3 + C_{10}H_8^{\div}Li^+$	1:10	THF	1	0·5	0·4	*11, 48*
$FeCl_3 + C_{10}H_8^{\div}Li^+$	1:25	THF	30–100	4	0·5[g]	*33*
$FeCl_3 + C_{10}H_8^{=}Li_2^{+}$	1:8	THF	ca. 110	4	0·82[g]	*33*
$[Cp_2TiH]_2 + C_{10}H_8^{\div}Li^+$	1:60	THF	200	1	0·99	*35*
$[Cp_2TiH]_2 + C_{10}H_8^{\div}Li^+$	1:3	1,2-DME	150	1	0·32	*35*
$[Cp_2TiCl_2] + C_{10}H_8^{\div}Li^+$	1:6	THF	1	16	0·96	*11, 49*
$[Cp_2TiCl_2] + C_{10}H_8^{\div}Na^+$	1:4·5	THF	1	2	0·52[h]	*91, 14*
$[CpTiCl_3] + C_{10}H_8^{\div}Na^+$	1:4·3	THF	1	0·5	0·90[i]	*39*
$[CpTiCl_2] + C_{10}H_8^{\div}Na^+$	1:3·1	THF	1	0·5	0·85[k]	*39*
$Ti(O{-}i{-}Pr)_4 + C_{10}H_8^{\div}Na^+$	not given	THF	1	0·5	1·3	*14, 50*
$[Cp_2ZrCl_2] + C_{10}H_8^{\div}Na^+$	not given	THF	1	not given	0·03	*32*
$[Cp_2NbCl_2] + C_{10}H_8^{\div}Na^+$	not given	THF	1	not given	0·03	*32*

[a] Unless especially noted the reactions were carried out at room temperature; [b] at 30°C; ca. 0·05 mole of N_2H_4 is also formed; [c] at 90°C; [d] also 0·10–0·15 moles of $PhNH_2$ (see Section III); [e] also 0·28 moles of N_2H_4; [f] also 0·06–0·07 moles of toluidines (see Section III); [g] at 60°C; [h] at −20°C; also 0·05 mole of N_2H_4 is formed; [i] also 0·015 moles of N_2H_4; [k] also 0·045 moles of N_2H_4.

THF = tetrahydrofuran; 1,2-DME = 1,2-dimethoxyethane.

ammonia per mole of $[Cp_2TiCl_2]$.[2,3,9,15] With increased N_2 pressure the yield of NH_3 could be brought up to 1 mole.

The activity of systems containing RLi, RMgX and R_3Al as reducing agents depends on the nature of the radical R bound to the metal. In the case of the system $CrCl_3$ + RMgX in ether, the yields of ammonia fall in the order:[3,15]

$$\text{n-Bu} \sim \text{n-Pr} > \text{Et} \gg \text{Me} \sim \text{Ph} \sim \text{PhCH}_2.$$

The same order holds for the reducing capacity of the organomagnesium reagents. A similar pattern is displayed by $[Cp_2TiCl_2]$ + RMgX in ether where the transition from EtMgBr to MeMgI or PhMgBr is accompanied by a sharp fall in ammonia yield (from 1 mole to 0·01–0·02 moles/Ti atom; RMgX/Ti = 9, $P_{N_2} \sim 100$ atm.). Interestingly, however, with PhLi, a stronger reducing agent than PhMgBr, $[Cp_2TiCl_2]$ forms a highly active N_2 reducing system (Section III).

Combinations of Ti, V, Cr, Mo or Mn compounds with aluminium alkyls, i.e. typical Ziegler catalysts for olefin polymerization, are also capable of reacting with N_2 at room temperature.[1,9,15,45] This can explain the inhibitory effect of dinitrogen earlier observed in the polymerization of ethylene by the system $TiCl_4$ + i-Bu_2AlH.[46] Hence in dealing with Ziegler catalysts one should not use gaseous dinitrogen as an inert atmosphere, otherwise the catalyst can partially or even completely lose its activity.

The use of metals as reducing agents substantially broadened the range of N_2-fixing systems. The first of these systems based on MgI_2-activated metallic magnesium in ether (or ether–benzene mixture) manifested considerable activity.[5,9,15,24] With $[Cp_2TiCl_2]$ and $[CpTiCl_3]$ up to 1 mole of ammonia per mole of titanium compound was formed (at 20°C and $P_{N_2} \sim 100$ atm.), and with $TiCl_4$ even 1·3 moles were produced. MgI_2 may be replaced by $MgBr_2$ in these reactions. In the absence of the magnesium halide the systems are inactive; MgI_2 is known to enhance the reducing capacity of Mg in ether solution.

Magnesium-containing systems $[TiCl_3(THF)_3]$ + Mg and $[VCl_3(THF)_3]$ + Mg in THF functioning efficiently at atmospheric pressure ($NH_3/MCl_3 = 1$) have been described by Yamamoto *et al*;[23,25] the addition of magnesium halides is not required here. Similar systems, $FeCl_3$ + Mg, $[TiCl_4(THF)_2]$ + Mg, VCl_4 + Mg and $CrCl_2$ + Mg in THF, have been studied.[27–30]

Various alkali and alkaline earth metals, Li, Na, K, Rb, Cs, Mg, Ca, as well as lanthanum and cerium have been employed by Bayer and Schurig.[22] In the presence of titanium and zirconium compounds ($[Cp_2TiCl_2]$, $[CpTiCl_3]$, $[Cp_2TiCl]_2$, $[(C_5H_5)_2(C_{10}H_8)Ti_2H_2]$, $[Cp_2TiBH_4]$, $TiCl_4$, $[Cp_2ZrCl_2]$) all these metals are also capable of reducing dinitrogen.

Van Tamelen *et al.*[14,47] reported the formation of ammonia (after hydrolysis) during electrolysis of a solution of Ti(O–i-Pr)$_4$ and $AlCl_3$ in 1,2-dimethoxyethane (DME) under N_2. As the result of passing a current through the solution for two days the yield of NH_3 was 0·1 mole/Ti atom.

Highly active nitrogen-fixing systems are formed when radical-anions or dianions of aromatic hydrocarbons are used as reducing agents.

Henrici-Olivé and Olivé[11,48,49] found that the products formed in the course of reductions of $TiCl_4$, VCl_3, $CrCl_3$ or $[Cp_2TiCl_2]$ by excess lithium naphthalide ($C_{10}H_8^{\dot{-}}\ Li^+$) in THF fix dinitrogen at room temperature. The reactions proceed very effectively. With $TiCl_4$ and $CrCl_3$ the ammonia yields exceed 1 mole per mole of salt ($P_{N_2} = 120$ atm.), while in the case of VCl_3 they attain a value of 2 moles, which is the highest yield of NH_3 observed up till now in stoichiometric nitrogen-fixing systems. The systems are active also at atmospheric pressure, particularly $[Cp_2TiCl_2] + C_{10}H_8^{\dot{-}}Li^+$ (NH_3/Ti = 1)[49] and $VCl_3 + C_{10}H_8^{\dot{-}}Li^+$ (NH_3/V = 0·9).[48]

Van Tamelen *et al.*[14,50,51] observed N_2 reduction at 20° and 1 atm. in the system Ti(O–i-Pr)$_4$ + sodium naphthalide ($C_{10}H_8^{\dot{-}}Na^+$) in THF or diglyme. Under optimal conditions the yield of ammonia was 1·3 mole/Ti atom. After decomposing the reaction products with alcohol and removal of the ammonia, the nitrogen-fixing activity of the system could be partly restored by adding fresh reducing agent. By repeating this procedure five times, 3·4 moles NH_3/Ti atom could be obtained.

On transition from titanium, vanadium and chromium to less active metals, the reductive power of the naphthalide ion turns out to be insufficient for high conversion of dinitrogen to nitrides. According to Bell and Brintzinger[33] in the case of the system $FeCl_3 + C_{10}H_8^{\dot{-}}Li^+$ in THF the yield of NH_3 does not exceed 0·5 mole even with a 25-fold excess of reducing agent and a temperature of 60°C ($P_{N_2} \sim 100$ atm.). The reaction efficiency increases if lithium naphthalide is replaced by stronger reducing agents such as dianions of naphthalene or of 2,6-dimethylnaphthalene. In the latter case up to 1 mole of NH_3 per mole of $FeCl_3$ was formed.

A similar effect was observed in our laboratory for the Na-adducts of diphenyl, naphthalene, *p*-terphenyl and anthracene in THF. On introducing these adducts into the reaction with N_2 in the presence of $CrCl_3$, $MnCl_2$, $FeCl_3$ (20°C, $P_{N_2} \sim 100$ atm.; $ArH^{\dot{-}}Na^+/M = 6:1$) the yields of ammonia gradually diminished with decreasing reducing potential of the radical-anion[52] in the order:

$$[\text{diphenyl}]^{\dot{-}} > [\text{naphthalene}]^{\dot{-}} > [p\text{-terphenyl}]^{\dot{-}} > [\text{anthracene}]^{\dot{-}}$$

In the case of $TiCl_4$ and VCl_4 the differences in the reductive power of $ArH^{\dot{-}}$ affected the reaction less. It is interesting that, whereas with anthracene–sodium ammonia-forming capacity was manifested only by $TiCl_4$, VCl_4 and $CrCl_3$, with anions of other hydrocarbons all the aforementioned salts (including $MnCl_2$ and $FeCl_3$) were active. Hence passing to stronger reducing agents widens the range of transition metals capable of inducing N_2 reduction to the level of ammonia.

Under certain conditions the reactions can lead to the formation of varying amounts of hydrazine along with ammonia after hydrolysis. Van Tamelen *et al.*[14,53] reported the formation of hydrazine at room temperature in the reaction of N_2 with $Ti(O{-}i\text{-}Pr)_4$ and sodium naphthalide in THF. Maximum yields of hydrazine (0·15–0·19 mole/Ti atom) were attained at $C_{10}H_8^{\dot{-}}Na^+/Ti$ = 5–6:1. The ratio $2NH_3/N_2H_4$ was 3·3–5·0. Increase in reaction time (above 90 min) and in the relative amount of reducing agent (e.g. up to 12:1) sharply decreased the yield of hydrazine due to further reduction of hydrazine precursor† to the ammonia level. When using $CoCl_2$, $MoCl_5$, WCl_6, $CrCl_3$ and $FeCl_3$ hydrazine did not form.

Shilov *et al.*[17,54,55] detected hydrazine when reacting dinitrogen at −60° with the systems $[Cp_2TiCl_2]([Cp_2TiCl])$ + RMgX (R = Et, i-Pr; X = Br, Cl) in ether. At temperatures above 0°C, hydrazine disappeared and ammonia was formed. Details of this study will be considered in Section I.B. The formation of small amounts of hydrazine was observed in our laboratory on heating dinitrogen with the complex $C_6H_6 \cdot TiCl_2 \cdot 2AlCl_3$ and in the system $TiCl_4$ + Al + $AlBr_3$.[43]

Subsequently hydrazine was found in the reactions of N_2 with the systems $MoCl_5$ + i-PrMgCl,[56] $[FeCl_3(Ph_3P)_2]$ + i-PrMgCl,[56,57] $FeCl_3$ + PhLi,[58,59] $[Cp_2TiCl]$ + MeMgI,[60,61] $[Cp_2TiCl_2]([Cp_2TiCl])$ + $C_{10}H_8^{\dot{-}}Na^+$,[55,91] $[CpTiCl_3]([CpTiCl_2])$ + $C_{10}H_8^{\dot{-}}Na^+$,[39] $[Cp_2TiR]$ + $C_{10}H_8^{\dot{-}}Na^+$,[21, 39, 91] $FeCl_3(CrCl_2)$ + Mg,[28, 30] $TiCl_4(FeCl_3)$ + $C_{10}H_8^{\dot{-}}Li^+$ + Ph_3P,[36] $TiCl_4$ + naphthalene + Li,[37,38] and also in N_2 fixation in protic media (Chapter 3), and in the reduction of the dinitrogen ligand in certain stable N_2 complexes (Chapters 6–9). In the case of $MoCl_5$ + i-PrMgCl, $[CpTiCl_3]([CpTiCl_2])$ + $C_{10}H_8^{\dot{-}}Na^+$, $[Cp_2TiR]$ + $C_{10}H_8^{\dot{-}}Na^+$ and $TiCl_4$ + naphthalene + Li it was shown that a hydrazine precursor is an intermediate in the formation of the nitride which hydrolyses to ammonia.

In order to compare the chemical nitrogen-fixing systems with enzymatic ones, the effect of nitrogenase inhibitors on the chemical fixation of N_2 was investigated. It is well known that biological fixation of N_2 is strongly inhibited by carbon monoxide, acetylenes, nitriles, isonitriles, CN^-, etc.[62] Experiments with the system $CrCl_3$ + EtMgBr in ether showed that carbon monoxide practically completely inhibits reduction of N_2. The same effect is exerted by acetylenes (e.g. tolane).[3,15] Hence, as in the enzymic reaction, carbon monoxide and acetylenes are effective inhibitors of dinitrogen reduction.

† Here and later, "hydrazine precursor" stands for a product that corresponds to the hydrazine level of N_2 reduction and yields hydrazine on hydrolysis. We use the term "diazene precursor" analogously.

A much weaker effect on the fixation of N_2 was exerted by olefins. With the presence of a large excess of hex-1-ene the yield of NH_3 in the system $CrCl_3$ + EtMgBr diminishes only 2·5 times, and the addition of *trans*-stilbene has no effect on the reaction at all.[3,15] Olefins, contrary to acetylenes, in general do not inhibit dinitrogen reduction in the course of biological fixation.[62]

A certain similarity between these two processes can be noted also in the way they are affected by molecular hydrogen. As is well known, in the enzymic reaction dihydrogen usually acts as an inhibitor of nitrogen fixation, but in artificial systems Clostridial ferredoxins and flavodoxins are capable of transferring electrons from H_2 via hydrogenase to various nitrogenases and further to N_2.[62] Dihydrogen affects chemical N_2 fixation in a similar fashion. Depending on the type of the system and on the reaction conditions, H_2 can either effectively inhibit nitrogen fixation[9,15,22,25,45,63] or take part in its reduction (Section II). The inhibiting effect of H_2 is the result of the competitive displacement of N_2 by the dihydrogen from the coordination sphere of the transition metal. The reactions of certain dinitrogen complexes with H_2 serve as a model for such competition, e.g.:[64]

$$[CoH(N_2)(PPh_3)_3] + H_2 \rightleftharpoons [CoH_3(PPh_3)_3] + N_2$$

Van Tamelen *et al.*[65] have investigated the reactions of a number of nitrogenase inhibitors (hex-1-yne, cyclohexylisonitrile, KCN) with the systems $[Mo(acac)_3] + C_{10}H_8^{\dot{-}}Na^+$, $[Fe(acac)_3] + C_{10}H_8^{\dot{-}}Na^+$, $MoCl_5$ + Mg, $FeCl_3$ + Mg, $TiCl_3$ + Mg and $[Cp_2TiCl_2] + C_{10}H_8^{\dot{-}}Na^+$ in THF. Nitrogenase is known to catalyse the reduction of acetylenes to olefins, of CN^- to CH_4 and NH_3 (giving also CH_3NH_2 and traces of C_2H_4 and C_2H_6), and of isonitriles (RNC) to RNH_2, CH_4 and higher hydrocarbons (e.g. C_2H_6 and C_2H_4).[62] It turned out the reactions of such substrates with the aforementioned non-enzymatic systems, while manifesting a number of significant differences from the corresponding enzymatic processes, also display common features. Thus hex-1-ene and only small amounts of hexane are formed from hex-1-yne ($FeCl_3$ + Mg), KCN yields CH_4, C_2H_4 and C_2H_6, whereas from cyclohexylisonitrile, CH_4 and traces of C_2H_4 and C_2H_6 are produced as well as cyclohexane ($FeCl_3$ + Mg; $TiCl_3$ + Mg). Finally all these nitrogenase inhibitors also inhibited the non-enzymatic reduction of N_2.

The effect of dioxygen was also investigated[3,14,15,50] and it was shown that although small amounts of O_2 have no strong influence on the reaction, large quantities inhibit nitrogen fixation up to its complete suppression. Continuing the comparison of the chemical systems with the enzymatic ones it should be noted that nitrogenase preparations isolated from cell-free extracts of nitrogen-fixing bacteria (both aerobic and anaerobic) are inactivated on contact with dioxygen, although the bacteria themselves are capable of binding dinitrogen directly from the atmosphere, i.e. in the presence of O_2. In all probability the cells of these bacteria contain some protective mechanism hindering the penetration of O_2 to the active centres of the enzyme so that the process of biological nitrogen fixation actually also proceeds in a dioxygen-free atmosphere.

Despite the high activity of the systems discussed above, they are not catalytic. Kinetic experiments show that the curves for the time dependence of the ammonia yields are similar for differing systems. In all cases increase in the amount of ammonia formed with time occurs only up to a certain limit after which it remains constant even in the presence of a large excess of reducing agent. The limiting yield of NH_3 depends upon the type of system employed, but never exceeds a value of 1 or 2 moles per atom of transition metal (M). The stoichiometric ratio $NH_3/M = 1$ may indicate that the dinitrogen reduction proceeds through an intermediate binuclear complex $[M \cdot N_2 \cdot M]$. This apparently is the case for the systems $[Cp_2TiCl_2]$ + RMgX in ether,[3,17] $[TiCl_3(THF)_3]$ + Mg in THF, $[VCl_3(THF)_3]$ + Mg in THF,[25] $[Cp_2TiCl_2]$ + $C_{10}H_8^{\dot{-}}Li^+$ in THF,[49] etc. If however, $NH_3/M = 2$ (as for VCl_3 + $C_{10}H_8^{\dot{-}}Li^+$ in THF[11,34,48]), one could visualize the formation of a mononuclear complex $[M \cdot N_2]$ (or of the corresponding dimer

$$\left[M \begin{smallmatrix} N_2 \\ N_2 \end{smallmatrix} M\right])$$

as a stage preceding dinitrogen reduction. Limiting NH_3 yields less than 1 mole or between 1 and 2 moles, could have different causes: irreversible loss of some of the metal in side reactions, a more complicated reaction stoichiometry, or (when $1 < NH_3/M < 2$) contribution by both the $[M \cdot N_2]$ and $[M \cdot N_2 \cdot M]$ species to the ammonia-forming process.

In some systems, e.g. $[Cp_2TiCl_2]$ + Mg + MgI_2 in ether,[15,24] $[Cp_2TiCl_2]$ or $[Cp_2TiPh_2]$ + PhLi in ether,[15] $[Cp_2TiCl_2]$ + Na in THF,[22] $[VCl_3(THF)_3]$ + Mg and $[TiCl_3(THF)_3]$ + Mg in THF,[25] dinitrogen absorption does not occur immediately after mixing the reagents, but only at the end of a certain induction period. Here the kinetic curves are S-shaped. It was found by experiment that the lag was due to reduction of the original transition metal compound to a species being able to complex and activate the N_2. Evidently this preliminary reduction stage is common to all nitrogen-fixing systems containing transition metals in their higher oxidation states.

Most often the active species is an intermediate rather than the end product of the reduction process. This is evidenced by experiments wherein the initial reacting system is held for varying periods under argon before introducing dinitrogen. The ammonia yields under such conditions are often greatly diminished. Typical examples are $[Cp_2TiCl_2]([Cp_2TiPh_2])$ + PhLi in ether, $CrCl_3$ + EtMgBr in ether, $[Cp_2TiCl_2]$ + Mg + MgI_2 in ether,[15,24] $[Cp_2TiCl_2]$ + $C_{10}H_8^{\dot{-}}Na^+$ in THF, Ti(O–i-Pr)$_4$ + $C_{10}H_8^{\dot{-}}Na^+$ in THF,[14,51,66] $[TiCl_3(THF)_3]$ +

Mg and $[VCl_3(THF)_3]$ + Mg in THF.[25] The system $[Cp_2TiCl_2]$ + EtMgBr in ether is more stable but it too, gradually loses activity.[67,68]

The loss in activity with time can be due to various causes. In the case of $[Cp_2TiCl_2]$ + EtMgBr one of the de-activating factors could be connected with formation of the dinitrogen-inert paramagnetic dihydride $[Cp_2TiH_2]^{\dot{-}}MgX^+$ characterized by a triplet signal in the ESR spectrum.[69,70] The hydrogen required for its formation originates from the ethyl radical of the Grignard reagent.[69] Similar hydrides displaying the same ESR spectral pattern are formed by $[Cp_2TiCl_2]$ + $C_{10}H_8^{\dot{-}}Li^+(Na^+)$ in THF,[11,49] $[Cp_2TiCl_2]$ + Li(Na, Mg) in THF,[22,69] $[CpTiCl_3]$ + Na(Mg) in THF,[22] $[CpTiCl_3]$ + Mg + MgX_2 in ether and $[Cp_2TiCl_2]$ + Mg + MgX_2 (X = I, Br) in ether.[24,71] Here the Cp rings are the hydrogen source for the hydride formation.[24,71] In the system $Ti(O{-}i{-}Pr)_4$ + $C_{10}H_8^{\dot{-}}Na^+$ the loss in activity is considered to be due to excessive reduction of the initial Ti(IV) alkoxide to the Ti(O) derivatives which are inert to N_2.[14] In the systems $[TiCl_3(THF)_3]$ + Mg and $[VCl_3(THF)_3]$ + Mg the activity diminishes because of reduction of Ti(III) and V(III) to a level below Ti(II) and V(II).[25]

Owing to the complexity of the N_2-fixing systems, detailed kinetic studies have been carried out only in isolated cases.

According to Maskill and Pratt,[67,68] the rate of N_2 reduction by $[Cp_2TiCl_2]$ + EtMgBr in ether ($P_{N_2} > 0{\cdot}67$ atm. and [EtMgBr] $> 0{\cdot}66$ M) obeys a second-order equation with respect to the Ti concentration and is independent of the partial N_2 pressure and EtMgBr concentration ($E_{act} \sim 4 \pm 1{\cdot}5$ kcal mol^{-1}). This circumstance and the fact that the limiting ammonia yield equals 1 mole (per mole of $[Cp_2TiCl_2]$)[3] means that here dinitrogen is activated by two titanium atoms in a binuclear complex $[Ti \cdot N_2 \cdot Ti]$. Subsequently, the formation of intermediate binuclear N_2 complexes in such systems was directly observed by Shilov *et al.* (Section I.B). The rate of N_2 reaction with the system $FeCl_3$ + PhLi in ether is also second-order with respect to the initial metal salt and independent of the dissolved dinitrogen and reducing agent concentrations, evidence of the formation of a binuclear N_2 complex under the reaction conditions.[58,59]

Yamamoto *et al.*[25] investigated the kinetics and stoichiometry of N_2 reduction by the systems $[TiCl_3(THF)_3]$ + Mg and $[VCl_3(THF)_3]$ + Mg in THF. The reactions proceed according to the equation:

$$[MCl_3(THF)_3] + \tfrac{5}{2}Mg + \tfrac{1}{2}N_2 \rightarrow [MNMg_2Cl_2(THF)] + \tfrac{1}{2}MgCl_2 \cdot 2THF, \quad M = Ti, V.$$

With the system $[TiCl_3(THF)_3]$ + Mg, which was studied in most detail, first-order kinetics with respect to the titanium, the amount of reacted magnesium and the N_2 pressure were observed ($E_{act} \sim 5$ kcal mol^{-1}). The proposed mechanism involves rapid and reversible reaction of N_2 with a dimeric titanium species (probably $[Ti(II)]_2$) to form the binuclear complex $[Ti(II) \cdot N_2 \cdot Ti(II)]$, followed by reduction of the N_2 ligand by magnesium in the rate-determining step. The kinetic data can be interpreted otherwise, but still necessitating the assumption of N_2 reduction in a binuclear complex.

B. Reaction mechanism

Despite the large variety of N_2 reducing systems in aprotic media the presently available data give grounds for the assumption that the basic features of the N_2 reduction mechanism are common to all reactions, whatever the transition metal compound or reducing agent used.

The following scheme depicts the common features of the nitrogen fixation mechanism:†

$$L_nM \xrightarrow[(1)]{[e]} L_mM \xrightarrow[(2)]{N_2} \left\{ \begin{matrix} L_mM \cdot N_2 \\ \text{or} \\ L_mM \cdot N_2 \cdot ML_m \end{matrix} \right\} \xrightarrow[(3)]{[e]} [{}^{1-}N{=}N^{1-}] \xrightarrow[(4)]{[e]} [{}^{2-}N{-}N^{2-}] \xrightarrow[(5)]{[e]} [N^{3-}]$$

$$\{L_mM \cdot N_2 \text{ or } L_mM \cdot N_2 \cdot ML_m\} \xrightarrow{(2')\ H^+} N_2 \quad \text{and} \quad NH_3 + N_2H_4$$

$$[{}^{1-}N{=}N^{1-}] \xrightarrow{(3')\ H^+} N_2H_2 \longrightarrow N_2 + H_2 \quad N_2H_4 + N_2 \quad NH_3 + N_2$$

$$[{}^{2-}N{-}N^{2-}] \xrightarrow{(4')\ H^+} N_2H_4 \qquad [N^{3-}] \xrightarrow{(5')\ H^+} NH_3$$

Stage (1)

In the first stage the initial transition metal compound (L_nM) is reduced to a species (L_mM) capable of forming a complex with dinitrogen. In systems based on organolithium, organomagnesium or organoaluminium compounds this process occurs via unstable σ-alkyl (or σ-aryl) derivatives of the transition metals.

The structure of the active species (L_mM) has been discussed in references *9, 11, 14, 17, 20–26, 31, 34, 35, 68, 74*. In the case of $[Cp_2TiCl_2]$ + RMgX in ether they are apparently monoalkyltitanocenes, $[Cp_2TiR]$ or $[Cp_2TiR]_2$ (see below), although contribution by more reduced forms cannot be excluded. In the systems based on $[Cp_2TiCl_2]$ and metals or naphthalides as reducing agents, the active species could be titanocene $[Cp_2Ti]_2$ or the corresponding monomer $[Cp_2Ti]$.[14,20,26,31,35] In all cases when cyclopentadienyl–metal complexes are used, one must bear in mind that one or both Cp-ligands may have to be eliminated before the active transition metal compound will appear in the solution.

---

† In the scheme the symbols $[{}^{1-}N{=}N^{1-}]$, $[{}^{2-}N{-}N^{2-}]$ and $[N^{3-}]$ designate the diazene, hydrazine and nitride levels of N_2 reduction, respectively.

In the reactions of N_2 with systems based on Ti(IV) alkoxides, it was assumed[14] that dinitrogen is activated by Ti(II) derivatives. The active species in the systems $[TiCl_3(THF)_3]$ + Mg and $[VCl_3(THF)_3]$ + Mg in THF are apparently Ti(II) and V(II).[25] Finally, in the system $VCl_4 + C_{10}H_8^{\dot{-}}Li^+$ in THF, ESR evidence is in favour of a contribution by V(I) and V(O) complexes to the N_2 activation.[11,34,72,73]

It should be noted, however, that despite the definite progress achieved here in latter years, the question of the structures of the species responsible for the N_2-fixing activity is in most cases still open.

Stage (*2*)

Next the active species reacts with the N_2 to form a mononuclear or binuclear complex.

The formation of dinitrogen complexes in such systems has been observed for many transition metals:[75] Ti, Zr, Nb, Cr, Mo, W, Fe, Ru, Os, Co, Rh, Ni (Table 2). The reactivity of the N_2 ligand in these compounds varies greatly and will be considered in Part 3. Here we shall discuss mainly those dinitrogen complexes that can participate directly as intermediates in the N_2 reduction process.

The assumption that dinitrogen reduction proceeds through the step of N_2 coordination with the transition metal compound was originally made in the first studies of molecular nitrogen fixation in solution.[1,3,9] Its experimental confirmation was obtained in a study of systems based on cyclopentadienyl derivatives of titanium.

Table 2. Examples of systems complexing dinitrogen in aprotic media

System	Solvent	Dinitrogen complex	Ref.
$[Cp_2TiCl]_2$ + i-PrMgCl	ether	$[(Cp_2TiPr\text{-}i)_2N_2]$	*17*
$[(\eta^5\text{-}C_5Me_5)_2ZrCl_2]$ + Na/Hg	toluene	$[\{(\eta^5\text{-}C_5Me_5)_2Zr(N_2)\}_2(N_2)]$	*76, 77*
$[NbCl_4(dmpe)_2]$ + Mg	THF	$[\{NbCl(dmpe)_2\}_2N_2]$	*75*
$CrCl_3$ + Mg + Me_3P	THF	*cis*-$[Cr(N_2)_2(Me_3P)_4]$	*78*
$[MoCl_4(PhMe_2P)_2]$ + Na + $PhMe_2P$	toluene	*cis*-$[Mo(N_2)_2(PhMe_2P)_4]$	*79*
$[WCl_4(PhMe_2P)_3]$ + Mg + $PhMe_2P$	THF	*cis*-$[W(N_2)_2(PhMe_2P)_4]$	*80*
$[RuHCl(Ph_3P)_3]$ + Et_3Al	ether	$[RuH_2(N_2)(Ph_3P)_3]$	*81*
$[Co(acac)_3]$ + i-Bu_3Al + Ph_3P	ether	$[CoH(N_2)(Ph_3P)_3]$	*64*
$RhCl_3 \cdot 3H_2O$ + Na/Hg + t-Bu_2PhP	THF	$[RhH(N_2)(t\text{-}Bu_2PhP)_2]$	*82*
$[NiBr_2(Cy_3P)_2]$ + Na	toluene	$[\{(Cy_3P)_2Ni\}_2N_2]$	*83*

dmpe = $Me_2PCH_2CH_2PMe_2$; Cy = cyclohexyl.

By reacting N_2 with the systems $[Cp_2TiCl_2]([Cp_2TiCl]) + i\text{-}PrMgCl$ in ether at low temperatures (−70° to −100°C) Shilov *et al.*[17,54,84–86] were able to observe the formation of a dark blue dinitrogen complex of the composition $[\{Cp_2Ti\text{–}i\text{-}Pr\}_2(N_2)]$, apparently an intermediate in N_2 reduction by these systems. The complex is unstable ($\Delta H = -5$ kcal mol^{-1}) and according to its IR spectra has a centrosymmetric bridge structure, probably

$$[Cp_2\overset{\diagup R}{Ti}\text{–}N{\equiv}N\text{–}\underset{R\diagup}{Ti}Cp_2]$$

(R = i-Pr). On addition of HCl, the complex liberates N_2, but with an excess of i-PrMgCl at −60°C it transforms into a product that is hydrolysed to hydrazine.[17,54,55,87] If the reduction is carried out at higher temperatures, ammonia is formed.

Similar, but more stable complexes $[\{Cp_2TiAr\}_2(N_2)]$ were obtained by Teuben *et al.*[88–91] by interaction of N_2 with $[Cp_2TiAr]$ in toluene at −78°C. When treated with sodium naphthalide, i-PrMgCl or n-BuLi followed by hydrolysis, the dinitrogen ligand in $[\{Cp_2TiAr\}_2(N_2)]$ is reduced to hydrazine and/or ammonia.[39,90–93]

The aforementioned data give grounds to assume that the active species in the systems $[Cp_2TiCl_2]$ + RMgX are the monoalkyltitanocenes in the form of the dimers $[Cp_2TiR]_2$ or the monomers $[Cp_2TiR]$. It should be noted, however, that $[Cp_2TiR]$ and $[Cp_2TiR]_2$ are probably not the only compounds that can activate dinitrogen in these systems. In fact, according to Maskill and Pratt[67,68] the activity of the system $[Cp_2TiCl_2]$ + EtMgBr in ether is lost only to a small extent when kept under argon at 20°C for 6·5 h. It is hard to conceive that under such conditions $[Cp_2TiEt]_2$ or $[Cp_2TiEt]$ would be preserved in the solution in any significant amounts after being held for so long. Evidently some of the more reduced products of $[Cp_2TiCl_2]$ produced by excess Grignard reagent are also able to activate dinitrogen, inducing its reduction.

On treating N_2 with titanocene $[Cp_2Ti]_2$ in toluene at −80°C Brintzinger *et al.*[35] isolated the dinitrogen complex $[\{Cp_2Ti\}_2(N_2)]$ that could form as an intermediate in N_2 fixation by systems based on $[Cp_2TiCl_2]$ and metals or naphthalides as reducing agents. The complex is also blue-coloured and judging from X-ray data[94] for the related decamethyl analogue $[\{(\eta^5\text{-}C_5Me_5)_2Ti\}_2(N_2)]$ probably has the end-on bridge structure $Cp_2Ti\text{–}N{\equiv}N\text{–}TiCp_2$. As in the case of $[\{Cp_2TiR\}_2(N_2)]$ treatment of the complex with HCl liberates N_2. With sodium naphthalide in THF followed by hydrolysis it gives ammonia. Apparently this same complex was observed in the work described in reference *26*.

According to Van Tamelen *et al.*[14,51] on prolonged (up to three weeks) contact with N_2 the so-called hydride isomer of titanocene $[(\eta^5$-

$C_5H_5)_2(C_{10}H_8)Ti_2H_2]$[95] also forms a dinitrogen complex formulated as $[C_{10}H_{10}TiN_2]_2$, wherein the N_2 ligand can be reduced to ammonia by sodium naphthalide. However the part played by this compound in N_2 fixation by systems based on $[Cp_2TiCl_2]$ should be insignificant because of the very low reaction rate of the hydride isomer of titanocene with N_2.

Stages (3)–(5)

As the result of the formation of a complex, a molecule of N_2 is activated and then reduced. The reaction proceeds in at least three stages. First there is formed a diazene intermediate, then a hydrazine intermediate and finally a nitride end-product which yields ammonia on hydrolysis. Products corresponding to all of these N_2 reduction stages have been observed.

The diazene level of N_2 reduction was approached by a complex isolated by Shilov *et al.* on reacting N_2 with a mixture of $[Cp_2TiCl]$ and MeMgI in ether.[17,60,61,96] At $-100°C$, in this system an unstable blue-coloured complex, evidently $[\{Cp_2TiMe\}_2(N_2)]$, is formed which, on raising the temperature to $-70°C$, gives a dark paramagnetic product formulated as $[Cp_2Ti]_2N_2$. The isolated compound differs in its properties from the blue complex of the same composition isolated by Brintzinger (see above) and on the basis of IR and magnetic data it is ascribed a structure close to $Cp_2Ti{-}N{=}N{-}TiCp_2$, i.e. the complex is regarded as a diazene derivative. Indeed, on treating the complex with methanolic HCl at $-60°C$ it yields equimolar quantities of N_2 and N_2H_4 (due to disproportionation of the liberated N_2H_2) and with HCl in ether it gives N_2 and NH_3.

The reaction of N_2 with $[Cp_2TiCl_2]$ and i-PrMgCl in ether at $-60°C$ leads to the formation of another paramagnetic complex formulated as $[(Cp_2Ti)_2N_2MgCl]$ where the dinitrogen is in a still more reduced state, approaching a hydrazine level of N_2 reduction.[87] On hydrolysis (HCl/EtOH; $-60°C$) hydrazine is formed in a yield $>80\%$. Finally, on treating $[(Cp_2Ti)_2N_2MgCl]$ with an excess of i-C_3H_7MgCl in ether at 20°C the nitride complex of composition $[Cp_2TiN(MgCl)_2]$ can be isolated; its hydrolysis gives ammonia.[17]

On the basis of these data the following mechanism of N_2 fixation by the systems $[Cp_2TiCl_2]([Cp_2TiCl]) + RMgX$ has been suggested:[17,87]

$$[Cp_2TiCl_2]\ (\text{or } [Cp_2TiCl]) \xrightarrow{RMgX} Cp_2TiR \xrightarrow{N_2} [Cp_2TiR]_2N_2 \searrow$$

$$[Cp_2TiN(MgX)_2] \xleftarrow{RMgX} [(Cp_2Ti)_2N_2MgX] \xleftarrow{RMgX} [Cp_2Ti]_2N_2$$

$$\downarrow H^+ \qquad\qquad \downarrow H^+ \qquad\qquad \downarrow H^+$$

$$NH_3 \qquad\qquad N_2H_4 \qquad\qquad N_2H_2$$

Stepwise transformation of the dinitrogen ligand was also observed by Van der Weij and Teuben[39,93] in the reaction of $[\{Cp_2TiAr\}_2(N_2)]$ with sodium naphthalide in THF. Although the intermediate complexes were not isolated, the results obtained are in good agreement with the mechanism involving the diazene and hydrazine stages of N_2 reduction (see also Chapter 9):

$$\text{(1)}\quad Cp_2Ti(Ar)\cdots N{\equiv}N\cdots Ti(Ar)Cp_2 + 2C_{10}H_8^{\dot{-}}Na^+ \longrightarrow (Cp)(Ar)Ti{-}N{=}N{-}Ti(Ar)(Cp) + 2C_{10}H_8 + 2CpNa$$

$$\text{(2)}\quad (Cp)(Ar)Ti{-}N{=}N{-}Ti(Ar)(Cp) + 2C_{10}H_8^{\dot{-}}Na^+ \longrightarrow \left[(Cp)(Ar)Ti{-}\bar{N}{-}\bar{N}{-}Ti(Ar)(Cp)\right]^{2-} 2Na^+ + 2C_{10}H_8$$

$$\text{(3)}\quad \left[(Cp)(Ar)Ti{-}\bar{N}{-}\bar{N}{-}Ti(Ar)(Cp)\right]^{2-} 2Na^+ \longrightarrow 2CpTi(Ar){=}NNa$$

It should be noted that the reaction of sodium naphthalide with $[\{Cp_2TiAr\}_2(N_2)]$ does not affect the Ti—Ar bonds, whereas two of the four Cp rings are eliminated. According to Teuben *et al.* fixation of N_2 by Cp_2TiCl_2 + RMgX is also accompanied by elimination of Cp-rings:[21,39]

$$[Cp_2TiCl_2] \xrightarrow{RMgX} Cp_2TiCl \xrightarrow{RMgX} Cp_2TiR \xrightarrow{N_2} [Cp_2TiR]_2N_2 \xrightarrow{RMgX} [CpTiR]_2N_2$$

$$[CpTiR]_2N_2 \xrightarrow{RMgX} [(CpTiR)_2N_2(MgX)_2] \longrightarrow [CpTiR(NMgX)]$$

$$[CpTiR]_2N_2 \xrightarrow{H+} N_2H_2; \quad [(CpTiR)_2N_2(MgX)_2] \xrightarrow{H+} N_2H_4; \quad [CpTiR(NMgX)] \xrightarrow{H+} NH_3$$

The diazene step of N_2 reduction was also observed in fixation of dinitrogen by the system $TiCl_4$ + lithium naphthalide in THF.[37,38] Here the reaction mechanism was studied by ^{15}N-isotope effects determined by means of precision mass spectrometry.

A dinitrogen reduction level intermediate between that of a diazene and a hydrazine is evidently realized in the binuclear nickel complex $[\{(PhLi)_3Ni\}_2N_2 \cdot 2Et_2O]_2$ ($r_{NN} = 1{\cdot}35$ Å) formed in the reaction of N_2 with the system Ni(CDT) + PhLi in ether at 0°C (CDT = *trans*-1,5,9-cyclodecatriene).[97,98] According to X-ray data the N_2-fragment in the complex is side-on coordinated to the nickel atoms:

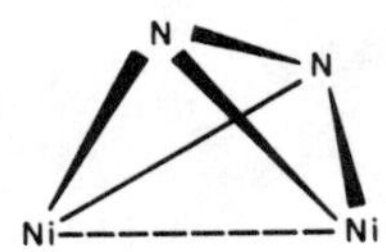

This structure shows that side-on dinitrogen complexes could be intermediates in N_2 reduction by these systems.

Recently the isolation of a paramagnetic zirconium dinitrogen complex $[Cp_2Zr(N_2)R][R = (Me_3Si)_2CH]$ from the reaction of N_2 with $[Cp_2Zr(Cl)R]$ and Na/Hg in THF at 20°C was reported.[99] It is assumed that in this, probably mononuclear, complex the N_2 ligand is also bound side-on to the metal. It is of interest that treatment of the complex with dry HCl liberates about 20% hydrazine together with traces of ammonia, an indication of considerable electron transfer from zirconium to the dinitrogen ligand. This also follows from ESR data.

In a number of cases the end products of N_2 reduction corresponding to complete rupture of the NN triple bond could be isolated. One such product—a nitride of the composition $[N(TiCl_2 \cdot 2AlCl_3)_3 \cdot C_6H_6]$—was obtained on heating dinitrogen with the complex $[C_6H_6 \cdot TiCl_2 \cdot 2AlCl_3]$[41–43] (Section II). By treating N_2 with the systems $[TiCl_3(THF)_3]$ + Mg and $[VCl_3(THF)_3]$ + Mg in THF, Yamamoto *et al.* isolated black diamagnetic nitrides formulated as $[MNMg_2Cl_2(THF)]$ (M = Ti, V). The nitride complex $[Cp_2TiN(MgCl)_2]$ has already been mentioned above. On hydrolysis all these compounds yield stoichiometric amounts of ammonia.

Stages (2′)–(5′)

The concluding stage of the reaction is hydrolysis of the products. When dinitrogen has been reduced to the nitride level this step is simple and requires no special commentary. As a rule the hydrolysis proceeds smoothly under the action of rather mild agents (alcohols, water, dilute acids), the final result usually being little dependent on which of these agents is taken as proton source for the hydrolysis.

The situation is more complicated when the products approach the diazene or hydrazine level of N_2 reduction. Hydrazine is readily decomposed by transition metal compounds, the rate and direction of the decomposition depending on the nature of the metal, the pH and other factors. To predict the behaviour of diazene is yet more difficult as, depending on the pH or on the character of the reaction media, it may decompose to a mixture of N_2 and H_2, N_2 and N_2H_4 and/or N_2 and NH_3. Consequently hydrolysis here becomes very sensitive to the conditions under which it is carried out, and what one would have expected to be minor differences in the nature of the hydrolysing agent used often lead to dramatic changes in the composition of the resultant nitrogen-containing products.

From the above discussion it also follows that the formation of ammonia or hydrazine on hydrolysis cannot by itself serve as a test for the presence of the corresponding nitride or of the hydrazine precursor in the mixture. Hydrazine and ammonia can be produced by disproportionation of diazene, and in addition ammonia can result from decomposition of hydrazine. Finally, as Chatt *et al.* have shown,[75] both ammonia and hydrazine can in principle be formed by protonation of the N_2 ligand in typical dinitrogen complexes, i.e. when the multiplicity of the NN bond does not differ considerably from three (Part 3).

II. Catalytic Nitrogen-Fixing Systems

As has already been mentioned, all the systems discussed above are non-catalytic. The absence of a catalytic effect could in some cases be due to instability of the lower-valent transition metal compound responsible for N_2 activation. However, this is evidently not a crucial factor since the catalytic process also cannot be achieved with relatively stable nitrogen-fixing systems.

A more general reason for the non-catalytic character of these reactions is that the transition metal compound activating dinitrogen is irreversibly lost in the form of the end nitride. The high affinity of the metal for nitrogen (favourable for scission of the N_2 molecule) hinders regeneration of the active species and consequently makes the N_2 reduction process non-catalytic. Thus to accomplish the catalytic cycle conditions must be established for splitting the transition metal–nitrogen bonds in the nitride products.

These considerations form the basis of our work on the development of catalytic nitrogen-fixing systems in solution. The N_2 reactions were carried out in aprotic solvents, strong aprotic acids such as aluminium bromide being used for breaking the nitride bonds, while metallic aluminium served as reducing agent and titanium compounds as the catalysts.[41–43] Some of the results obtained are given in Table 3, from which one can immediately see the sharp difference between these systems and those treated in the previous section.

When dinitrogen is heated with a mixture of Al and $AlBr_3$ in the presence of $TiCl_4$, $TiBr_4$ or $Ti(OBu)_4$, the yields of ammonia formed by hydrolysis constantly increase with increase in the amount of Al and $AlBr_3$ and can be brought up to 200 moles and more per mole of titanium compound. In the absence of TiX_4, neither aluminium nor its mixture with $AlBr_3$ react with dinitrogen. Hence, titanium compounds are catalysts for the reduction of N_2 by aluminium.

Interestingly, the halides of other transition metals ($ZrCl_4$, VCl_3, $CrCl_3$, $MoCl_5$, WCl_6, $MnCl_2$, $FeCl_3$, $FeBr_3$, $CoCl_2$, $NiBr_2$) are practically inactive under these conditions.

Benzene was found to be the best solvent for this reaction. Ethers, giving stable complexes with aluminium bromide, sharply decrease the activity of the systems. The reaction can also be successfully carried out in the absence of solvent, i.e. in molten aluminium bromide (m.p. 97·5°C).

Table 3. Catalytic reduction of N_2 by aluminium and lithium–aluminium hydride[41–43]†

System	Molar ratio of reagents	Reaction time (h)	Ammonia yield	
			(mol/g-at. of Ti)	referred to reducing agent (%)
$TiCl_4 + Al + AlBr_3$	1:6:0	8	0·02	0·4
	1:6:2	8	1·3	23
	1:12:33	14	10·7	92
	1:150:200	30	95	64
	1:600:1000	30	286	48
$TiBr_4 + Al + AlBr_3$	1:116:370	45	74	64
$Ti(OBu)_4 + Al + AlBr_3$	1:150:250	36	105	70
$C_6H_6 \cdot TiCl_2 \cdot 2AlCl_3 + Al + AlBr_3$	1:150:250	30	115	77
$TiCl_4 + LiAlH_4 + AlBr_3$	1:4:44	14	3·7	74
	1:8:44	24	9·8	95
	1:16:90	14	17·8	85
	1:16:90	40	19·2	92
	1:16:120	40	19·8	94
	1:300:1500	40	125	31
$Ti(OBu)_4 + LiAlH_4 + AlBr_3$	1:16:90	40	17·6	84
$TiCl_4 + LiAlH_4 + AlBr_3 + AlCl_3$	1:300:1500:690	35	100	25

† $P_{N_2} \sim$ 100 atm; with $TiCl_4 + LiAlH_4 + AlBr_3 + AlCl_3$ the reaction was carried at 60–70°C; and in the other cases at 130°C.

Both the rate of the reaction and the maximum amount of fixed dinitrogen are strongly dependent on the amount of aluminium bromide. Complete conversion of the initial aluminium into nitrides can be achieved only with a sufficient excess of $AlBr_3$ over Al (~3:1). The necessity for large amounts of $AlBr_3$ results from its consumption in the rupture of the metal–nitride bonds. It should be noted, however, that in the absence of $AlBr_3$, the systems not only become non-catalytic, but in general cease to reduce dinitrogen. Evidently $AlBr_3$ is necessary also in other stages of the process.

The mechanism of the reaction apparently involves, first, reduction of Ti(IV) to Ti(II) which is the actual N_2 reduction catalyst. This is supported by the fact that the complex $[C_6H_6 \cdot TiCl_2 \cdot 2AlCl_3]$, formed under similar conditions on heating $TiCl_4$ with a mixture of Al and $AlCl_3$ in benzene,[100,101] is capable of reacting with N_2 to give a Ti(III) nitride of the composition $[N(TiCl_2 \cdot 2AlCl_3)_3 \cdot C_6H_6]$.[41–43] Hydrolysis of this compound yields a stoichiometric amount of ammonia. On the other hand the same complex $[C_6H_6 \cdot TiCl_2 \cdot 2AlCl_3]$ when added in small amounts to a mixture of aluminium and aluminium bromide gives rise to the catalytic reduction of N_2 (Table 3).[41–43] It is likely that complexes of the type $[C_6H_6 \cdot TiX_2 \cdot 2AlBr_3]$ act as N_2-reduction catalysts when the reaction is carried out in benzene. In the

$AlBr_3$ melt the catalyst function is evidently assumed by titanium (II) halide in the form of an aluminium bromide adduct: $TiX_2 \cdot 2AlBr_3$. The formation of such adducts on heating $TiCl_4$ (or $TiCl_3$) with Al in the presence of aluminium halide has been demonstrated in ref. *102*.

These results, when combined, lead to the following scheme for the reaction mechanism:† stage (1) formation of the catalyst; stage (2) complexing of dinitrogen; stage (3a) reduction of the coordinated N_2 molecule by Ti(II) and aluminium or (3b) only by Ti(II); stage (4) cleavage of the titanium–nitride bonds by $AlBr_3$ to give Ti(III) halide and the end nitride (yielding ammonia on hydrolysis) and stage (5) regeneration of the catalyst.

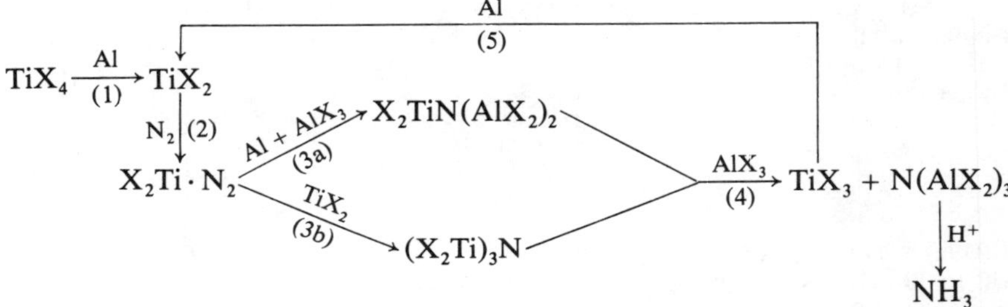

Reduction of dinitrogen at stages (3a) and (3b) proceeds stepwise as is substantiated by identification of hydrazine among the hydrolysis products when the reaction is carried out at lowered temperatures.

The overall process can be expressed by the following equation:

$$N_2 + 2Al + 4AlX_3 \xrightarrow{Ti(II)} 2[N(AlX_2)_3].$$

Effective catalytic systems can be obtained on substituting aluminium by $LiAlH_4$ as reducing agent[43] (Table 3).

Reduction of N_2 by lithium–aluminium hydride to ammonia (after hydrolysis) is also catalysed by titanium compounds. Here, too, a prerequisite for quantitative yields of ammonia (with reference to $LiAlH_4$) is the presence of a sufficient excess of $AlBr_3$ (up to 5–6 moles per mole of $LiAlH_4$). Interestingly, $LiAlH_4$ behaves in these reactions as donor of four electrons, i.e. the stoichiometric ratio $LiAlH_4 : N_2$ equals 3 : 2. In the absence of $AlBr_3$ the yields of NH_3 do not exceed 1 mol/Ti atom even in the presence of large amounts of lithium–aluminium hydride, and in the absence of titanium compounds, ammonia is not formed at all.

Temperatures above 100°C are necessary because below this temperature the aluminium bromide melt (which serves as the reaction medium) solidifies.

† In the scheme the reaction mechanism is illustrated by the case of N_2 activation in a mononuclear complex; for the sake of simplicity $AlBr_3$ is not shown in the products.

However if $AlBr_3$ is used in a eutectic mixture with $AlCl_3$ ($AlBr_3:AlCl_3 = 2{\cdot}17:1$ m.p. 67°C) the system will operate efficiently at 60–70°C giving up to 100 moles of NH_3 per mole of catalyst (Table 3), and with the ternary low melting mixture $AlBr_3$–$AlCl_3$–benzene the reaction can be carried out even at 30°C ($NH_3/Ti = 5$). A similar procedure can be employed for lowering the temperature of catalytic N_2 reduction by aluminium.

Van Tamelen *et al.* have described a catalytic system using electric current as reducing agent.[14,103] The system comprised Ti(O–i-Pr)$_4$, Al(O–i-Pr)$_3$ and naphthalene in 1,2-DME. When the N_2 reaction was carried out in the electrolytic cell with an aluminium anode and nichrome cathode, up to 3 moles of NH_3 per mole of titanium compound were formed in the course of 11 days (20°C). Probably in this reaction naphthalene plays the role of electron carrier and Al(O–i-Pr)$_3$, that of a Lewis acid. In the analogous, but non-electrochemical system Ti(O–i-Pr)$_4$ + Na + Al(O–i-Pr)$_3$ + naphthalene in 1,2-DME the yield of ammonia was 2·8 mole/Ti atom. It is assumed that the catalyst in these reactions is the alkoxy derivative of Ti(II).

The data presented show that in the presence of a sufficiently strong aprotic acid a number of systems acquire the ability to reduce dinitrogen catalytically. The efficiency of these reactions is determined by the nature of the transition metal compound, of the Lewis acid and by other factors. Under optimal conditions the yields of the N_2 reduction products can attain a value of hundreds of moles per mole of catalyst and the conversion of the reducing agent approaches 100%.

In further studies Shilov *et al.* were able to find catalytic nitrogen-fixing systems effective in protic media (Chapter 5). The catalysts here are molybdenum compounds and the reducing agents are Ti(III) hydroxide or sodium amalgam. One might suppose that the part played by the protons in these reactions is also that of splitting the metal–nitrogen bonds in the reduction products and that the basic features of the catalytic mechanism are in general the same for both types of systems.

One of the important problems in the present stage of development of this field is the accomplishment of low temperature catalytic reactions that can utilize molecular hydrogen for the reduction of N_2.

In the previous section it was mentioned that H_2 has an inhibiting effect on nitrogen fixation. First noted in reference *45*, it was subsequently observed for a large number of systems. However, this is not a general property of dihydrogen, and systems are now known where dinitrogen reduction is markedly augmented in the presence of H_2.[15,44,45] Thus in the case of the non-catalytic, homogeneous system $Ti(OEt)_4$ + i-Bu_3Al in toluene, the yield of ammonia increases by about twofold when the reaction with N_2 is carried out in the presence of dihydrogen ($N_2:H_2 = 3:1$).[45] A still more marked effect is exerted by dihydrogen on the systems $Ti(OBu)_4$ + i-Bu_3Al in toluene and n-heptane, and $Ti(OEt)_4$ + i-Bu_3Al in cyclohexane, where replacement of N_2 by a 5:1 N_2—H_2 mixture increases the yield of NH_3 by 3 to 5 times.[104] Evidently reduction of dinitrogen by the systems $Ti(OR)_4$ + i-Bu_3Al + H_2 (R = Et, Bu) proceeds at least partially with participation of molecular hydrogen, as is suggested by the detection of the N—D bonds in the final products when the reaction is carried out in the presence of D_2.[104]

It is of interest that although addition of $AlBr_3$ to the systems TiX_4 + i-Bu_3Al (X = Cl, BuO) does not make them catalytic even at 130°C, they become so when dihydrogen is introduced in addition to $AlBr_3$ (NH_3 yields up to 10–15 moles per mole of TiX_4).[43] Experiments on nitrogen fixation from a N_2—D_2 atmosphere here again reveal the presence of N—D bonds in the end products. Consequently in the catalytic reaction dihydrogen also participates in the reduction of N_2.

In recent years it has been shown that the systems discussed in this chapter under certain conditions can catalyse gaseous phase ammonia synthesis from N_2 and H_2.[105–118] Systems using the phthalocyanines, acetylacetonates and halides of Fe, Ru, Os, Mo, Co and Ti,[105,107,118] laminar compounds of graphite with Fe, Os, $FeCl_3$, $RuCl_3$, $OsCl_3$ and $MoCl_5$,[106,115–117] and metallic Ru, Os, Fe, Co, Re and Mo on various supports,[108–114] have been reported. Alkali metals and their naphthalides and also organolithium and organomagnesium compounds serve as reducing agents. Hydrogenation of dinitrogen usually proceeds at temperatures above 200°C; however, some of these systems (for instance, Ru/C + K[108,109] and iron phthalocyanine + Na[105,107]) are active even at 100–150°C. Investigation of the mechanism of catalysis leads to the conclusion that ammonia synthesis occurs here on negatively charged atoms of the transition metal stabilized by interaction with the conjugated ligand or with the support. In the reaction of N_2 with one of these catalytic systems (Ru/Al_2O_3 + K) at 170–350°C the formation of a surface dinitrogen complex was observed (ν_{NN} = 2020 cm^{-1})[114] and on heating N_2 with Ru and K at 250–450°C the product $[KRuN_2]_n$ was obtained, which hydrolyses to form 10% hydrazine.[119] These studies are laying a bridge between the homogeneous and heterogeneous reactions of N_2.

III. Nitrogen Fixation with the Formation of Nitrogen-containing Organic Compounds

The reactions of dinitrogen to give products with N—C bonds form yet another promising direction which arose in studying N_2 fixation in aprotic media. In prospect, such conversions could open a fundamentally new way to the synthesis of nitrogen-containing organic compounds—directly from molecular nitrogen.

The first example of such reaction was observed in 1966, when it was found that in the reaction of N_2 with $[Cp_2TiCl_2]$ (or $[Cp_2TiPh_2]$) and excess PhLi in ether the products after hydrolysis contain noticeable amounts of aniline along with ammonia.[120] The formation of aniline in this reaction proceeds even at room temperature and atmospheric pressure ($PhNH_2/Ti = 0{\cdot}03$; $NH_3/Ti = 0{\cdot}17$). An increase in the N_2 pressure markedly enhances the yields of both aniline and ammonia ($PhNH_2/Ti = 0{\cdot}10$–$0{\cdot}15$; $NH_3/Ti = 0{\cdot}65$; $P_{N_2} =$ 80–100 atm.). Subsequently it was shown that approximately the same amounts of aniline are obtained with other titanium compounds: $[CpTiCl_3]$, $Ti(OBu)_4$, $TiCl_4$.[121] Chlorides of V(IV), Cr(III), Mo(V), and W(VI) are poorly effective in the formation of aniline although ammonia can be produced here in rather large quantities (e.g. up to 0·4–0·5 moles in the case of VCl_4). The halides of Zr(IV), Mn(II), Fe(III), Co(II) and Ni(II) are practically inactive in the formation of either aniline or ammonia.

The reaction of dinitrogen with $[Cp_2TiCl_2]$ and *p*-tolyllithium in ether affords *p*-toluidine and only traces of the *m*-isomer. Under the same conditions *m*-tolyllithium gives *m*-toluidine and traces of the *p*-isomer.[121] Hence dinotrogen adds almost exclusively to that carbon atom which is bound to the metal in the original organolithium compound. The picture differs with *o*-substituted lithium aryls, where rearrangements take place resulting in a mixture of isomeric amines. Thus, from *o*-tolyllithium and N_2 a mixture of *o*- and *m*-toluidines with a small amount of *p*-isomer is formed; *o*-lithiumdiphenyl gives *o*-, *m*- and *p*-aminodiphenyls, and α-naphthyllithium gives α- and β-naphthylamines. The mechanism of these rearrangements is discussed in references *15* and *121*.

The replacement of ArLi in these systems by ferrocenyllithium leads to the formation of ferrocenylamine.[122] However, with aliphatic organolithium and organomagnesium compounds amines are either not formed at all or are formed only in trace amounts.[121] Arylmagnesium halides in combination with $[Cp_2TiAr_2]$, $[Cp_2TiCl_2]$, $TiCl_4$ or $Ti(OBu)_4$ in ether yield small amounts of aromatic amines (not more than 0·01 mole) but only on heating (80–100°C).

Marked amine-forming capacity is also displayed by systems formed on treating diaryltitanocenes, $[Cp_2TiAr_2]$, with some alkali and alkaline earth metals such as Li, Na and Mg.[123] The reactions proceed at room temperature in ethereal solutions (THF, diethyl ether, etc.) and give rise to aniline from $[Cp_2TiPh_2]$, *p*-toluidine from $[Cp_2Ti(p\text{-}MeC_6H_4)_2]$, and *m*-toluidine from $[Cp_2Ti(m\text{-}MeC_6H_4)_2]$. Besides the amines quite large amounts of ammonia are also obtained. The highest activity in amine production is displayed by the system $[Cp_2TiPh_2]$ + Li in THF where 0·10–0·15 moles of aniline (per mole of $[Cp_2TiPh_2]$) are formed with ~1·1 moles of ammonia (20°C, $P_{N_2} = 100$ atm.). Change in the dinitrogen pressure affects the yields of $PhNH_2$ and NH_3 differently. On lowering the N_2 pressure to atmospheric the yield of aniline falls sharply (down to 0·01–0·03 moles), whereas the amount of ammonia remains constant or even increases somewhat. Raising the temperature similarly affects the yield of $PhNH_2$.

The formation of amines in these reactions possibly proceeds via insertion of dinitrogen into the Ti–Ar bond of some lower-valent titanium derivative arising during reduction of the initial titanium compound by the excess ArLi, Li, etc.[120,121]

$$L_mTi{-}Ar + N_2 \longrightarrow L_mTi{<}^{Ar}_{N_2} \longrightarrow L_mTi{-}N{=}N{-}Ar \xrightarrow[(2)\ H^+]{(1)\ [e]} ArNH_2 + NH_3$$

or

$$2L_mTi{-}Ar + N_2 \longrightarrow L_m\overset{Ar}{Ti}\cdot N_2 \cdot \underset{Ar}{Ti}L_m \longrightarrow L_mTi{-}\overset{Ar}{N}{-}\underset{Ar}{N}{-}TiL_m \xrightarrow[(2)\ H^+]{(1)\ [e]} 2ArNH_2$$

The resultant arylazo or arylhydrazo derivative then undergoes reductive splitting.

Another possible mechanism proposed earlier for explaining the formation of aniline in the systems $[Cp_2TiCl_2]([Cp_2TiPh_2])$ + PhLi involves direct nucleophilic attack on the N_2-ligand in the intermediate dinitrogen complex by an arylcarbanion (from ArLi) followed by reduction of the ArN=N group bound to the titanium.[120] The possibility of such nucleophilic attack has been demonstrated recently by Sellmann *et al.*[124,125] for the manganese dinitrogen complex $[CpMn(CO)_2(N_2)]$:

$$[Cp(CO)_2Mn{-}N{\equiv}N] \xrightarrow[-30^\circ]{RLi} [Cp(CO)_2Mn{-}\underset{R}{N}{=}N]^-Li^+$$

$$[Cp(CO)_2Mn{-}\underset{R}{N}{=}N]^-Li^+ \xrightarrow[0^\circ]{H^+} [Cp(CO)_2Mn{-}\underset{R}{N}{=}NH] \quad R = Ph$$

$$[Cp(CO)_2Mn{-}\underset{R}{N}{=}N]^-Li^+ \xrightarrow[-30^\circ]{R_3O^+} [Cp(CO)_2Mn{-}\underset{R}{N}{=}\overset{R}{N}] \quad R = Me$$

Formation of amines was also observed on reaction of N_2 with σ-aryl titanium derivatives alone, that is in the absence of a specially added reducing agent.[126–129]

The thermal decomposition of $[Cp_2TiPh_2]$ in ether or in aromatic hydrocarbons at 80–100°C under pressure of N_2 gives products which liberate ammonia and small amounts of aniline on hydrolysis.[44,126–128] It turned out that in this reaction aniline formation proceeds differently from that in the above reactions of N_2 with the systems $[CpTiPh_2]([Cp_2TiCl_2])$ + PhLi and $[Cp_2TiPh_2]$ + Li. Investigation of the behaviour of the isomeric ditolyl-titanocenes, $[Cp_2Ti(p\text{-}MeC_6H_4)_2]$, $[Cp_2Ti(m\text{-}MeC_6H_4)_2]$ and $[Cp_2Ti(o\text{-}MeC_6H_4)_2]$, under conditions of thermolysis has shown that they all react with N_2 to form mixtures of isomeric toluidines: *p*- and *m*-toluidines (~35:65) from both $[Cp_2Ti(p\text{-}MeC_6H_4)_2]$ and $[Cp_2Ti(m\text{-}MeC_6H_4)_2]$, and *o*- and *m*-toluidines (46:54) from $[Cp_2Ti(o\text{-}MeC_6H_4)_2]$.[127–129]

The proposed mechanism involves a stage wherein N_2 reacts with a benzyne complex of titanocene (**A**) according to the scheme:

$$[Cp_2TiPh_2] \xrightarrow[-PhH]{80\text{–}100°C} [Cp_2Ti(C_6H_4)]\ (\mathbf{A}) \xrightarrow{N_2} [Cp_2Ti(C_6H_4)(N_2)] \longrightarrow [Cp_2Ti(C_6H_4N{=}N)]\ (\mathbf{B}) \xrightarrow[(2)\ H^+]{(1)\ [e]} PhNH_2 + NH_3$$

The formation of (**A**) during the thermolysis of $[Cp_2TiPh_2]$ has been shown by Dvorak *et al.*[130] This complex is capable of reacting with tolane and CO_2 to give cyclic products similar to the postulated (**B**).† [131,132]

Apparently dinitrogen reacts analogously with the system $[Cp_2TiCl_2]$ + Mg + *o*-fluorobromobenzene in THF, also giving rise to aniline and ammonia, but at room temperature.[128] The reaction of *o*-fluorobromobenzene with Mg, as is well known, generates benzyne[133] and the reaction of $[Cp_2TiCl_2]$ with Mg can yield Cp_2Ti. The conditions are thus at hand for the formation of the benzyne complex (**A**), and for its reaction with N_2. In this connection it is noteworthy that replacement of *o*-fluorobromobenzene with the corresponding *o*-dibromo and *o*-chlorobromo derivatives (which are less efficient sources of benzyne[133]) lowers the yield of $PhNH_2$. Of significance is also the fact that if the products from the reaction of N_2 with $[Cp_2TiCl_2]$, Mg and *o*-fluorobromobenzene are

† Reductive splitting of the N=N bond in (**B**) leading to the formation of aniline can be brought about by lower-valent titanium complexes forming at the final stages of $[Cp_2TiPh_2]$ decomposition.[136]

decomposed by D_2O, *o*-deuteroaniline is formed. Similar treatment in the case of $[Cp_2TiCl_2]$ + PhLi yields aniline not containing deuterium in the aromatic ring.

The conclusion that the systems $[Cp_2TiCl_2]([Cp_2TiPh_2])$ + PhLi and $[Cp_2TiCl_2]$ + Mg + *o*-FC_6H_4Br produce aniline by different mechanisms is in accord with the data on the ^{15}N isotopic effects as determined by precision mass spectrometry.[128] In the reactions of N_2 with $[Cp_2TiCl_2]$ + PhLi and $[Cp_2TiPh_2]$ + PhLi a kinetic ^{15}N isotope effect (similar for NH_3 and $PhNH_2$) is manifested, resulting in a lower ^{15}N content in both products than in the initial N_2. In the case of the "benzyne" system $[Cp_2TiCl_2]$ + Mg + *o*-FC_6H_4Br the ^{15}N isotope effects for $PhNH_2$ and NH_3 differ significantly from each other, aniline in contrast to ammonia being more enriched with ^{15}N than the initial N_2. Evidently aniline, and at least a major part of the ammonia, originate here from different N_2 complexes.

The possibility of synthesizing amines directly from molecular nitrogen and aromatic hydrocarbons has also been investigated. On reacting N_2 at room temperature with a mixture of $TiCl_4$, naphthalene and excess lithium (or sodium) in THF in addition to ammonia, the following amines were found: α-naphthylamine, 5,8-dihydro-α-naphthylamine, 5,6,7,8-tetrahydro-α-naphthylamine and β-naphthylamine. Under the same conditions *o*- and *p*-aminodiphenyls were obtained from dinitrogen and diphenyl.[134,135]

The formation of small amounts of amines from benzene and toluene was observed in the aforementioned reactions of N_2 with diaryltitanocenes.[128] It appears that if the thermolysis of $[Cp_2Ti(p\text{-}MeC_6H_4)_2]$ is carried out in benzene under N_2 pressure, aniline, as well as *p*- and *m*-toluidine, is formed. Similarly among the reaction products of N_2 with $[Cp_2TiPh_2]$ in toluene, *p*- and *m*-toluidines (1 : 1) are detected along with aniline. Apparently the lower-valent titanium complexes, formed in the thermolysis of $[Cp_2TiAr_2]$,[136] split the C—H bonds of the aromatic solvents with the formation of σ-arylhydride derivatives of titanium. This is substantiated by the fact that $[Cp_2TiAr_2]$ (Ar = Ph, *p*-MeC_6H_4) and $[Cp_2TiMe_2]$ are able to induce isotopic exchange of aromatic hydrocarbons (benzene, toluene, anisole, fluorobenzene) with D_2 at 100–130°C.[137]

According to Shiina,[138] if the reaction of trimethylchlorosilane with lithium in THF is carried out in a N_2 atmosphere in the presence of certain transition metal halides, considerable amounts of tris-(trimethylsilyl)amine are formed. With $TiCl_4$, $[Cp_2TiCl_2]$, VCl_3, $MnCl_2$, $CoCl_2$ and $MoCl_5$, the yields of amine are from 0·8 to 1·2 moles per mole of initial halide, and in the case of $FeCl_3$ and $CrCl_3$ they even reach a value of 2·3 and 4·5 moles, respectively, i.e. the reaction becomes catalytic. Analogously, tris-(triethylsilyl)amine was obtained from Et_3SiCl. Although the mechanism of amine formation is obscure, one may assume that the catalytic effect results from the action of trialkylchlorosilane as an aprotic acid that splits the nitride bonds in the reaction products, thereby regenerating the catalyst.

Van Tamelen and Rudler[14,139] have investigated the possibility of utilizing nitrides formed in N_2 fixation in aprotic media for preparing nitrogenous organic compounds. The dinitrogen reduction products after reaction of N_2 with $[Cp_2TiCl_2]$ and excess magnesium (or sodium naphthalide) in THF were not hydrolysed but treated with excess diethylketone. This led to the slow formation of 3-pentyl- and di(3-pentyl)amines in a yield of 25–50%. Under the same conditions di-n-butylketone gave 5-nonyl- and di-(5-nonyl)amines; cyclohexanone gave cyclohexyl- and dicyclohexylamines; benzaldehyde gave benzylamine and small amount of dibenzylamine and benzoyl chloride gave benzonitrile.

Analogous reactions were subsequently explored by Dormond *et al.*[122] On reacting ethyl phenylglyoxylate with the N_2 fixation products they observed the formation of small amounts of ethylphenylalanine.

Sobota *et al.*[29] found that the nitrides formed in the reduction of N_2 by the systems $[Cp_2TiCl_2]$ + Mg and MCl_4 + Mg (M = Ti, V) in THF, readily react with CO_2 to give isocyanate complexes of the composition $[Cp_2Ti(NCO)]$ and $[(THF)_3Cl_2Mg_2OM(NCO)]$, respectively. On treating these complexes with acids, CO_2 and NH_3 are evolved, and with CH_3I methylisocyanate is obtained.

IV. Conclusion

Transition metal compounds are capable of inducing multifarious reactions of molecular nitrogen in aprotic media. The dinitrogen can be reduced at room temperature to the level of ammonia, it can form hydrazine, and it can also react with organometallic compounds to form amines. Evidently all these reactions proceed through a stage of dinitrogen coordination with the transition metal compound.

Comparison of the chemical nitrogen-fixing systems with the enzymatic ones show that they have many features in common. In both, a transition metal is involved in the active centre: Mo or V in the enzymatic reaction; Ti, V, Mo and some other metals in the chemical reaction. Both types effect N_2 reduction at potentials which are as a rule more negative than those for H_2, although this is not required from a thermodynamic standpoint. There is also much in common in inhibitor action on these systems.

In both the enzymatic and the corresponding chemical reactions dinitrogen is ultimately reduced to ammonia. Naturally, when the reactions are carried out in aprotic media, hydrolysis is usually required for ammonia liberation. It is possible, however, that in the case of biological nitrogen fixation the complexation and reduction of N_2 also takes place in a peculiar aprotic medium created by the hydrophobic environment of the active centre. If so, then the processes of electron transfer to the activated dinitrogen and of proton transport to its reduction products can be time-divided in this case as they are in the nonenzymatic reactions of N_2 in aprotic media.

The achievements in the study of nitrogen fixation in solution and of model reactions of the individual dinitrogen complexes have served as a firm basis for the modern structure of the low-temperature chemistry of dinitrogen.

Acknowledgment

The authors are grateful to Dr. G. Peck for editing the English translation of this chapter.

References

1. M. E. Vol'pin and V. B. Shur, *Dokl. Akad. Nauk SSSR* (1964) **156**, 1102.
2. M. E. Vol'pin, V. B. Shur and M. A. Ilatovskaya, *Izvest. Akad. Nauk SSSR, Ser. Khim.* (1964) 1728.
3. M. E. Vol'pin and V. B. Shur, *Vestn. Akad. Nauk SSSR* (1965) No. 1, 51.
4. M. E. Vol'pin, V. B. Shur and L. P. Bichin, *Izvest. Akad. Nauk SSSR, Ser. Khim.* (1965) 720.
5. M. E. Vol'pin, A. A. Belyi and V. B. Shur, *Izvest. Akad. Nauk SSSR, Ser. Khim.* (1965) 2225.
6. M. E. Vol'pin and V. B. Shur, *Nature* (1966) **209**, 1236.
7. N. T. Denisov, V. F. Shuvalov, N. I. Shuvalova, A. K. Shilova and A. E. Shilov, *Kinet. Katal.* (1970) **11**, 813.
8. A. D. Allen and C. V. Senoff, *Chem. Commun.* (1965) 621.
9. M. E. Vol'pin and V. B. Shur, *Zh. Vses. Khim. Obshchestva im D. I. Mendeleeva* (1967) **12**, No. 1, 31.
10. R. Murray and D. C. Smith, *Coord. Chem. Rev.* (1968) **3**, 429.
11. G. Henrici-Olivé and S. Olivé, *Angew. Chem.* (*Internat. Edn.*) (1969) **8**, 650.
12. K. Kuchynka, *Catal. Rev.* (1969) **3**, 111.
13. W. Büttner, *Z. Chem.* (1969) **9**, 219.
14. E. E. van Tamelen, *Acc. Chem. Res.* (1970) **3**, 361.
15. M. E. Vol'pin and V. B. Shur. *In* "Organometallic Reactions" (Ed. E. Becker and M. Tsutsui), Vol. I., p. 55. Wiley, New York (1970).
16. B. Jeżowska-Trzebiatowska and P. Sobota, *Wiadomośći Chem.* (1972) **26**, 229.
17. A. E. Shilov, *Uspekhi Khim.* (1974) **43**, 863.
18. R. Juza, *Adv. Inorg. Chem. Radiochem.* (1966) **9**, 81.
19. M. E. Vol'pin, N. K. Chapovskaya and V. B. Shur, *Izvest. Akad.Nauk SSSR, Ser. Khim.* (1966) 1083.
20. H. H. Brintzinger, "Proc. 1st Internat. Symp. Nitrogen Fixation, Pullman, Washington, 1974", Vol. I, p. 33. Pullman, Washington (1976).
21. F. W. van der Weij, H. Scholtens and J. H. Teuben, *J. Organomet. Chem.* (1977) **127**, 299.
22. E. Bayer and V. Schurig, *Chem. Ber.* (1969) **102**, 3378.
23. A. Yamamoto, M. Ookawa and S. Ikeda, *Chem. Commun.* (1969) 841.

24. M. E. Vol'pin, A. A. Belyi, V. B. Shur, Yu. I. Lyakhovetsky, R. V. Kudryavtsev and N. N. Bubnov, *Dokl. Akad. Nauk SSSR* (1970) **194**, 577.
25. A. Yamamoto, S. Go, M. Ookawa, M. Takahashi, S. Ikeda and T. Ken, *Bull. Chem. Soc. Japan* (1972) **45**, 3110.
26. E. E. Van Tamelen, W. Cretney, N. Klaentschi and J. S. Miller, *Chem. Commun.* (1972) 481.
27. B. Jeżowska-Trzebiatowska, P. Sobota, H. Kozłowski and A. Jezierski, *Bull. Acad. Polon. Sci., Ser. Sci. Chim.* (1972) **2**, 194.
28. B. Jeżowska-Trzebiatowska and P. Sobota, *J. Organometallic Chem.* (1972) **46**, 339.
29. P. Sobota, B. Jeżowska-Trzebiatowska and Z. Janas, *J. Organometallic Chem.* (1976) **118**, 253.
30. P. Sobota and B. Jeżowska-Trzebiatowska, *J. Organometallic Chem.* (1977) **131**, 341.
31. C. Ungurenasu and S. Streba, *J. Inorg. Nucl. Chem.* (1972) **34**, 3753.
32. D. R. Gray and C. H. Brubaker, *Chem. Commun.* (1969) 1239.
33. L. G. Bell and H. H. Brintzinger, *J. Amer. Chem. Soc.* (1970) **92**, 4464.
34. T. P. M. Beelen and W. Van Erk, *Rec. Trav. Chim.* (1971) **90**, 1197.
35. J. E. Bercaw, R. H. Marvich, L. G. Bell and H. H. Brintzinger, *J. Amer. Chem. Soc.* (1972) **94**, 1219.
36. B. Lorenz, G. Möbius, S. Rummel and M. Wahren, *Z. Chem.* (1975) **15**, No. 6, 242.
37. S. Rummel, *Z. Chem.* (1976) **16**, No. 7, 288.
38. M. Wahren, B. Bayerl, B. Lorenz, G. Möbius, S. Rummel and K. Schmidt. *In* "Stable Isotopes in the Life Sciences", p. 281. IAEA-CMEA-Technical Committee Meeting, Leipzig 1977, IAEA, Vienna (1977).
39. F. W. Van der Weij, Dissertation, Rijksuniversiteit te Groningen (1977).
40. B. Akermark and M. Almemark, *Acta Chem. Scand.* (1975) **29A**, No. 1, 155.
41. M. E. Vol'pin, M. A. Ilatovskaya, L. V. Kosyakova and V. B. Shur, *Dokl. Akad. Nauk SSSR* (1968) **180**, 103.
42. M. E. Vol'pin, M. A. Ilatovskaya, L. V. Kosyakova and V. B. Shur, *Chem. Commun.* (1968) 1074.
43. M. E. Vol'pin, M. A. Ilatovskaya and V. B. Shur, *Kinet. Katal.* (1970) **11**, 333.
44. M. E. Vol'pin, V. B. Shur, V. N. Latyaeva, L. I. Vyshinskaya and L. A. Shulgaitser, *Izvest. Akad. Nauk SSSR, Ser. Khim* (1966) 385.
45. M. E. Vol'pin, M. A. Ilatovskaya, E. I. Larikov, M. L. Khidekel', Yu. A. Shvetsov and V. B. Shur, *Dokl. Akad. Nauk SSSR* (1965) **164**, 331.
46. A. Schindler, *J. Polymer Sci.* (1963) **C4**, 81.
47. E. E. van Tamelen and B. Akermark, *J. Amer. Chem. Soc.* (1968) **90**, 4492.
48. G. Henrici-Olivé and S. Olivé, *Angew. Chem.* (*Internat. Edn.*) (1967) **6**, 873.
49. G. Henrici-Olivé and S. Olivé, *Angew. Chem.* (*Internat. Edn.*) (1968) **7**, 386.
50. E. E. van Tamelen, G. Boche and R. G. Greeley, *J. Amer.Chem. Soc.* (1968) **90**, 1677.
51. E. E. van Tamelen, R. B. Fechter, S. W. Schneller, G. Boche, R. H. Greely and B. Akermark, *J. Amer. Chem. Soc.* (1969) **91**, 1551.
52. A. Streitwieser, Jr., "Molecular Orbital Theory for Organic Chemists". Wiley, New York and London (1962).
53. E. E. van Tamelen, R. B. Fechter and S. W. Schneller, *J. Amer. Chem. Soc.* (1969) **91**, 7196.
54. A. E. Shilov, A. K. Shilova and E. F. Kvashina, *Kinet. Katal.* (1969) **10**, 1402.

55. A. E. Shilov and A. K. Shilova, *Zhur. Fiz. Khim.* (1970) **44**, 288.
56. M. O. Broitman, N. T. Denisov, N. I. Shuvalova and A. E. Shilov, *Kinet. Katal.* (1971) **12**, 504.
57. Yu. G. Borod'ko, M. O. Broitman, L. M. Kachapina, A. E. Shilov and L. Yu. Ukhin, *Chem. Commun.* (1971) 1185.
58. M. O. Broitman, T. A. Vorontsova and A. E. Shilov, *Kinet. Katal.* (1972) **13**, 61.
59. B. Tchoubar, A. E. Shilov and A. K. Shilova, *Kinet. Katal.* (1975) **16**, 179.
60. Yu. G. Borod'ko, I. N. Ivleva, L. M. Kachapina, S. I. Salienko, A. K. Shilova and A. E. Shilov, *Chem. Commun.* (1972) 1178.
61. Yu. G. Borod'ko, I. N. Ivleva, S. I. Salienko, A. K. Shilova and A. E. Shilov, *Zhur. Strukt. Khim.* (1973) **14**, 1112.
62. R. W. F. Hardy, R. C. Burns and G. W. Parshall. *In* "Inorganic Biochemistry" (Ed. G. B. Eichhorn), Vol. 2. Elsevier, Amsterdam, Oxford, New York (1973).
63. M. L. Khidekel' and Yu. B. Grebenshchikov, *Izvest. Akad. Nauk SSSR, Ser. Khim.* (1965) 761.
64. A. Yamamoto, S. Kitazume, L. S. Pu and S. Ikeda, *J. Amer. Chem. Soc.* (1971) **93**, 371.
65. E. E. van Tamelen, H. Rudler and C. Bjorklund, *J. Amer. Chem. Soc.* (1971) **93**, 3526.
66. E. E. van Tamelen, D. Seeley, S. Schneller, H. Rudler and W. Cretney, *J. Amer. Chem. Soc.* (1970) **92**, 5251.
67. R. Maskill and J. M. Pratt, *Chem. Commun.* (1967) 950.
68. R. Maskill and J. M. Pratt, *J. Chem. Soc. (A)* (1968) 1914.
69. H. Brintzinger, *J. Amer. Chem. Soc.* (1967) **89**, 6871.
70. G. Kenworthy, J. Myatt and M. C. R. Symons, *J. Chem. Soc. (A)* (1971) 1020.
71. M. E. Vol'pin, A. A. Belyi, V. B. Shur, Yu. I. Lyakhovetsky, R. V. Kudryavtsev and N. N. Bubnov, *J. Organometallic Chem.* (1971) **27**, C 5.
72. G. Henrici-Olivé and S. Olivé, *J. Organometallic Chem.* (1967) **9**, 325.
73. G. Henrici-Olivé and S. Olivé, *J. Amer. Chem. Soc.* (1970) **92**, 4831.
74. G. N. Nechiporenko, G. M. Tabrina, A. K. Shilova and A. E. Shilov, *Dokl. Akad. Nauk SSSR* (1965) **164**, 1062.
75. J. Chatt, J. R. Dilworth and R. L. Richards, *Chem. Rev.* (1978) **78**, 589.
76. J. Manriquez and J. E. Bercaw, *J. Amer. Chem. Soc.* (1974) **96**, 6229.
77. J. Manriquez, R. D. Sanner, R. E. Marsh and J. E. Bercaw, *J. Amer. Chem. Soc.* (1976) **98**, 8351.
78. H. H. Karsch, *Angew. Chem. (Internat. Edn.)* (1977) **16**, 56.
79. M. Aresta and A. Sacco, *Gazz. Chim. Ital.* (1972) **102**, 755.
80. J. Chatt, G. A. Heath and R. L. Richards, *J. Chem. Soc., Dalton Trans.* (1974) 2074.
81. W. H. Knoth, *J. Amer. Chem. Soc.* (1972) **94**, 104.
82. P. R. Hofmann, T. Yoshida, T. Okano, S. Otsuka and J. A. Ibers, *Inorg. Chem.* (1976) **15**, 2462.
83. M. Aresta, C. F. Nobile and A. Sacco, *Inorg.Chim. Acta* (1975) **12**, 167.
84. A. E. Shilov, A. K. Shilova, E. F. Kvashina and T. A. Vorontsova, *Chem. Commun.* (1971) 1590.
85. Yu. G. Borod'ko, E. F. Kvashina, V. B. Panov and A. E. Shilov, *Kinet. Katal.* (1973) **14**, 255.
86. V. B. Panov, Yu. M. Shul'ga, E. F. Kvashina and Yu. G. Borod'ko, *Kinet. Katal.* (1974) **15**, 518.

87. Yu. G. Borod'ko, I. N. Ivleva, L. N. Kachapina, E. F. Kvashina, A. K. Shilova and A. E. Shilov, *Chem. Commun.* (1973) 169.
88. J. H. Teuben and H. J. de Liefde Meijer, *Rec. Trav. Chim.* (1971) **90**, 360.
89. J. H. Teuben and H. J. de Liefde Meijer, *J. Organometallic Chem.* (1972) **46**, 313.
90. J. H. Teuben, *J. Organometallic Chem.* (1973) **57**, 159.
91. J. H. Teuben, Dissertation, Rijksuniversiteit te Groningen (1973).
92. F. W. van der Weij and J. H. Teuben, *J. Organometallic Chem.* (1976) **105**, 203.
93. F. W. van der Weij and J. H. Teuben, *J. Organometallic Chem.* (1976) **120**, 223.
94. R. D. Sanner, D. M. Duggan, T. C. McKenzie, R. E. Marsh and J. E. Bercaw, *J. Amer. Chem. Soc.* (1976) **98**, 8358.
95. H. H. Brintzinger and J. E. Bercaw, *J. Amer. Chem. Soc.* (1970) **92**, 6182.
96. I. N. Ivleva, A. K. Shilova, S. I. Salienko and Yu. G. Borod'ko, *Dokl. Akad. Nauk. SSSR* (1973) **213**, 116.
97. K. Jonas, *Angew. Chem.* (Internat. Edn.) (1973) **12**, 997.
98. C. Kruger and Y.-H. Tsay, *Angew. Chem.* (*Internat. Edn.*) (1973) **12**, 998.
99. M. J. S. Gynane, J. Jeffery and M. F. Lappert, *J. Chem. Soc. Chem. Commun.* (1978) 34.
100. G. Natta, G. Mazzanti and G. Pregaglia, *Gazz. Chim. Ital.* (1959) **89**, No. 9, 2065.
101. H. Martin and F. Vohwinkel, *Chem. Ber.* (1961) **94**, 2416.
102. J. Brynestad, S. von Winbush, H. L. Yakel and G. P. Smith, *Inorg. Nucl. Chem. Lett.* (1970) **6**, 889.
103. E. E. van Tamelen and D. A. Seeley, *J. Amer. Chem. Soc.* (1969) **91**, 5194.
104. Ya. A. Shvetsov, L. G. Korabliova, G. N. Nechiporenko, M. A. Ilatovskaya and M. L. Khidekel', *Dokl. Akad. Nauk SSSR* (1968) **178**, 400.
105. M. Sudo, M. Ichikawa, M. Soma, T. Onishi and K. Tamaru, *J. Phys. Chem.* (1969) **73**, 1174.
106. M. Ichikawa, T. Kondo, K. Kawase, M. Sudo, T. Onishi and K. Tamaru, *Chem. Commun.* (1972) 176.
107. K. Tamaru, *Catalysis Rev.* (1970) **4**, 161.
108. A. Ozaki, K. Aika and H. Hori, *Bull. Chem. Soc. Japan* (1971) **44**, 3216.
109. K. Aika, H. Hori and A. Ozaki, *J. Catalysis* (1972) **27**, 424.
110. K. Urabe, K. Aika and A. Ozaki, *J. Catalysis* (1974) **32**, 108.
111. K. Aika and A. Ozaki *J. Catalysis* (1974) **35**, 61.
112. K. Urabe, K. Aika and A. Ozaki, *J. Catalysis* (1975) **38**, 430.
113. M. Ishizuka, K. Aika and A. Ozaki, J. Catalysis (1975) **38**, 189.
114. M. Oh-Kita, K. Aika, K. Urabe and A. Ozaki, *Chem. Commun.* (1975) 147.
115. V. A. Postnikov, L. M. Dmitrienko, R. F. Ivanova, N. L. Dobrolubova, M. A. Golubeva, T. I. Gapeeva, Yu. N. Novikov, V. B. Shur and M. E. Vol'pin, *Izv. Akad. Nauk SSSR* (1975) 2642.
116. A. V. Nefed'ev, R. A. Stukan, V. A. Postnikov, V. B. Shur, Yu. N. Novikov and M. E. Vol'pin, *Izv. Akad. Nauk SSSR* (1976) 2376.
117. M. E. Vol'pin, Yu. N. Novikov, V. A. Postnikov, V. B. Shur, B. Bayerl, L. Kaden, M. Wahren, L. M. Dmitrienko, R. A. Stukan and A. V. Nefed'ev, *Z. Anorg. Allg. Chem.* (1977) **428**, 231.
118. V. V. Karpov and G. I. Kozub, *Kinet. Katal.* (1976) **17**, 1334.
119. A. Ohya, K. Urabe and A. Ozaki, *Chem. Lett.* (1978) **3**, 233.
120. M. E. Vol'pin and V. B. Shur, *Izvest. Akad. Nauk SSSR*, *Ser. Khim.* (1966) 1873.

121. M. E. Vol'pin, V. B. Shur, R. V. Kudryavtsev and L. A. Prodayko, *Chem. Commun.* (1968) 1038.
122. A. Dormond, J. C. Leblanc, F. Le Moigne and J. Tirouflet, *C.R. Acad. Sci. Paris, Ser. C* (1972) **274**, 1707.
123. V. B. Shur, E. G. Berkovitch, S. M. Yunusov, Sh. M. Mamedov, I. I. Nizker and M. E. Vol'pin, *Izvest. Akad. Nauk SSSR Ser. Khim.* (1977) 2841.
124. D. Sellmann and W. Weiss, *Angew. Chem.* (*Internat. Edn.*) (1977) **16**, 880.
125. D. Sellman and W. Weiss, *Angew. Chem.* (*Internat. Edn.*) (1978) **17**, 269.
126. V. B. Shur, E. G. Berkovitch and M. E. Vol'pin, *Izvest. Akad. Nauk SSSR, Ser. Khim.* (1971) 2358.
127. V. B. Shur, E. G. Berkovitch, L. B. Vasiljeva, R. V. Kudryavtsev and M. E. Vol'pin, *J. Organometallic Chem.* (1974) **78**, 127.
128. E. G. Berkovitch, V. B. Shur, M. E. Vol'pin, B. Lorenz, S. Rummel and M. Wahren, *Chem. Ber.*, (1980) **113**, 70.
129. M. Kh. Grigorjan, I. S. Kolomnikov, E. G. Berkovitch, T. V. Lysjak, V. B. Shur and M. E. Vol'pin, *Izvest. Akad. Nauk SSSR, Ser. Khim.* (1978) 1177.
130. J. Dvorak, R. J. O'Brien and W. Santo, *Chem. Commun.* (1970) 411.
131. H. Masai, K. Sonogashira and N. Nagihara, *Bull. Chem. Soc. Japan.* (1968) **41**, 750.
132. I. S. Kolomnikov, T. S. Lobeeva, V. V. Gorbachevskaja, G. G. Alexandrov, Yu. T. Struchkov and M. E. Vol'pin, *Chem. Commun.* (1971) 972.
133. G. Wittig, *Angew. Chem.* (1957) **69**, 245.
134. M. E. Vol'pin, A. A. Belyj, N. A. Katkov and V. B. Shur, *Izvest. Akad. Nauk SSSR, Ser. Khim.* (1969) 2858.
135. M. E. Vol'pin, A. A. Belyj, V. B. Shur, N. A. Katkov, I. M. Nekaeva and R. V. Kudryavtsev, *Chem. Commun.* (1971) 246.
136. C. P. Boekel, J. H. Teuben and H. J. De Liefde Meijer, *J. Organometallic Chem.* (1974) **81**, 371.
137. V. B. Shur, R. G. Berkovitch, E. I. Mysov and M. E. Vol'pin, *Izvest. Akad. Nauk SSSR, Ser. Khim.* (1975) 1908.
138. K. Shiina, *J. Amer. Chem. Soc.* (1972) **94**, 9267.
139. E. E. van Tamelen and H. Rudler, *J. Am. Chem. Soc.* (1970) **92**, 5253.

Part 2

Nitrogen Reduction and Fixation in Protic Media

4

Studies of the Mechanism of Biological Nitrogen Fixation with Functional Model Systems

G. N. SCHRAUZER

University of California at San Diego, California, USA

THE reduction of molecular nitrogen in a protic solvent poses the major problem of finding a powerful reducing agent that will react with traces of dissolved nitrogen in preference to the protons of the medium. Although a number of metals are known to combine with dinitrogen at remarkably low temperatures, those that form salt-like nitrides that can be hydrolysed to yield ammonia are also strong nucleophiles and react with protic solvents violently to produce dihydrogen, irrespective of whether the reaction is conducted in the presence or absence of dinitrogen. The fact that close to 2×10^8 tons of nitrogen are converted to ammonia annually near the surface of the earth by soil microorganisms indicates that there exist biogenic catalysts for nitrogen reduction that operate in protic environments. Therefore, the desire to understand biological nitrogen fixation and to duplicate the natural process with the enzyme *in vitro*, or preferably, with artificial catalysts, has intrigued chemists throughout this century. By the mid-Sixties, nitrogenase, the enzyme responsible for the natural process of nitrogen fixation, had been isolated and methods for studying the enzymatic process in the laboratory had been found. However, in spite of extensive efforts, the chemical mechanism of biological nitrogen fixation remained obscure. Only during the past decade has progress been made toward the understanding of the mechanism of biological nitrogen fixation, and then only after the substrate reactions of the enzyme had been duplicated under non-enzymatic conditions with artificial functional model systems. In addition, a number of abiological chemical systems were discovered with which the reduction of dinitrogen to ammonia and hydrazine could be demonstrated in protic media under mild conditions. It can now be stated with assurance that the reduction of dinitrogen in aqueous solution at ambient temperature and atmospheric pressure is no longer the exclusive domain of certain microorganisms and that the biological process of nitrogen fixation is now better understood than many other enzymatic reactions. The present account focuses on these newer developments. For historical and other reasons it is indispensable to describe the biological process prior to the discussion of the mechanism of the reaction. This will be done only briefly and with special emphasis on the chemistry rather than the biology of the system.

I. Biological Nitrogen Fixation: Properties and Reactions of Nitrogenase

A. The dinitrogen-reducing enzymes

Cell extracts from various nitrogen-fixing organisms have been shown to contain two proteins, a MoFe-protein and an Fe-protein, which, if combined in the presence of cofactors and a reductant, reduce molecular nitrogen to ammonia. Both proteins have been obtained in crystalline form and are chemically well characterized.[1] The MoFe-proteins from several bacterial sources have molecular weights of 210 000–270 000 and all contain iron, molybdenum and labile sulphide, but their compositions are somewhat variable and dependent on their source: MoFe-protein from *Azotobacter vinelandii*, for example, contains, Mo, Fe and S^{2-} in the atomic ratios of 2:32:25; *Clostridium pasteurianum* MoFe-protein contains Mo, Fe and S^{-2} in the ratios of 2:11–12:10–15; MoFe-protein from *Klebsiella pneumoniae* contains Mo and Fe in the ratios of $2:33 \pm 3$. This protein was shown[2] to be tetrameric, consisting of two each of two types of subunits of molecular weights of 50 000 and 66 000. It may be suspected that the differences in composition are a consequence of the lability of the non-haeme iron components of the protein and due to partial decomposition during the isolation. The nitrogenase Fe-proteins from different nitrogen fixing-organisms are typical ferredoxins, varying in molecular weight between 40 000 and 60 000 and containing one Fe_4S_4 cluster per molecule.[1]

Optimal nitrogen reducing activity of mixtures of MoFe- and Fe-protein is observed at approximately equimolar ratios. The *in vitro* assays are conducted in pH 6–8 buffered solution, usually with $Na_2S_2O_4$, occasionally with KBH_4 as the reductant. All substrate reactions of nitrogenase are ATP-dependent and require the presence of a bivalent cation, usually Mg^{2+}. For each electron pair transferred to substrate, 2–5 ATP are hydrolysed to ADP and inorganic phosphate. To prevent the build-up of inhibitory ADP, most *in vitro* experiments are conducted in the presence of a phosphatidokinase ATP generating system. It has been established recently that MoFe- and Fe-protein associate and dissociate during each turnover.[3] Accordingly, the suggestion has been made to name the MoFe-protein "nitrogenase" and the Fe-protein "nitrogenase reductase".[3] It is more convenient for our present purpose to retain the older nomenclature. Hence, if reference is made to nitrogenase, this is to designate the functional mixture of the MoFe- and Fe-protein.

B. Substrate reactions of nitrogenase

Nitrogenase exhibits a remarkable substrate non-specificity. The enzyme reduces not only dinitrogen but in addition numerous organic and inorganic unsaturated compounds. The reductions are invariably accompanied by protonations; in some cases C=N bonds are cleaved and new C–C bonds are formed. In the absence of substrates, ATP is still consumed and the electrons from the external reductant are used to discharge protons of the medium. The reduction of all substrates is inhibited by CO, the ATP-dependent evolution of H_2 is not.

C. The reduction of dinitrogen

The reduction of dinitrogen by nitrogenase is unusual as it is accompanied by a simultaneous evolution of dihydrogen even under normal field conditions. The electrons from the external reductant are partitioned only to 75% for the reduction of N_2, the remainder is used for dihydrogen production.[4,5] The stoichiometry of biological nitrogen fixation should accordingly be expressed in terms of equation (1):

$$N_2 + 8e + 8H^+ \rightarrow 2NH_3 + H_2 \quad (1)$$

and not by the customary equation (2):

$$N_2 + 6e + 6H^+ \rightarrow 2NH_3. \quad (2)$$

It is necessary to stress this point because the simultaneous dihydrogen evolution during biological nitrogen fixation is of mechanistic significance. That dihydrogen is evolved during dinitrogen reduction is in itself interesting since dihydrogen is also a competitive inhibitor of N_2 reduction. Moreover, dinitrogen is the only substrate of nitrogenase which stimulates the enzyme-mediated exchange reaction between gaseous D_2 and solvent protons (equation (*3*)):[5,6]

$$D_2 + H^+ \xrightarrow{N_2,\ \text{MgATP, reductant, nitrogenase}} DH + D^+. \quad (3)$$

The exchange reaction (viz. equation (3)) occurs even at very low partial pressures of N_2,[1,32] a fact which is again of significance for the mechanism of biological dinitrogen reduction. The K_m (Michaelis–Menten constant) for N_2 is 0·1 mM, but may vary between 0·03 and 0·12 mM, depending on the quality of the enzyme preparation. Turnover numbers of around 50 (moles of N_2 reduced per mole of Mo per minute) have been observed with the most active enzyme preparations *in vitro*, but much lower activities are often seen due to variations of the activity of the enzyme.

D. Early mechanistic hypotheses

All attempts to explain nitrogenase action prior to 1960 were highly speculative and will not be reviewed in any detail. Although H. Wieland suggested that the overall pathway of nitrogen fixation is reductive as early as in 1922, leading investigators of the period and as late as 1963 favoured mechanisms of fixation involving hydroxylamine as the intermediate. In almost all of the early proposals, iron was given a central role. And since the stoichiometry of N_2 reduction to NH_3 demands six electrons and six protons it was assumed that several iron atoms were present at the active site at which the binding and reduction of N_2 was thought to occur.[6]

A later version[7] of the mechanism suggested a bimetallic site containing iron and molybdenum. The iron atom was assumed to bind dinitrogen, while molybdenum was considered to reduce it. All mechanistic proposals which rely exclusively on the results derived from studies with the enzyme suffer from the "black box" effect and even the application of sophisticated methods of structural analysis cannot provide insights into the mechanism of a complex biochemical reaction if the basic chemical processes have never been investigated in simpler abiological systems.

II. The Mechanism of Biological Nitrogen Fixation as Deduced from Studies with Functional Model Systems

A. Simulation of nitrogenase reactions under non-enzymatic conditions

The limitations of the classical biochemical approaches as applied to nitrogenase became apparent in 1968. By this time extensive studies had been performed with the enzyme and with numerous substrates, but due to the lack of information on the chemistry of iron and molybdenum, chemically plausible mechanisms of the reactions of the enzyme could not be formulated. It also appeared at that time, to us at least, that a study of reactions of isolable dinitrogen complexes would not produce answers relevant to the mechanism of biological nitrogen fixation.

It was decided at this point to attack the problem by means of systematic model studies. The first aim of these studies was to simulate the substrate reactions of the enzyme with simpler catalysts. Since this was the first complicated enzyme system to be investigated in this fashion, care was taken to employ a consistent approach, in accord with established methods of model theory. The first models consisted of mixtures of a thiol, a metal salt and a reducing agent and were examined with acetylene as the substrate.[8,9] They immediately favoured molybdenum as the catalysts over all elements of the periodic system. Molybdenum thus became a logical choice for the subsequent detailed investigations of the reactions of all other nitrogenase substrates, but since nitrogenase also contains iron, all reactions were routinely studied with iron complexes as well.

The first models were devised to fulfil a "principle of maximum simplicity" and had to contain a minimum number of essential components.[8,10] When it

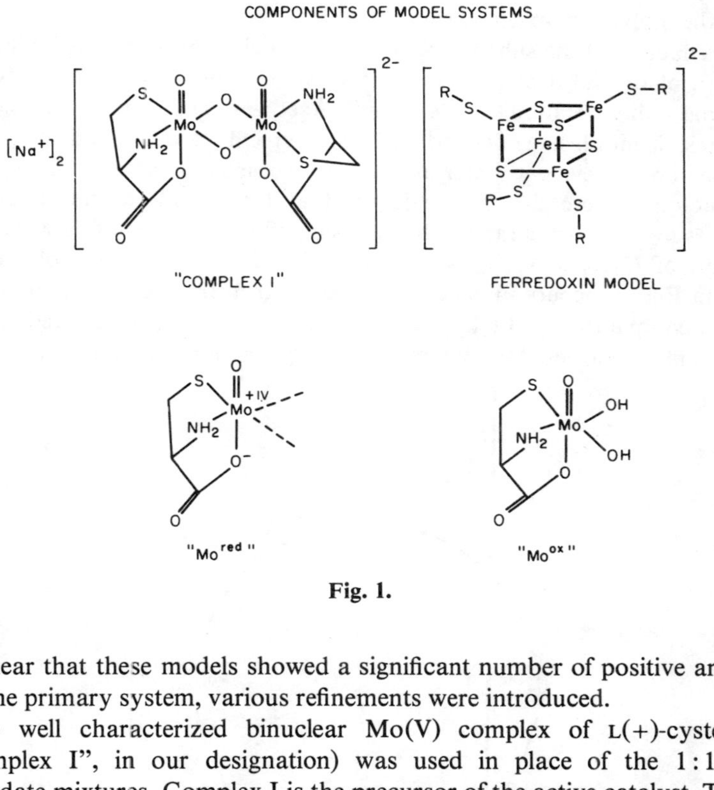

COMPONENTS OF MODEL SYSTEMS

"COMPLEX I"

FERREDOXIN MODEL

"Mo^{red}"

"Mo^{ox}"

Fig. 1.

was clear that these models showed a significant number of positive analogies with the primary system, various refinements were introduced.

The well characterized binuclear Mo(V) complex of L(+)-cysteine[11,12] ("Complex I", in our designation) was used in place of the 1 : 1 thiol–molybdate mixtures. Complex I is the precursor of the active catalyst. The iron cofactor was first added in the form of simple iron salts; in later studies, ferredoxin model compounds[13] were employed as cocatalysts (Fig. 1). Since it appeared that the active reduced form of the catalyst contained molybdenum in the +4 oxidation state, a number of defined Mo(IV) complexes were also investigated.[14–16]

B. Mechanism of nitrogenase action as derived from model studies

Since the details of the model work on nitrogenase have been outlined elsewhere,[10] only the key conclusions necessary for the understanding of nitrogenase action will be summarized here.

First and foremost, the model studies established inconvertibly that the *substrate chemistry of nitrogenase is molybdenum chemistry.* All substrate reactions are furthermore characteristic of a *mononuclear molybdenum active site* located in a sterically partially obstructed or "pocketed" environment. The molybdenum active site is accessible to the protons of the medium as well as to substrates other than dinitrogen. ATP hydrolysis is associated with the reduction of the molybdenum active site.

The reduction of the substrates occurs in two-electron steps, the binding and reduction of the substrates occurs at the molybdenum active site. Substrates are bound either "end-on" or "side-on". Acetylene and nitriles are "side-on" substrates, isonitriles, azide and N_2O are typical "end-on" substrates. The molybdenum site evidently can bind up to two substrates "end-on", but only one "side-on". There also is no evidence that it can bind one end-on and one side-on substrate simultaneously. A simplified scheme of the catalytic reduction of C_2H_2 to C_2H_4 is shown in Fig. 2. The reduction of CH_3CN, shown in Fig. 3, occurs in successive two-electron transfer protonation steps and is accompanied by the hydrolysis of the C=N bond. In the reduction of cyanide and isonitriles, Mo—C bond insertion reactions can take place, giving

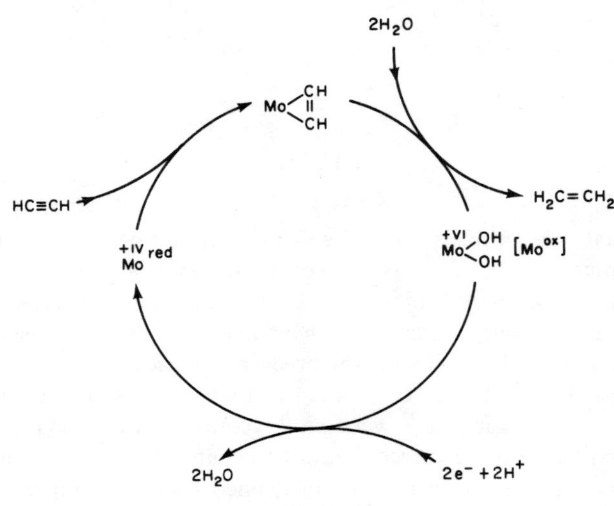

Fig. 2.

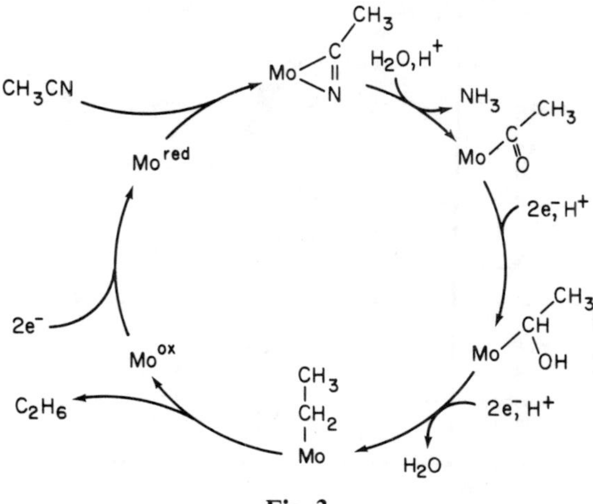

Fig. 3.

rise to C_2 and C_3 hydrocarbons. All reactions with carbon-containing substrates take place via organomolybdenum intermediates. The properties of the Mo–C bonds in such intermediates have recently been investigated in isolable model compounds.[16]

The non-haeme iron components of nitrogenase function as electron storage compartments and as catalysts of the transfer of electrons from the external reductant to the molybdenum active site. If this process is to occur efficiently, non-haeme iron clusters must be located in close vicinity of the molybdenum active site, but in a region which is not as readily accessible by the substrates, especially CN^- which is known to have a destructive effect on non-haeme iron clusters. The reduction of the FeMo-protein is ATP-dependent. Model studies[10] support the idea that ATP is hydrolysed at the molybdenum active site, since molybdate is a specific catalyst of ATP hydrolysis to ADP and inorganic phosphate.[17] The hydrolysis is assumed to occur in a complex of molybdate with the terminal phosphate group of ATP.[17] The same reaction is envisaged to occur at the molybdenum active site and is assumed to generate an activated, dehydroxylated modification of the site which is more rapidly reduced. A similar role for ATP was proposed in principle by Hardy, Parshall and Knight.[18,19] Magnesium appears to participate in this process, possibly by assisting in the removal of ADP or phosphate from the molybdenum active site. Protons generated during ATP hydrolysis may produce a locally acidic environment at the active site, facilitating substrate reductions. A diagrammatic representation of nitrogenase is shown in Fig. 4. MgATP, the Fe-protein and the MoFe-protein are pictured as building blocks which are arranged in a specific fashion and interact in a prescribed manner.

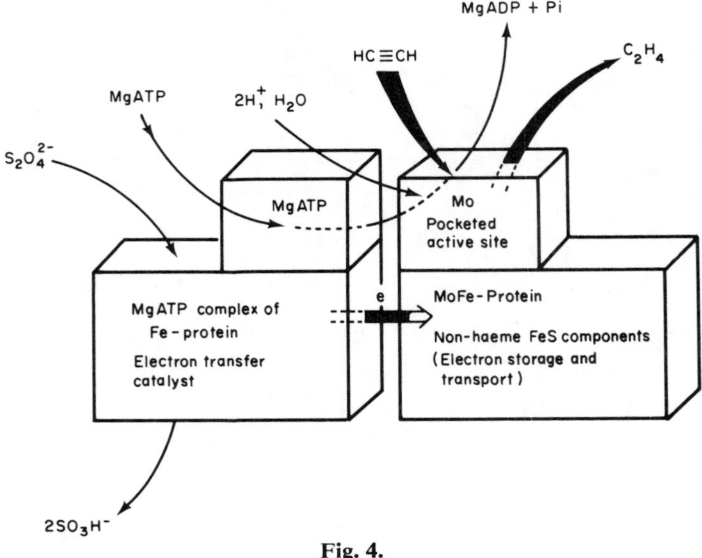

Fig. 4.

C. The mechanism of dinitrogen reduction

The stereospecific reduction of C_2H_2 to *cis*-1,2-dideuterioethylene by nitrogenase in D_2O by analogy suggests that N_2 should be reduced to *cis*-diazene as the initial product.

Diazene (or diimide) is a highly reactive, unstable compound.[20,21] Although it can be isolated and stored for some time at low temperatures, it has a great tendency to decompose into the elements or to disproportionate into N_2 and N_2H_4. Both reactions are competitive and were known years before the existence of diazene was firmly established. F. Raschig[22] reported its apparent decomposition on May 20, 1910, at a meeting of the German Chemical Society, only one day after T. Curtius[23] had proposed that it should disproportionate, expressing himself as follows: "Würden wir das Diimid wenigstens vorübergehend stationär machen, es eine Zeitlang als Individuum beobachten können, so würden wir es sehr wahrscheinlich in Stickstoff und Hydrazin zerfallen sehen". However, Raschig's observation of the decomposition of diazene into its elements sufficiently confused the issue and made diazene a controversial compound for the next five decades. The competitive nature of the decomposition and disproportionation reactions of diazene were recognized clearly only as late as 1960 by organic chemists[24–27] who also discovered its utility in organic synthesis.

Diazene has been postulated by several workers as an intermediate in biological nitrogen fixation, but never in sufficient detail or consistently with its

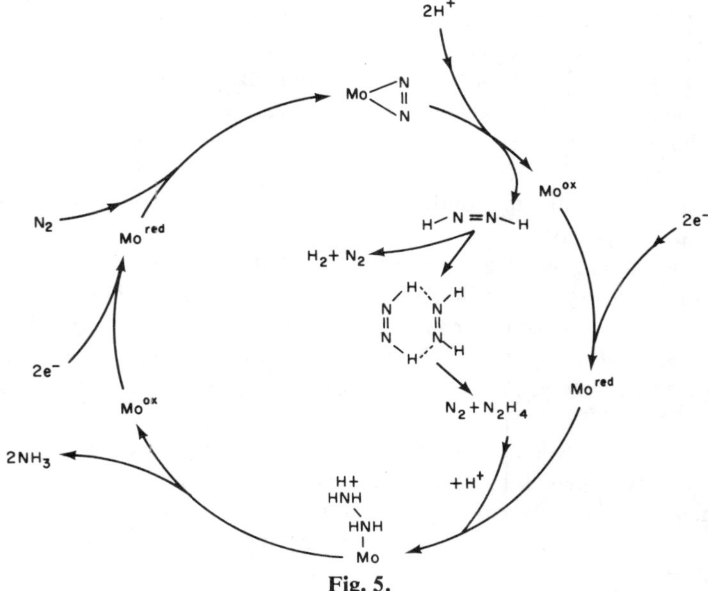

Fig. 5.

chemical properties. In 1974, we formulated the first comprehensive diazene mechanism of biological nitrogen fixation and supported it with model experiments.[28] This mechanism is outlined schematically in Fig. 5. Dinitrogen is assumed to interact with the reduced molybdenum catalyst in the side-on fashion just like acetylene. The resulting nitride-type intermediate complex undergoes hydrolysis to yield *cis*-diazene. The latter *is not a substrate* and accumulates in the vicinity of the active site, possibly within the "pocket" and either disproportionates into hydrazine and dinitrogen or decomposes into dihydrogen and dinitrogen without leaving the pocket. Once formed, hydrazine is reduced to ammonia at the molybdenum active site. The mechanism in Fig. 5 was substantiated by model studies with molybdothiol catalysts. These clearly reduce dinitrogen to diazene but because there is no protecting environment or protein pocket, virtually all the diazene decomposes into its elements and very little disproportionates. We concluded from this that the protein in the vicinity of the molybdenum active site of nitrogenase has a "diazene protecting effect".

D. Extended model systems

To simulate the pocketed environment of the nitrogenase active site at least to a degree, we considered the replacement of the low molecular weight thiols in the "molybdothiol" model systems by peptides as a means of producing a sterically partially obstructed environment for the molybdenum catalysts.[29] In the first of these extended model systems the peptide components of bovine insulin were used as the ligands. "Chain A" of bovine insulin is composed of 21 aminoacids of which four are cysteine. "Chain B" consists of 30 aminoacids and contains two cysteines and two histidines. We found that the cys-SH residues have a natural binding affinity for molybdenum and form catalytically active complexes. The imidazole groups of the two histidines also bind molybdate under reducing conditions, but the resulting complexes have only weak catalytic activity.[29]

The molybdoinsulin model systems reduce C_2H_2 at rates approaching 8% of the rate of nitrogenase, with $NaBH_4$ as the reductant. They thus are equally as active as the so-called "iron molybdenum cofactor"[30] (FeMo-co) of nitrogenase,[31] but not quite as substrate-specific. For example, they still reduce 2-butyne, albeit slowly, although this alkyne is not reduced by nitrogenase at all.

Nitriles make excellent molecular probes for nitrogenase because they are reduced to easily detectable hydrocarbons and ammonia. Their reduction by nitrogenase was chiefly studied by Hardy *et al.*[1] For six saturated and unsaturated nitriles, the observed reactivity sequence is: $CH_2{=}CHCN \gg$ *cis*-$CH_3CH{=}CHCN > CH_3CN \simeq$ *trans*-$CH_3CH{=}CHCN > CH_2{=}C(CH_3)CN \simeq C_2H_5CN$. With the simple molybdocysteine model systems, a completely different reactivity sequence is observed, since in the absence of steric obstruction of the catalyst, the reduction of the nitriles is controlled by electronic factors. With molybdoinsulin catalysts, however, the above reactivity sequence could be very closely reproduced.[29] It thus follows that the steric obstruction at the molybdenum active site of nitrogenase can be partly simulated by a peptide chain containing an internal cys-SH residue.

The molybdoinsulin systems were also found to reduce dinitrogen to ammonia with clearly enhanced efficiency. With $NaBH_4$ as the reductant, rates of N_2 reduction approaching up to 5% of nitrogenase were observed.[29] We attribute the superior performance of the molybdoinsulin systems in the reduction of N_2 to a diazene-protecting effect of the insulin peptide chains.

By attaching Fe_4S_4 clusters to some of the cys-S^- residues of the insulin peptides it may become possible to obtain models of the MoFe-protein that could operate efficiently with $Na_2S_2O_4$ and not just with $NaBH_4$ as the reductant. The use of "molybdenum only" model systems, however, offers the advantage of separating molybdenum reactions from those of iron, and thus provides the essential mechanistic information that iron is not required for the simulation of the reduction of the substrates by nitrogenase.

E. Catalysis of D_2–H^+ exchange

The stimulatory effect of N_2 on the exchange between D_2 in the gas phase and H^+ in solutions of functional nitrogenase is most plausibly attributed to a process involving the homogeneous activation of D_2 at the nitrogenase active site, followed by the formation of diazene-D_1 and its decomposition into HD and N_2. The fact that this exchange occurs even at very low partial pressures of nitrogen[1,32] is not surprising in view of the greater tendency of diazene to decompose at high dilution. The D_2–H^+ exchange catalysis by N_2 thus provides a telling argument in favour of the diazene mechanism of N_2 reduction.

The effect of N_2 on D_2–H^+ exchange has been fully duplicated with the molybdothiol systems.[28]

F. Recent developments concerning the nature of the molybdenum active site

There is presently some argument as to whether molybdenum is part of a "molybdenum–iron cluster", or, as we proposed in 1972,[33] is just closely located to at least one Fe_4S_4 cluster component of the MoFe-protein. EXAFS-measurements have been interpreted to suggest a Mo–Fe cluster,[34] but these conclusions must be treated with caution if only because this new experimental technique is not sufficiently rigorous to allow a decision between competing structural hypotheses. Two Mo, Fe, S clusters with different numbers of thiol ligands have been prepared and structurally characterized[35,36] Our experiments indicate that they do not possess stability and/or catalytic activity in the reduction of nitrogenase substrates until or unless they are decomposed first.[37] The EXAFS measurements also led to the conclusion that the molybdenum active site does not contain a Mo=O bond.[34] This may be true for certain reduced forms of the MoFe-protein but the situation is probably entirely different for the molybdenum site in actively turning over nitrogenase. Our work clearly shows that the reactions of nitrogenase substrates with molybdenum catalysts require oxygen in the coordination sphere of molybdenum, and a strongly hydrophilic environment. Cluster complexes of molybdenum with sulphide ligands are all decidedly hydrophobic in character and thus lack some of the main prerequisites for activity.

G. Stepwise N_2 reduction by nitrogenase

Until about 1974, the enzymological evidence apparently disfavoured mechanisms involving diazene and hydrazine as the intermediates.[6] However, hydrazine has since been shown to be a substrate of nitrogenase, and recently published results of quenching experiments indicate that the reduction of N_2 occurs in a stepwise manner via a "bound N_2H_x species", detectable as hydrazine.[38] These experimental results are fully consistent with the diazene mechanism and rule out a concerted mechanism.

III. Reductions of Nitrogen in Other Protic Systems

A. Reduction by $V(OH)_2$–$Mg(OH)_2$

In 1970, Shilov *et al.*[39,40] discovered that precipitates of $Mg(OH)_2$ containing $V(OH)_2$ reduce molecular nitrogen at room temperature to hydrazine. This abiological nitrogen-fixing system is of interest for the present discussion in that its study in our laboratory[41,42] provided additional evidence for the diazene mechanism of N_2 reduction. As in the work with the molybdothiol model systems, we first investigated the reduction of C_2H_2 in this system. This was necessary because it was assumed by the originators that V^{2+} acted as a one-electron reductant and that nitrogen was reduced in aggregates of $4V^{2+}$ ions directly to hydrazine. We first demonstrated that V^{2+} in $Mg(OH)_2$ acts as a two-electron reductant and subsequently showed that the reduction of C_2H_2 to $C_2H_2D_2$ occurs exclusively *cis*, just as in the molybdenum systems. It thus seemed plausible to expect that V^{2+} would also reduce N_2 in a two-electron step, affording *cis*-diazene, from which N_2H_4 could form by way of disproportionation. Further chemical and kinetic studies, which also included the monitoring of H_2 evolved during the reactions, led to the mechanism shown in Fig. 6. A key feature of this mechanism is that the formation and disproportionation of diazene occurs *inside* the $Mg(OH)_2$ lattice, since N_2H_2 would otherwise undergo rapid decomposition if generated in the strongly

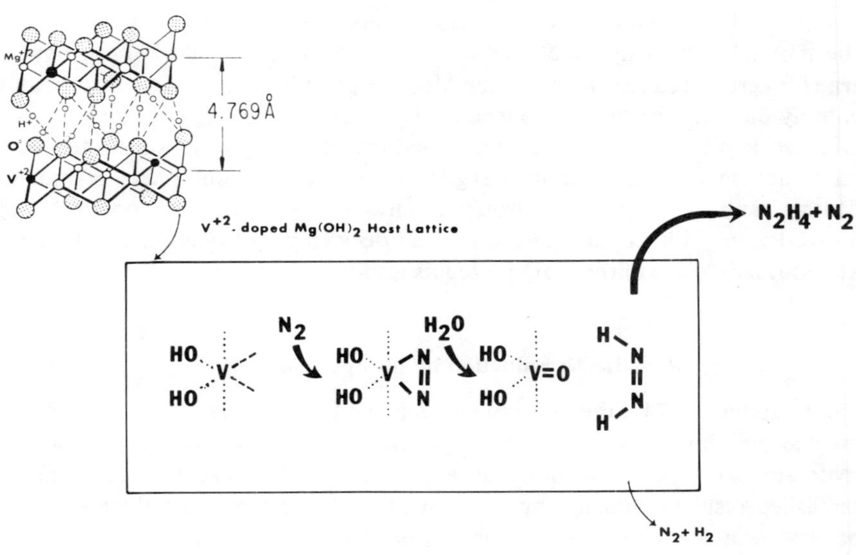

Fig. 6.

alkaline aqueous phase. The $Mg(OH)_2$ layer structure in which Mg^{2+} ions are isomorphously substituted by V^{2+} allows the entrance of N_2 *but not of free OH^- ions*. The $Mg(OH)_2$ lattice thus exerts a "diazene protecting effect", allowing diazene to accumulate and to disproportionate without undergoing base-catalysed decomposition into the elements. The presence of the $Mg(OH)_2$ lattice is thus essential. Only very few solids can replace $Mg(OH)_2$ in this function; $ZrO_2 \cdot aq$ is one such solid,[41] but perhaps due to structural differences is slightly less effective than $Mg(OH)_2$.

Freshly precipitated $V(OH)_2 \cdot aq$ also reacts with N_2, but in the absence of a protective lattice, the N_2H_2 generated decomposes quantitatively into its elements. This gives rise to a *dinitrogen-stimulated* H_2 evolution. The intermediate formation of diazene in the reactions of $V(OH)_2$–$Mg(OH)_2$ with dinitrogen was also confirmed through trapping experiments with allyl alcohol, which is reduced to propanol to a greater extent under N_2 than if the same reaction is run in an argon atmosphere. Finally, the diazene mechanism is supported by the observed[42] dependence of the yields of N_2H_4 on $p_{N_2}^2$, which is fully consistent with the stepwise reduction of N_2 and inconsistent with a direct mechanism.

V. Reduction of N_2 in other systems and some theoretical considerations

Diazene is an endothermic compound ($\Delta H = 49$ kcal mol^{-1}) whose reduction potential in water is believed to be about $-1 \cdot 1$ V.[43] It has been argued that because such strong reducing agents cannot exist in water, a reduction of dinitrogen to diazene is impossible.[43] This argument can be countered by the fact that reducing agents exist which are *metastable* in water; diazene itself is an example of this: it can be generated in protic solvents and used for the reduction of olefinic substrates, under appropriate conditions without serious losses due to decomposition into its elements. In the absence of valid thermodynamic reasons against the diazene mechanism of nitrogen reduction, diazene must be suspected as an intermediate whenever dinitrogen reacts with a sufficiently strong reducing agent in a protic medium, or, in the hydrolysis of certain dinitrogen complexes of transition metals. Thus, diazene has been postulated as the product of the hydrolysis of the complex formed on reacting cp_2TiCl_2 with N_2 in the presence of CH_3MgCl.[44] The acid hydrolysis of dinitrogen–phosphine complexes of Mo(0), yields only 2/3 of the theoretical amount of NH_3. This has been attributed to the possible formation and disproportionation of a N_2H_2 (isodiazene-) intermediate.[45] Neither this work nor the results of other studies with complexes of nitrogen with zerovalent transition metals will be discussed here as these are treated elsewhere (Part 3) and because in these cases the important reaction steps (formation of the N_2

complex) require aprotic conditions It must be emphasized, however, that N_2 can be reduced by several different mechanisms and that the stepwise diazene mechanism described above is but one of them. To illustrate this, the reduction of N_2 in yet another system will be described. It has been known for many years[46] that suspensions of $Fe(OH)_2$ decompose under certain conditions with H_2 evolution. A recent study in our laboratory[47] suggested that this is because $Fe(OH)_2$ may disproportionate into Fe° and mixed Fe^{2+}–Fe^{3+} hydroxides. The Fe° reacts with protons of the medium to yield H_2 but is reactive enough to combine with N_2 to produce NH_3 and traces of N_2H_4. The disproportionation of $Fe(OH)_2$ is accelerated by light. It has also been found that the Fe° generated in this fashion reacts with C_2H_2 to yield a mixture of CH_4, C_2H_6 and C_2H_4. Interestingly, this work has recently been criticized and the experimental results were considered to be in error.[48] It was argued that the thermodynamic potentials of $Fe(OH)_2$ and Fe° are insufficient to effect a reduction of N_2. This, however, ignores the well known dependence of the thermodynamic potentials on the degree of dispersion. Generated in the $Fe(OH)_2$ system, the Fe° is initially probably atomic and therefore indeed sufficiently reactive. The reactions of atomic iron and of low molecular weight aggregates of it confirm this. According to two recent reports,[49,50] an "Fe_2" species reacts with N_2 in a frozen N_2 matrix to form nitrides; it even reacts with CH_4 at low temperatures to yield an organoiron complex.

It is not inconceivable that other metals generated in sufficiently highly dispersed form will react with dinitrogen even if suspended in a protic solvent.

While multielectron reductions are to be expected in reactions of N_2 with zerovalent metals or aggregates of them, the reduction of N_2 by concerted four- or six-electron transfer in homogeneous protic solutions still remains to be demonstrated. For several homogeneous N_2 reducing systems that have been described as occurring via four-electron transfer steps, the evidence that hydrazine is formed directly is not convincing and the available experimental data do not rule out the diazene mechanism.

IV. Concluding Remarks

The present account illustrates, in part with historical details, the evolution of a mechanism of biological nitrogen fixation and of other aspects of the reduction of nitrogen in protic media. During the past 10 years of our endeavours in this exciting field of chemistry, a *stepwise mechanism* at a mononuclear molybdenum centre via diazene and hydrazine has been developed and substantiated by extensive studies in abiological model systems. Of all past and present mechanisms of biological nitrogen fixation that have been proposed, the diazene mechanism can best account for the available experimental evidence.

Acknowledgments

The work described in this chapter was supported by grants from the National Science Foundation (USA), by Climax Molybdenum Company, and by Mitsubishi Chemical Industries Ltd. The following students have been, or are presently, engaged in research related to the topic of this account: P. A. Doemeny, J. M. Doemeny, R. Frazier, Jr., J. H. Grate, L. Hughes, T. D. Guth, G. W. Kiefer, Dr. H. Kisch, E. L. Moorehead, J. G. Palmer, M. R. Palmer, P. R. Robinson, G. Schlesinger, N. Strampach, K. Tano, B. Ufkes, Dr. T. M. Vickrey, B. J. Weathers and S. I. Zones.

References

1. R. C. Burns and R. W. F. Hardy, *Molecular Biology, Biochemistry & Biophysics* (1975) **21** and references cited therein.
2. B. E. Smith, R. N. F. Thorneley, M. G. Yates, R. R. Eady and J. R. Postgate *Proc. 1st Internat. Symposium on Nitrogen Fixation* (Ed. W. E. Newton and C. J. Nyman), p. 150. Washington State University Press (1976).
3. (a) R. V. Hageman and R. H. Burris, *Biochemistry* (1978) 4117.
   (b) R. V. Hageman and R. H. Burris, *Proc. Natl. Acad. Sci., USA* (1978) **75**, 2699.
4. (a) R. Silverstein and W. A. Bulen, *Biochemistry* (1970) **9**, 3809.
   (b) R. C. Burns and R. W. F. Hardy, *Methods in Enzymology* (1972) **24**, 482.
5. (a) G. L. Turner andf F. J. Bergersen, *Biochem. J.* (1969) **115**, 529.
   (b) E. K. Jackson, G. W. Parshall and R. W. F. Hardy, *J. Biol. Chem.* (1968) **243**, 4952.
6. (a) J. R. Postgate (Ed.), "The Chemistry and Biochemistry of Nitrogen Fixation". Plenum Press, London and New York (1971) and references cited therein.
   (b) E. N. Mishustin and V. K. Shil'nikova, "Biological Fixation of Atmospheric Nitrogen". Macmillan, London and Basingstoke (1971) and references therein.
7. R. W. F. Hardy, R. C. Burns and C. W. Parshall, *Advances in Chemistry Series* (Bioorganic Chemistry) (1971) **100**, 219.
8. G. N. Schrauzer, *Advances in Chemistry Series* (Bioinorganic Chemistry) (1971) **100**, 1.
9. G. N. Schrauzer and G. Schlesinger *J. Amer. Chem. Soc* (1970) **92**, 1808.
10. G. N. Schrauzer, *Angew. Chem.* (1975) **87**, 579; *Angew. Chem.* (*Internat. Edn. Engl.*) (1975) **14**, 514.
11. A. Kay and P. C. H. Mitchell, *Nature* (*London*) (1967) **219**, 267.
12. J. R. Knox and C. K. Prout, *Chem. Commun.* (1968) 1227.
13. R. H. Holm, *Endeavour* (1975) **34**, 1, and references therein.
14. P. R. Robinson, E. L. Moorehead, B. J. Weathers, E. A. Ufkes, T. M. Vickrey and G. N. Schrauzer, *J. Amer. Chem. Soc.* (1977) **99**, 3657.
15. G. N. Schrauzer, P. R. Robinson, E. L. Moorehead and T. M. Vickrey, *J. Amer. Chem. Soc.* (1975) **97**, 7069.

16. (a) E. L. Moorehead, B. J. Weathers, E. A. Ufkes, P. R. Robinson and G. N. Schrauzer, *J. Amer. Chem. Soc.* (1977) **99**, 6089.
(b) G. N. Schrauzer, E. L. Moorehead, J. H. Grate and L. Hughes, *J. Amer. Chem. Soc.* (1978) **100**, 4760.
17. H. Weil-Malherbe and R. H. Green, *J. Biol. Chem.* (1951) **49**, 286.
18. R. W. F. Hardy and E. Knight, Jr. *Bacteriol. Proc.* (1967) 112.
19. G. W. Parshall, *J. Amer. Chem. Soc.* (1967) **89**, 1822.
20. N. Wiberg, G. Fischer and H. Bachhuber, *Angew. Chem.* (1976) **88**, 386; *Angew. Chem.* (*Internat. Edn. Engl.*) (1976) **15**, 385.
21. C. Willis, R. A. Back and J. G. Purdon, *Internat. J. Chem. Kinetics* (1977) **IX**, 787, and references cited therein.
22. F. Raschig, *Z. Angew Chem.* (1910) **23**, 972.
23. T. Curtius, *Z. Angew. Chem.* (1911) **24**, 2.
24. S. Hünig, H. R. Müller and W. Thier, *Angew. Chem.* (1965) **77**, 368; *Angew. Chem.* (*Internat. Edn. Engl.*) (1965) **4**, 271.
25. S. Hünig, H. R. Müller and W. Thier, *Tetrahedron Letters*, (1961) 353.
26. E. J. Corey, W. L. Mock and D. J. Pasto *Tetrahedron Letters*, (1961) 347.
27. E. E. van Tamelen and R. S. Dewey, *J. Amer. Chem. Soc.* (1961) **83**, 3729.
28. G. N. Schrauzer, G. W. Kiefer, K. Tano and P. A. Doemeny, *J. Amer. Chem. Soc.* (1974) **96**, 641.
29. B. J. Weathers, J. H. Grate, N. Strampach and G. N. Schrauzer, *J. Amer. Chem. Soc.* (1979) **101**, 925
30. P. T. Pienkos, V. K. Shah and W. J. Brill, *Proc. Natl. Acad. Sci., USA* (1977) **74**, 5468.
31. V. K. Shah, J. R. Chisnell and W. J. Brill, *Biochem. Biophys. Res. Commun.* (1978) **81**, 232.
32. K. L. Hadfield, Ph.D. Thesis, Brigham Young University (1970), cited in ref. 1, p. 127.
33. G. N. Schrauzer, P. A. Doemeny, G. W. Kiefer and R. H. Frazier, *J. Amer. Chem. Soc.* (1972) **94**, 3604.
34. S. P. Cramer, K. O. Hodgson, W. O. Gillum and L. E. Mortenson *J. Amer. Chem. Soc.* (1978) **100**, 3398.
35. T. E. Wolff, J. M. Berg, C. Warrick, K. Hodgson, R. H. Holm and R. B. Frankel, *J. Amer. Chem. Soc.* (1978) **100**, 4630.
36. (a) G. Christou, C. D. Garner, F. E. Mabbs and T. J. King, *J.C.S. Chem. Commun.* (1978) **740**.
(b) G. Christou, C. D. Garner and F. E. Mabbs, *Inorg. Chim. Acta* (1978) **28**, L189.
37. Unpublished results, 1978.
38. R. N. F. Thorneley, R. R. Eady and D. J. Lowe, *Nature* (*London*), (1978) **272**, 557.
39. N. T. Denisov, O. N. Efimov, N. I. Shuvalova, A. K. Shilova and A. E. Shilov, *Zh. Fiz. Khim.* (1970) **44**, 2694.
40. A. E. Shilov, N. T. Denisov, O. N. Efimov, N. F. Shuvalov, N. I. Shuvalova and E. Shilova, *Nature* (*London*) (1971) **231**, 460.
41. S. I. Zones, T. M. Vickrey, J. G. Palmer and G. N. Schrauzer, *J. Amer. Chem. Soc.* (1976) **98**, 7289.
42. S. I. Zones, M. R. Palmer, J. G. Palmer, J. M. Doemeny and G. N. Schrauzer *J. Amer. Chem. Soc.* (1978) **100**, 2113.
43. A. E. Shilov, *Uspekhi Khim.* (1974) **43**, 863.

44. Y. G. Borodko, I. N. Ivleva, L. M. Kachapina, S. I. Salienko, A. K. Shilova and A. E. Shilov, *J.C.S. Chem. Commun.* (1972) 1178.
45. J. Chatt, A. J. Pearman and R. L. Richards, *J.C.S.* (*Dalton*) (1977) 1853.
46. G. Schikorr, *Z. Elektrochem.* (1929) **35**, 65.
47. G. N. Schrauzer and T. D. Guth, *J. Amer. Chem. Soc.* (1976) **98**, 3508.
48. N. T. Denisov, A. F. Zueva and A., K. Shilova, *Kinet. Katal.* (1977) **19**, 267.
49. P. H. Barrett and P. A. Montano, *J.C.S.* (*Faraday*) (1977) **2**, 72, 378.
50. P. H. Barrett, M. Pasternak and R. G. Pearson, *J. Amer.Chem. Soc.* (1979) **101**, 222.

5

Comparison of Biological Nitrogen Fixation with Its Chemical Analogues

A. E. SHILOV

Institute of Chemical Physics, USSR Academy of Sciences, USSR

I. Introduction

THE primary process of molecular nitrogen fixation in industry is the catalytic synthesis of ammonia from dinitrogen and dihydrogen:

$$N_2 + 3H_2 \xrightarrow{(Fe)} 2NH_3.$$

There is also the natural process of dinitrogen fixation, which is presently the main source of bound nitrogen on our planet, and is carried out by a number of bacteria and algae. In this case, the first product of dinitrogen transformation is also ammonia and this first stage can be schematically presented as taking place with the participation of the protons of water:

$$N_2 + 6H^+ + 6e^- \rightarrow 2NH_3.$$

Despite all the obvious outward resemblance of these two reactions, chemists have always been amazed by the differences in the conditions under which they proceed. Due to the well known chemical inertness of dinitrogen, industrial synthesis of ammonia must be conducted at high temperatures and pressures, whereas the biological fixation proceeds effectively directly from the air, under natural environmental conditions. The existence of the natural process has led to numerous attempts at implementing the reduction of dinitrogen in the laboratory, in the same conditions under which the biological reaction takes place. Until recently such attempts have ended in failure. Reports appearing from time to time in the literature on the formation of small amounts of ammonia from dinitrogen in aqueous solutions were not subsequently verified.

It is only in the last nine years that reproducible results have been obtained demonstrating the reduction of dinitrogen to hydrazine and ammonia in protic media. The present paper reviews the results obtained in this area and brings up to date an earlier review.[1]

The way was paved for the discovery of N_2 reduction in protic media by the results, obtained mainly in the sixties (see below), on the biological fixation of dinitrogen, and on the chemical reduction of dinitrogen in aprotic media. The latter can be briefly summarized as follows (for details see Reference *2*).

Starting with the work of Vol'pin and Shur in 1964,[3] it became known that dinitrogen could be reduced to ammonia derivatives (usually nitrides) in the presence of compounds of transition metals (Ti, Zr, V, Nb, Cr, Mo, W, Mn, Fe, Co) under the action of such reducing agents as free metals (Li, Mg, Al), hydrides of metals ($LiAlH_4$), and various organometallic compounds. The role of transition metals in low oxidation states in these systems is from all indications associated with their ability to form dinitrogen complexes in which the N_2 molecule is activated. In 1965 Allen and Senoff discovered the first such N_2 complex, which was a Ru(II) compound.[4] This complex was obtained in an indirect way by the action of hydrazine on $RuCl_3$. Then, in 1966, the work of Shilov, Shilova and Borod'ko[5] demonstrated the feasibility of the formation of similar complexes directly from dinitrogen.

A large number of N_2 complexes is now known (see references *6–8*), and their formulae are: L_nMN_2, $L_nMN_2ML_n$, or $L_nM(N_2)_2$, where M is the transition metal, and L a ligand. They usually have a linear fragment MNN or MNNM. At first, investigators failed to observe complexes with dinitrogen in which the N_2 would enter into chemical reactions (e.g., those involving its reduction). The first intermediate complexes containing N_2 bound to a transition metal were observed in 1969,[10] and it was shown then for the first time that N_2 in the complex can be reduced further to a hydrazine derivative, forming N_2H_4 under the action of acid.[10,11] Subsequently, several such complexes involving titanium,[12] iron[13,14] and zirconium[15] were found and, in some cases, isolated. The complexes are binuclear (L_nMNNML_n). They were formed in aprotic media with the participation of strong reducing agents, but it was noted that, in the iron N_2 complexes,[13,14] the reduction of N_2 to hydrazine took place only under the action of acid protons (in the presence of alcohol, the dinitrogen in the complex is fully converted into free dinitrogen). This was explained by protonation of N_2 in the coordination sphere of the metal with resulting facilitation of further reduction.

More recently Chatt and co-workers[9,16] succeeded in showing that, with the protonation of nitrogen in mononuclear bis-dinitrogen complexes of W(0) and Mo(0), one N_2 molecule was being reduced finally to hydrazine and ammonia.

II. Theoretical Considerations

A. On the possibility of one-step multielectronic reduction of dinitrogen

Reduction of dinitrogen to hydrazine and ammonia requires the transfer of several electrons to the N_2 molecule and, in principle, this can proceed in several stages. In this connection, we shall examine the thermodynamics of the successive stages of reduction of the N_2 molecule. We first note that the reduction of dinitrogen leads to the increase of electron density on N atoms and, correspondingly, to the enhancement of its basic properties. This results in ammonia stabilization, as well as stabilization of the intermediate products of N_2 reduction in protic media as compared with reaction in the gaseous phase or in aprotic media, due to the formation of hydrogen bonds with the protons of the solvent. For example, the enthalpy of hydrazine formation from H_2 and N_2 is $+22{\cdot}8$ kcal mol^{-1} in the gaseous phase and only $+8{\cdot}2$ kcal mol^{-1} in aqueous solutions. Although the entropy of N_2H_4 formation is also reduced (because of the "freezing" of the solvent molecules close to N_2H_4), the overall effect of the change in free energy of formation, with the transfer of N_2H_4 from the gaseous phase to the aqueous solution at 25°C, amounts to $-7{\cdot}5$ kcal mol^{-1}, which corresponds to the equilibrium constant of the reaction

$$2H_2 + N_2 \rightleftharpoons N_2H_4$$

increasing 3×10^5 times. It can be assumed that the rate constant for N_2 reduction in complexes of transition metal compounds will also increase in protic media as compared with aprotic solutions, all other conditions being equal, due to hydrogen bonds with the solvent being formed in the transition state. Thus, with the same reducing agent strength, protic media conditions are more favourable for reduction to take place.

Energy expenditures of the stages of successive reduction of N_2 by 1, 2, 4 and 6 electrons are distributed in an extremely non-uniform way (Fig. 1). One can see that, if the reaction proceeds through N_2H, N_2H_2 or N_2H_4, a reducing agent stronger than H_2 is required for the first stage, which for aqueous or aqueous-alcohol media means that the reaction should be conducted in conditions for which H_2 evolution is more advantageous. This results from an especially high energy required to split the first bond in N_2; this energy ($\sim$130 kcal mol^{-1}) amounts to more than half the energy of the $N\equiv N$ triple bond (225 kcal mol^{-1}). This is why both one- and two-electron reductions of N_2 with the formation of N_2H_2 require a very strong reducing agent.

This distinguishes dinitrogen from the other compounds with multiple bonds, e.g., acetylene where for the two-electron reduction to ethylene a reducing

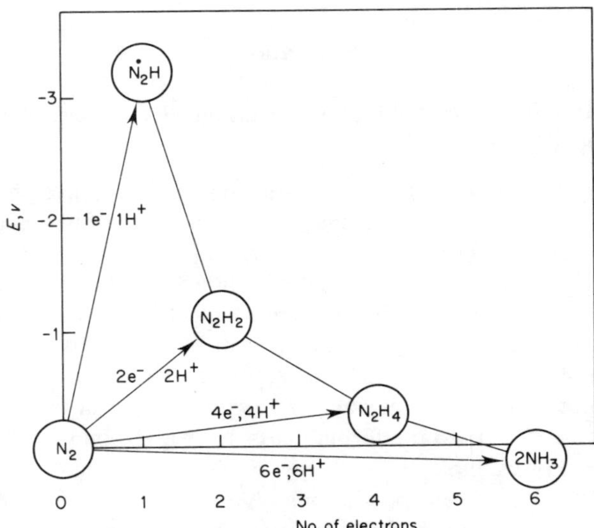

Fig. 1. Oxidation reduction potentials of 1, 2, 4 and 6 electron reduction reactions of dinitrogen to corresponding compounds.

agent with a potential of +0·37 V is sufficient. Thus, in this case, dihydrogen itself is a sufficiently strong reducing agent.

The high reduction potentials† of the one- and two-electron reductions of N_2 explain why it is difficult to reduce dinitrogen in protic media, although, as noted above, the energy requirements for N_2 reduction in protic media are smaller than in aprotic media. For reduction to N_2H_2 to be accomplished, a reducing agent with a potential greater than −1·1 V at pH 0, −1·5 V at pH 7, or −1·9 V at pH 14, must be stable in the protic media (free metals are among the reducers of this strength, but their inertness in water is associated with the formation of a protective film of oxides which will interfere with the reaction with dinitrogen).

At the same time, the existence of biological dinitrogen fixation with a moderate value of the reduction potential indicates that a mechanism exists for which this thermodynamic limitation is avoided. The study of the properties of transition metal N_2 complexes made it possible, as early as 1969, to propose a mechanism in which the N_2H_2 molecule did not appear at an intermediate stage. The reaction could proceed with the transformation of coordinated

† In almost all cases reductants stronger than H_2 will be considered here. Their reduction potentials (E_0) (oxidation–reduction potential related to standard H_2 electrode at pH = 0) are thus negative. The potentials will be considered "higher" or "lower" according to their absolute values.

dinitrogen directly into a derivative of hydrazine, for which a much less powerful reducing agent is needed.[6] Various specific cases of this mechanism are considered in a review in reference *2*. At present, the mechanism considered most probable by the author for protic media can be summarized as follows:

$$M^{n+} \cdots N{\equiv}N \cdots M^{n+} \rightarrow$$

$$M{=}N{-}N{=}M \ (\text{or } M{=}N{-}N{=}M) \xrightarrow{H^+} N_2H_4 + 2M^{n+2},$$

or with the prior participation of the solvent protons (taking into account the enhancement of the basic properties of nitrogen in the intermediate complex) as:

$$M^{n+} \cdots N{\equiv}N \cdots M^{n+} \xrightarrow{H^+} M{=}N{-}\overset{+}{N}H{=}M \xrightarrow{H^+} N_2H_4 + 2M^{n+2}.$$

In this process the oxidation level of every M must in the end increase by two units. If we assume that there is in the system next to every M at least one M_R ion capable of donating one electron to an M ion simultaneously with the reduction of dinitrogen, then the formation of hydrazine in the presence of such a four-nucleus catalyst leads to an increase in the oxidation level of every M and M_R by 1 unit:

$$M_R^{m+} \cdots M^{n+} \cdots N{\equiv}N \cdots M^{n+} \cdots M_R^{m+} \xrightarrow{H^+}$$

$$M_R \cdots M{=}N{-}\underset{H}{\overset{+}{N}}{=}M \cdots M_R \xrightarrow{H^+} N_2H_4 + 2M^{n+1} + 2M_R^{m+1}.$$

Such a mechanism obviously makes it possible to utilize, in a catalytic process, a one-electron reducing agent (Red) (in a specific case Red = M_R), for which the reduction potential of the process

$$\text{Red} \rightarrow \text{Ox} + e^-$$

is only slightly higher than the potential of the reaction

$$N_2 + 4e^- + 4H^+ \rightarrow N_2H_4.$$

Thus a reducing agent which is much weaker than one requires for the formation of diazene can be used.

We should note that for an optimum four-nucleus catalyst $(M^n)_2\,(M_R^m)_2$, the mean oxidation–reduction potential of the processes

$$M^{n+} \rightarrow M^{(n+1)} + e^-$$

and

$$M_R^{m+} \rightarrow M_R^{(m+1)+} + e^-$$

must lie between E_0 for Red and E_0 for N_2H_4. In this case when a polynuclear catalyst includes six and more metal ions one can visualize a further reaction with complete disruption of the N–N bond and the formation of ammonia derivatives:

$$\begin{matrix} M_R^{m+} & & & & M_R^{m+} \\ & \vdots M^{n+} \ldots N{\equiv}N \ldots M^{n+} \vdots & & & \\ M_R^{m+} & & & & M_R^{m+} \end{matrix} \xrightarrow{H^+} \begin{matrix} M_R & & & & & M_R \\ & \vdots M{=}N{-}\overset{+}{N}(H){=}M \vdots & & & & \\ M_R & & & & & M_R \end{matrix} \xrightarrow{H^+}$$

$$\qquad\qquad (1) \qquad\qquad\qquad\qquad\qquad\qquad (2)$$

$$\longrightarrow \begin{matrix} M_R & \\ & \vdots M{=}NH \\ M_R & \end{matrix} \qquad \begin{matrix} & M_R \\ NH{=}M \vdots & \\ & M_R \end{matrix} \xrightarrow{H^+} 2NH_3 + 4M_R^{m+1} + 2M^{n+1}$$

$$(3)$$

Utilizing this mechanism it is thus possible to exclude intermediate formation not only of diazene but also of hydrazine and conduct the reaction directly to derivatives of ammonia using a reducing agent which is even weaker than hydrazine. In this case the intermediate or transitional forms must certainly be stabilized in an appropriate way; this seems to be assured by the conjugation in the intermediate complexes (**1**) and (**2**).

B. On the nature of the intermediate complex in the reduction of dinitrogen in protic media

Thermodynamic difficulties encountered in the reduction of dinitrogen in protic media, even if the possibility of multielectron reduction of N_2 is taken into account, require special properties for the intermediate complex with dinitrogen. In fact, the requirement of a certain reduction potential for the M complexing agent is supplemented by the requirements of a certain stability and a certain structure for the intermediate complex. The stability must be optimal (it is felt that excessive stability will impede further reactions of coordinated dinitrogen) and it is obviously useful for the intermediate complex to be as close as possible to the reaction products. At present we have numerous data on various complexes of N_2 with transition metal compounds and can draw some conclusions on the nature of the optimum catalyst for the reduction of dinitrogen (certainly taking into account the experimental data on reactions in protic media already available).

As already noted, both the mono- and binuclear complexes whose structures have been determined usually have a linear fragment M–N≡N or

M—N≡N—M. It is only for a complex of low-valent nickel that a non-linear structure with a fragment:

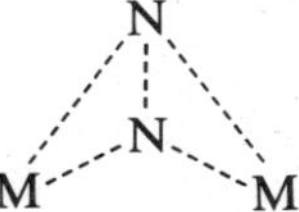

has been demonstrated.[17,18] The linearity of the M—N≡N—M fragment is evidently determined by the possibility of maximum "conjugation of bonds" (i.e. maximum overlapping of the d orbitals of M and the p orbitals of nitrogen). The transition into hydrazine derivatives corresponds to the formation of the planar configurations

M=N—N=M or M=N—N=M,

which is again determined by the best overlap of d and p orbitals. The order of the filling of molecular orbitals in a linear binuclear complex has been considered.[19,20] For the same M in a binuclear complex made up of two octahedra, a simplified picture of the molecular orbitals can be represented as:

$$1e_u < 1e_g < b_{2g} \approx b_{1u} < 2e_u < 2e_g$$

(Fig. 2). The bonding of dinitrogen in the M—N≡N—M complex is mainly effected by the $1e_g$ orbital, constructed from the d_{xy} and d_{yz} orbitals of M and

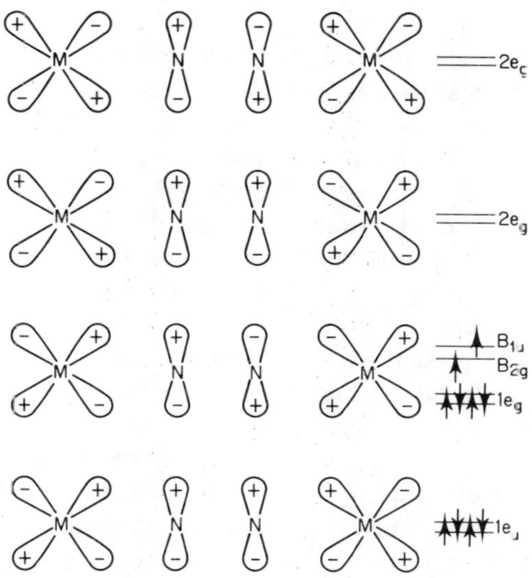

Fig. 2. The molecular orbitals for a linear binuclear $L_nM—N≡N—ML_n$ complex (D_{4h} symmetry).

the antibonding $1\pi_g$ orbital of N_2 (the lower doubly-degenerate $1e_u$ orbital accommodates four electrons of the π bonds of the N_2 molecule). The filling of the $1e_g$ orbital ensures electron transfer from M onto the N_2 molecule and the weakening of the N–N bond. Addition of four electrons to this doubly-degenerate orbital, i.e., with a change in electronic configuration from d^0 to d^2, will, other things being equal, enhance the weakening of the N–N bond. Further increase of the number of d electrons on M will first lead to the filling of the non-bonding orbitals b_{2g} and b_{1u} (d^3 and d^4); then the $2e_u$ orbitals (d^5 and d^6) will be filled which will weaken the M–N bonds and stabilize the N–N bond; and finally, the filling of the $2e_g$ orbital (d^7 and d^8) will correspond to the weakening of both the M–N bonds and the N–N bond.

On the whole, the experimental data for binuclear complexes conform to this picture. For example, in linear M–N≡N–M binuclear complexes the N–N bond for a d^6 metal system is no weaker than in the mononuclear d^6 complexes. For d^1 complexes $[\{(\pi\text{-}C_5H_5)_2Ti(III)R\}_2(N_2)]$ the reduction of N_2 proceeds only when there is a further reduction of titanium (to Ti(II)) by some reducing agent, and only N_2 is evolved when the complexes are decomposed by HCl.[12] Thus two d electrons are not sufficient to reduce N_2. It is for binuclear d^2 complexes (for Ti(II) and Zr(II)[15]) with four d electrons that the N–N bond is characterized by considerable weakening and hydrazine is formed under the action of HCl. From the foregoing considerations, it can be assumed that the catalytic properties of M in binuclear complexes must be more nearly optimal than in mononuclear complexes†. Further, in a series of M_2N_2 complexes with the same reducing properties for M, the optimum conditions for catalysis in transforming a complex with linear configuration into one with planar configuration will be realized when the N–N bond is maximally weakened and the charge of the N atoms is maximally negative.

From this it follows that with the same oxidation–reduction potential for M and a linear intermediate complex M–N≡N–M, the catalytic properties of M will improve from d^0 to d^2 and deteriorate from d^5 to d^{10}.

Electronic configurations d^3 and d^4 deserve special consideration. The filling of each of the non-bonding b_{2g} and b_{1u} orbitals with one electron will prevent the possibility of electron transfer from the $1e_g$ orbital to the b_{2g} and b_{1u} orbitals (with a small difference in levels, this transition, for a d^2 configuration complying with Hund's rule, will decrease the weakening of the N–N bond and the negative charge on the N atoms).

The change from a d^2 to d^3 configuration will thus apparently improve the catalytic properties of M. The further filling of b_{2g} and b_{1u} orbitals with two electrons (to configuration d^4) is undesirable from the point of view of creating optimum catalytic properties, as these very orbitals will be the highest

---

† Since less change is necessary in the oxidation state of M during reduction of N_2.

filled in the complex and, with equal reduction potentials for M(d^4) and M(d^3) (or equal ionization potentials), electron transfer onto the bonded N_2 molecule will be reduced for a d^4 system. The same conclusion may be arrived at if we consider that in octahedral complexes with d^4 configuration one electron is at a higher energy level, due to the ligand-field splitting, and this must lead to a greater difference in reduction potentials when there is a change from one- to two-electron processes (e.g. Cr^{2+}, while being a strong one-electron reducing agent, is far less capable of transferring two electrons). For equal reduction potentials in a one-electron process, the transfer of two electrons to the N_2 molecule in a complex must therefore be greater for a d^3 configuration than for d^4.

Thus in the series of transition elements, the d^3 configuration corresponds to the optimum properties for the catalytic reduction of dinitrogen in protic media following the mechanism discussed above, which seems to correspond to the most economical energy expenditure. For systems with the participation of strong reducing agents (e.g., for complexes obtained in aprotic media), the linear binuclear M—N≡N—M′ complex may not be a necessary intermediate in the reduction of N_2. The participation of M with d^5–d^{10} electronic configurations can then prove suitable for the activation of dinitrogen. As shown by the results of Chatt and co-workers[9,16] the reduction of N_2 can take place via protonation of coordinated N_2 molecules in previously prepared mononuclear W(0) and Mo(0) complexes.

III. Systems Fixing Nitrogen in Protic Media

A. Nitrogenase

In the enzymatic reduction of dinitrogen, the protons of water take part in the formation of ammonia which is the first detectable product. Taking into account that hydrophobic areas exist in enzymes, it is not known to what extent one can speak of a "protic medium", in the strict sense of the word, as a solvent in which the enzymatic process of dinitrogen fixation is developing. In any case the active centre proves to be accessible to the water protons which, in the absence of N_2, are reduced to dihydrogen. It is thus necessary to consider here the advantages and limitations resulting from the reaction being conducted with the participation of protons, which justifies our considering this process along with the non-biological reactions for the reduction of dinitrogen in protic media.

The structure and mechanism of action of nitrogenase have been studied with special intensity from the early sixties, when crude extracts which fixed nitrogen, and then the enzyme itself, were successfully isolated. Various aspects of the biological fixation of dinitrogen have been considered in a number of reviews (e.g., see references. *21*, *22*), and we shall only briefly formulate the principal results here. The enzyme isolated from microorganisms is a combination of two proteins, one of which ("MoFe-protein") contains two molybdenum atoms and 24–32 iron atoms, as well as sulphide and thiol groups; the other ("Fe-protein") contains 4 iron and 4 sulphur atoms.

The reaction of dinitrogen reduction *in vitro* can take place under the action of specially added sodium dithionite. No intermediate products (e.g., N_2H_4) have been observed in the process of NH_3 formation. However, additions of acid or alkali during N_2 fixation revealed formation of some of N_2H_4,[23] indicating intermediate formation of a substance with a single N–N bond present. Apparently, this is not free hydrazine since its reduction by nitrogenase is slow and with the concentrations of the intermediate observed could not account for the rate of NH_3 production.

A large role in the reaction mechanism belongs to ATP; its hydrolysis, coupled with electron transfer from the reducing agent to the dinitrogen molecule bonded in the active centre, evidently raises the potential of the reduction system. The reduction of iron seems to take place first in the Fe-protein, then the electrons pass from the Fe-protein to the MoFe-protein, and it is at this stage that ATP takes part. It is felt[2] that the mechanism of ATP action includes the non-equilibrium protonation of one of the MoFe-protein ligands during the process of ATP hydrolysis, which imparts the capacity to accept electrons from the Fe-protein. With subsequent equilibrium deprotonation of the ligand, the MoFe-protein potential increases above the initial potential of the reduced Fe-protein; this actually means a transfer of the electron from the Fe- to the MoFe-protein against the thermodynamic potential. According to the data of Likhtenshtein and co-workers,[24] it is essential that all iron atoms in the enzymes are bonded with each other, evidently forming a cluster capable of transferring electrons to the active centre. Electron transfer in conjunction with ATP hydrolysis is the rate-controlling stage of the process. The reactions of NH_3 formation from N_2, and of H_2 formation from the protons of water, compete for the electrons of the reducing agent, the yield of NH_3 increasing and that of H_2 decreasing with the growth of dinitrogen pressure from zero to atmospheric. However, even at elevated dinitrogen pressures, the yield of H_2 does not drop below 25% of the transferred electrons of the reducing agent, 75% accounting for formation of ammonia from dinitrogen. The stoichiometry of the reaction for dinitrogen at sufficiently high pressures of N_2 can thus be represented:

$$N_2 + 8e^- + 8H^+ \rightarrow 2NH_3 + H_2.$$

It is obvious that, along with the independent reaction of H_2 formation which is retarded by the introduction of dinitrogen, there exists a coupled reaction of hydrogen formation, which consumes 25% of the electrons at any dinitrogen pressure. Besides dinitrogen, other molecules with a triple bond, such as acetylene, CN^-, N_3^-, N_2O, CH_3CN, and CH_3NC are capable of being reduced in the presence of nitrogenase.

Carbon monoxide inhibits N_2 reduction, but at the same time has no effect on dihydrogen formation. It is of interest that acetylene is reduced only to ethylene, which is not reduced further to ethane and does not inhibit either the reduction of dinitrogen or the evolution of dihydrogen. The reduction of C_2D_2 produces *cis*-deuteroethylene.

The reduction rate of dinitrogen amounts to approximately 0·1–1 molecule of N_2 per one active centre in 1 s; the activation energy (pertaining to the process of electron transfer) is 14 kcal at temperatures above 20°C.

Molybdenum evidently plays a significant part in the activation of N_2 and the other substrates of nitrogenase, and seems to participate in the formation of the intermediate complex.

Application of X-ray absorbtion spectroscopy has revealed[25] that molybdenum in nitrogenase is surrounded by sulphur atoms and has an atom of iron in the neighbourhood. Taking into account that iron atoms are known to form sulphur-containing clusters, one can suggest that the electron-transferring chain of iron–sulphur clusters has at the end a molybdenum atom bound to the iron cluster by sulphur bridges.

B. Hydroxides of Ti(III), Cr(II), and V(II), with Mo(III) Participation

These systems were discovered in 1970[26–29] based on previous assumptions for the mechanism of dinitrogen reduction.[6] They were the first systems that reproducibly reduced molecular nitrogen in aqueous alkaline and aqueous alcoholic alkaline media with appreciable yields of hydrazine and ammonia as products. The behaviour of these systems was subsequently described in a number of publications (see reference *1*).

Table 1 presents some of the results obtained. $Ti(OH)_3$ and $Cr(OH)_2$ are practically inactive toward dinitrogen without the Mo(III) under the experimental conditions. Mo(III) also activates vanadium(II) hydroxide preserved in the absence of N_2 (reduction of dinitrogen with freshly prepared vanadium(II)–magnesium hydroxide is described later). A Mo(III) compound is formed when the initially added Mo compound (e.g., $MoOCl_3$ or MoO_4^{2-}) is reduced by a reducing agent *in situ*. The oxidation state of molybdenum (Mo(III) with electronic configuration d^3) was confirmed by the analysis of ESCA data.[30] $Mo(OH)_3$ does not by itself reduce dinitrogen, and thus in the case of $Ti(OH)_3$ and $Cr(OH)_2$, only the combined action of the reducing agent and Mo(III) ensures N_2 reduction. Addition of magnesium salts to acid solutions of $Ti(OH)_3$ activates the $Ti(OH)_3$–Mo(III) system at elevated

Table 1. N_2 reduction in $(TiOH)_3$–Mo(III) and $Cr(OH)_2$–Mo(III) systems

Mo/Red	P_{N_2} (atm)	T (°C)	Time (min)	N_2H_4/Mo
Red = $Ti(OH)_3$	Medium, $CH_3OH + H_2O$ (mole ratio 6·5 : 1) KOH = 1 M			
$2{\cdot}0 \times 10^{-5}$	100	85	15	87·5
$2{\cdot}7 \times 10^{-4}$	120	95	15	46·2
1×10^{-2}	100	85	10	2
Red = $Cr(OH)_2$	Medium, $CH_3OH + H_2O$ (mole ratio 5·5 : 1) KOH = 0·4 M			
1×10^{-3}	40	90	15	0·8
1×10^{-3}	40	24	180	0·35

temperatures (above 80°C). Catalytic yields of N_2H_4 and NH_3 with respect to molybdenum are formed, and the stoichiometry of the reaction for N_2H_4 formation can be written as follows:

$$4Ti(OH)_3 + N_2 + 2H_2O \xrightarrow{Mo(III)} N_2H_4 + 4TiO_2.$$

In the absence of dinitrogen, the system forms dihydrogen from water, the reaction proceeding even without Mo(III) and being retarded by $Mg(OH)_2$ formed together with $Ti(OH)_3$. This explains the increased yields of N_2H_4 and NH_3 in the presence of $Mg(OH)_2$, though the activation energy for N_2 reduction turned out to be even lower in the absence of $Mg(OH)_2$ (7 kcal mol^{-1}) than with $Mg(OH)_2$ (13 kcal mol^{-1}). Analysing kinetic results, the authors[31] concluded that the catalytic action of the system was associated with the transfer of electrons along a polymer chain of Ti^{3+} ions in the hydroxide surrounding the complex, to a N_2 complex of Mo(III), which is an electron trap. Investigation of the electrical conductivity of dried Ti(III) hydroxide, which is a semiconductor, supports this conclusion. Additions of molybdenum compounds decrease the electrical conductivity of the system, and this explains the decrease of the catalytic effect of Mo with increase of its concentration in the hydroxide. Increase of electroconductivity with the temperature corresponds to virtually the same activation energies as those for reaction with N_2 (~13 and ~7 kcal mol^{-1}, respectively, for the $Ti(OH)_3$–Mo(III) system with and without $Mg(OH)_2$),[32,33] supporting the view that electron transfer through the hydroxide is an important stage in N_2 reduction. In contrast to $Ti(OH)_3$, the hydroxides of Cr(II) and V(II) which apparently have no pronounced electrical conductivity do not give catalytic yields of products. However, the activity of these systems with respect to dinitrogen remains sufficiently high and hydrazine, for example, is observed to form at an appreciable rate in the $Cr(OH)_2$–Mo(III) system even at room temperature.

Carbon monoxide inhibits the reduction of dinitrogen in the presence of Mo(III). In a CO atmosphere the same yellow carbonyl complex of Mo passes into solution in all three systems,[34] this complex undergoing equilibrium dissociation in the solution forming free CO. The complex isolated from the solution after the removal of one CO molecule is also a carbonyl derivative of Mo and according to X-ray analysis[35] contains four $Mo(CO)_3$ moieties bound by oxygen bridges. The oxidation state of molybdenum in the complex corresponds to Mo(0). Thus further reduction of Mo(III) proceeds in the presence of CO which stabilizes the low valency state of Mo. The carbonyl complex formed is again capable of catalysing hydrazine formation when added to $Ti(OH)_3$ or $Cr(OH)_2$. The reduction of N_2 apparently proceeds only when carbon monoxide molecules in the coordination sphere of the complex are reduced (mainly to CH_3OH) on the surface of Ti(III) hydroxide. The activity of the Mo cluster towards N_2 serves as indirect confirmation of a non-monomeric structure of the active Mo(III) complex in the hydroxide systems with respect to dinitrogen fixation.

$Ti(OH)_3$ in alkaline media is a stronger reducing agent than H_2, thus the $Ti(OH)_3$–Mo(III) system cannot be a catalyst for hydrogenation of N_2 with N_2H_4 formation. However, the Ti(IV) formed can again be reduced by H_2 (e.g., with participation of a Pt catalyst) after addition of acid. Therefore, H_2 can be used as a reducing agent for N_2, with accumulation of N_2H_4 successively changing the pH; in alkaline media N_2H_4 is formed from N_2, while in acid media Ti(III) is again regenerated from Ti(IV). The energy needed by the H_2 in order to form N_2H_4 from N_2 and H_2 is thus provided by the neutralization reaction. The authors of this work[36,37] suggest that the acid action in this case is in effect similar to that of ATP in biological N_2 fixation, though naturally "acidification" in enzymatic ATP hydrolysis proceeds specifically only in the reaction centre, and this is so far unattainable in model experiments.

Other substrates of nitrogenase do not require the joint presence of the reducing agent and Mo(III) for the reduction to take place. Acetylene, for example, is reduced both under the action of $Ti(OH)_3$, $Cr(OH)_2$, or $V(OH)_2$, in the absence of Mo(III), and also under the action of $Mo(OH)_3$ itself,[38] ethylene and ethane being observed in the reaction products. Since ethylene is not by itself reduced in these systems, ethane formation indicates that the reaction with the participation of four electrons of a reducing agent, e.g.,

$$4Cr(OH)_2 + C_2H_2 + 4H_2O \rightarrow 4Cr(OH)_3 + C_2H_6$$

proceeds without release of free C_2H_4 from the coordination sphere of the complex.

C. Amalgams of alkaline metals as reducing agents of dinitrogen

Following the mechanism for the reduction of dinitrogen in the $Ti(OH)_3$–Mo(III) system, it was assumed[39] that, if electron transfer to the activated dinitrogen molecule were conducted by the lattice of titanium hydroxide, the use of a stronger reducing agent (e.g., sodium or potassium amalgam) should make it possible to transform the $Ti(OH)_3$–Mo(III) system into a catalyst for dinitrogen reduction under the action of the amalgam. The presence of alkaline metal amalgam does, in fact, increase the activity of the $Ti(OH)_3$–Mo(III) system toward dinitrogen.

An interesting feature of this system is the concurrent formation of hydrazine and ammonia (Fig. 3), which shows that free hydrazine is not the intermediate product in the formation of the majority of the ammonia.[40] This conclusion was verified by adding labelled hydrazine, $^{15}N_2H_4$, to the system. Both in the absence and the presence of dinitrogen, hydrazine is slowly reduced

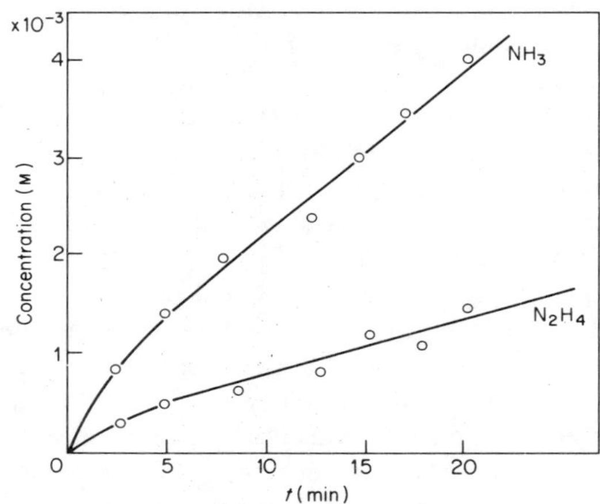

Fig. 3. Kinetics of NH_3 and N_2H_4 formation in the Na(Hg)–$Ti(OH)_3$–Mo(III) system in aq. CH_3OH; $T = 20°C$, $P = 100$ atm N_2.

at room temperature but its reduction rate cannot account for the observed rates of ammonia formation. This fact suggested that two types of active centres exist.[40] On one centre dinitrogen is reduced to hydrazine, while on the other type of active centre the reduction proceeds directly to ammonia without any release of hydrazine from the coordination sphere of the catalytic complex.

If titanium(III) hydroxide is the carrier of electrons from the alkaline metal amalgam to the Mo(III)-activated dinitrogen molecule, the reduction of N_2 should proceed without $Ti(OH)_3$ by direct contact of the alkaline metal amalgam with the Mo(III) complex, and the catalytic reduction of dinitrogen by sodium amalgam in the presence of Mo complexes in methanol was actually observed.[34,41] The reduction product was hydrazine, the reaction proceeding at room temperature with every Mo complex catalysing formation of a number of hydrazine molecules. The reaction, however, has to be carried out with very small concentrations of molybdenum (10^{-5}–10^{-7} M), because with higher concentrations the catalytic formation of dihydrogen proceeds very rapidly. Reproducible results were obtained when the Mo(III) complex was prepared by the reduction of Mo(V) with salts of Mg which apparently take part in the formation of a catalytic particle.[41]

In the Na(Hg)–Mo(III) system, azide is catalytically reduced to N_2 and NH_3, acetylene to ethylene and ethane, and ethylene to ethane. These results, especially the capacity to reduce ethylene, are evidently the consequence of the high reduction potential of the amalgam ($E_0 = -1{\cdot}85$ V).

D. Electrochemical reduction of dinitrogen in protic media

The possibility of using amalgams of alkaline metals as reducing agents for dinitrogen shows that a mercury cathode can be used to reduce N_2 at least by the amalgam formed electrochemically. Accordingly, considerable quantities of hydrazine and ammonia were observed to form in the Na(Hg)–$Ti(OH)_3$–Mo(III) system with the amalgam obtained from the electrolysis of an aqueous-alcohol suspension of $Ti(OH)_3$–Mo(III) on a mercury cathode at room temperature.[39] Hydrazine and ammonia could be obtained in this system as a result of electrolysis in the presence of N_2.[42]

In homogeneous solutions of Mo complexes only small amounts of N_2H_4 were found in electrochemical reduction of N_2 on a mercury cathode in water solutions,[43] but with methylates of Ti(III) and Mo(III) in methyl alcohol considerable and reproducible yields of hydrazine are obtained during electrolysis with controlled potential of a mercury cathode which at its optimal value turned out to be lower than that for amalgam formation.[44]

Acetylene is readily reduced electrochemically with good yields to ethylene and ethane in the presence of various dissolved Mo(III) complexes.[45–48] Ethane, for the most part, is again formed when acetylene is reduced in the coordination sphere of Mo(III) without intermediate formation of free ethylene. The Mo(III) complexes can reduce acetylene catalytically only in contact with the cathode which is the source of electrons; the complexes alone practically do not react with acetylene.

E. The $V(OH)_2$–$Mg(OH)_2$ system

If dinitrogen is passed into an aqueous or aqueous-alcohol suspension of freshly prepared V(II)–Mg(II) hydroxides, formed by adding alkali to a solution of a mixture of VCl_2 and $MgCl_2$, hydrazine can be formed even without added molybdenum. In this simple system discovered in 1970[27] the reaction proceeds at high rates, even at room and lower temperatures, resulting

Table 2. N_2 reduction via the $V(OH)_2$–$Mg(OH)_2$ system[a]

Mg^{2+}/V^{2+}	T (°C)	P_{N_2} (atm)	Time (h)	Yield (%)	
				N_2H_4	NH_3
20	−20	1	8·1	14·4	traces
6	−20	100	0·16	40	traces
6	−20	100	2·0	62	traces
20	+20	1	1·0	5	44

[a] Aqueous CH_3OH (20%), KOH = 6 M

in considerable yields of hydrazine and ammonia (Table 2). The excess of Mg^{2+} ions in the hydroxide is essential since the reduction hardly takes place without them. Depending on the temperature, dinitrogen pressure and the solvent, the reaction can proceed mainly either to hydrazine (high pressure, high pH values) or ammonia. The following reactions can be carried out:

$$4V(OH)_2 + N_2 + 4H_2O \xrightarrow{Mg(OH)_2} 4V(OH)_3 + N_2H_4$$

$$6V(OH)_2 + N_2 + 6H_2O \xrightarrow{Mg(OH)_2} 6V(OH)_3 + 2NH_3.$$

Kinetic studies show that NH_3 in this system is formed as a result of further N_2H_4 reduction. The activity of the mixed vanadium–magnesium hydroxide towards dinitrogen is demonstrated by the fact that it reduces N_2 directly from the air, with the yield being about half that from pure dinitrogen. Moreover, carbon monoxide does not in this case much inhibit the reaction,[40] being reduced to methyl alcohol.[50]

Acetylene is reduced by the hydroxide of divalent vanadium to ethylene and ethane, additions of magnesium salts not being necessary in this case. The mixed vanadium–magnesium hydroxide, as distinct from pure $V(OH)_2$, acquires the capability of also reducing ethylene to ethane.[38]

A quantitative study of the inhibition of dinitrogen reduction by adding V^{3+} salts led to the conclusion[51] that four V^{2+} ions are present in the active centre (Fig. 4). The role of magnesium ions becomes somewhat clearer as a result of studying the magnetic susceptibility of the system; in the presence of Mg^{2+} ions the effective magnetic moment of the hydroxide increases, which indicates that the spin exchange among V^{2+} ions forming the hydroxide decreases.[52]

Schrauzer *et al.*[53,54] came to a completely different conclusion about the mechanism of $V(OH)_2$–$Mg(OH)_2$ reaction with N_2 and other substrates suggesting that V^{2+} reacts in all the cases as a single ion oxidizing to V^{4+}. For N_2 reduction, diazene, N_2H_2, is proposed as an intermediate. However, new evidence has been obtained[55] for the multinuclear structure of an active centre for N_2 reduction to N_2H_4 and C_2H_2 reduction directly to C_2H_6 (without

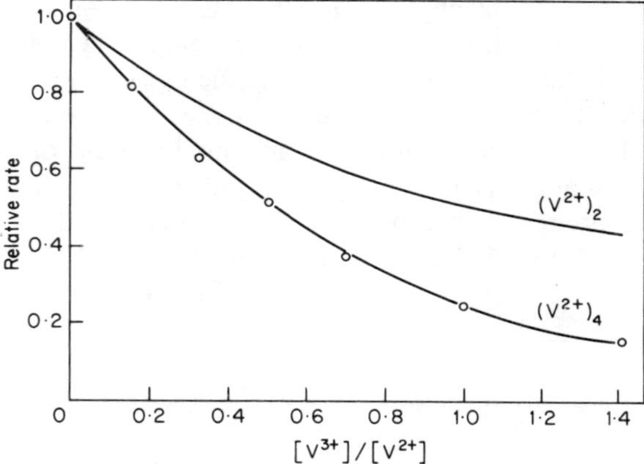

Fig. 4. Inhibiting effect of V(III) additions on N_2H_4 formation from N_2 in the $V(OH)_2$–$Mg(OH)_2$ system. Curves 1 and 2 are calculated for $(V^{2+})_2$ and $(V^{2+})_4$ systems, respectively, on the grounds of equal probability of V^{2+} and V^{3+} entering the system. The points correspond to experimental data.

intermediate formation of C_2H_4). According to reference *55* ions of V^{2+} in the mixed Mg^{2+}–V^{2+} hydroxide diffuse to the surface of the hydroxide and form clusters capable of reducing N_2 and C_2H_2.

The following mechanism of N_2 reduction via the V(II)–Mg(II) mixed hydroxide may be suggested:

$$>V^{II}\langle^{O}_{O}\rangle V^{II}-N{\equiv}N-V^{II}\langle^{O}_{O}\rangle V^{II}< \quad (V^{II}-O-Mg-O-V^{II}) \xrightarrow{H_2O}$$

$$OH^-$$

$$>V^{III}\langle^{O}_{O}\rangle V^{III}{=}N-\overset{+}{N}H{=}V^{III}\langle^{O}_{O}\rangle V^{III}< \quad (V^{III}-O-Mg-O-V^{III}) \xrightarrow{H_2O} N_2H_4$$

F. Soluble complexes of V(II) with catechol and its analogues

In 1972 and later, it was shown that dinitrogen can be reduced with good yields to ammonia in homogeneous aqueous and alcoholic solutions of V(II) complexes with catechol and some other aromatic diols.[56,57,58] For catechol in aqueous media the reaction proceeds in the pH range 8·5–13·5, the maximum yields being observed at pH $\approx$ 10. The reduction of N_2 takes place at room and lower temperatures. In alcoholic solutions at room temperature and atmospheric pressure of dinitrogen, the ammonia yield reached 60% with respect to the V(II) expenditure; in aqueous solutions such yields are observed only at elevated pressures. In the absence of dinitrogen, a parallel oxidation reaction of V(II) by the solvent proceeds with H_2 evolution. The yield of NH_3 increases with increase of dinitrogen pressure due to the inhibition of the competing reaction of dihydrogen formation, but at no pressure does the yield of ammonia exceed 75% of the reducing agent (Fig. 5); 25% of the V(II)

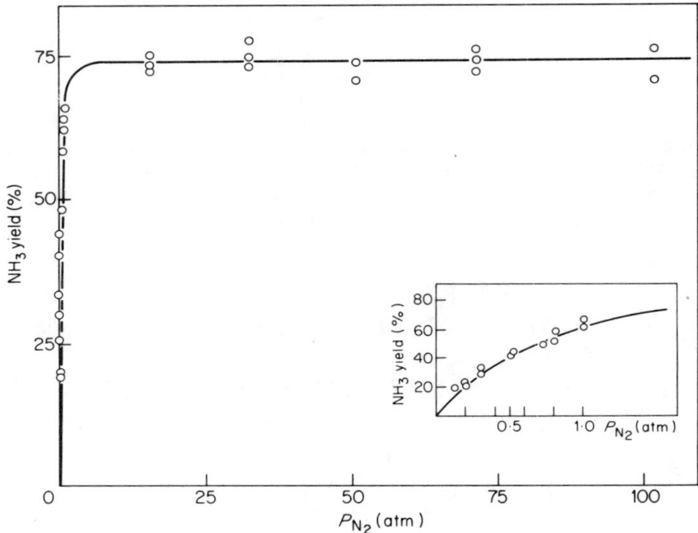

Fig. 5. Dependence of NH_3 yield on P_{N_2} for N_2 reduction by the V(II) catechol complex in methanol.

electrons account for the hydrogen formed concurrently. This residual process of hydrogen evolution is coupled with the reduction of N_2, since kinetic measurements show that both the rate of dihydrogen formation (at dinitrogen pressures corresponding to the limiting yields of ammonia) and the rate of ammonia formation grow in proportion to the dinitrogen pressure. Thus, in a manner similar to the biological fixation of dinitrogen, the stoichiometry of the reaction corresponds to the following equation:

$$8V^{2+} + N_2 + 8H_2O \rightarrow 8V^{3+} + 2NH_3 + H_2 + 8OH^-.$$

Again, as in biological dinitrogen fixation, carbon monoxide inhibits the formation of ammonia but not the formation of dihydrogen. In fact, the rate of

H_2 evolution is even increased in the presence of CO.[58] In CH_3OH, however, the yield of H_2 in the presence of CO is much lower than that required according to stoichiometric oxidation of V^{2+} by protons. Investigation of this phenomenon in the presence of ^{14}CO in CH_3OH has shown that carbon monoxide is reduced mainly to methyl alcohol. Some reduction of CO to CH_3OH was shown later also to take place in water solutions.[50] Acetylene is quantitatively reduced to ethylene and, in the case of C_2D_2 in H_2O, *cis*-deuteroethylene is formed. The range of pH for acetylene reduction is much broader than that for dinitrogen (from pH 5 to concentrated alkali).

Kinetic studies on the oxidation of the V(II)–catechol complex in a dinitrogen atmosphere (Fig. 6) showed[59] the rate equation to be of the form:

$$-d[V(II)]/dt = k_1[V(II)]^2 N_2 + k_2[V(II)]^{1/2}.$$

The first term pertains to the reduction of dinitrogen to ammonia, and the second to the parallel (and independent) reaction of H_2 evolution from the solvent protons. Hydrazine which could be the intermediate reaction product was found to be easily reduced to ammonia in V(II)–catechol solutions. When the dinitrogen reduction is abruptly stopped in its initial stages by adding excess acid or oxidant ($VOSO_4$), a small quantity of hydrazine is found which may confirm its intermediate formation.[60] A kinetic study of hydrazine reduction with the help of a similar procedure, the reaction being stopped by

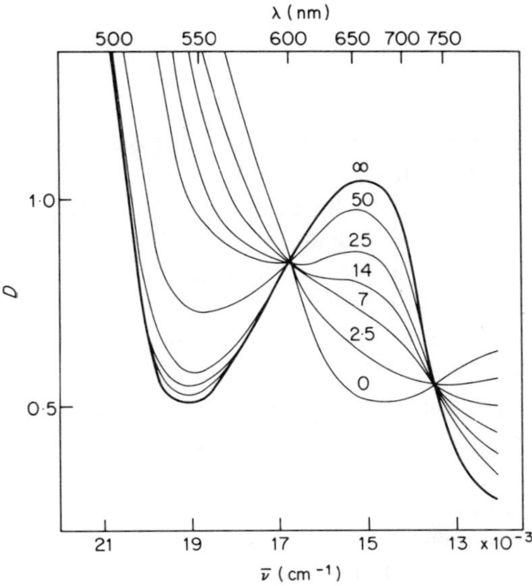

Fig. 6. Electronic spectra of V(II) catechol system under N_2 in methanol. Figures on the curves correspond to time of the reaction in minutes.

adding acid or V(IV), has, however, shown the rate constant to be at least two orders of magnitude smaller than it would have been had hydrazine, found in the system when dinitrogen is being reduced, been the only intermediate product during ammonia formation. From this it can be inferred that if the main pathway of the reaction proceeds via hydrazine, this molecule must be in a different state from free hydrazine. This implies that the reducible hydrazine does not leave the coordination sphere of the vanadium complex. As already mentioned, similar behaviour was observed for nitrogenase, where hydrazine is formed upon decomposition of the acting enzyme by acid or alkali.[23] In both cases (enzyme and V(II)–catechol complex) the hydrazine derivative is apparently an intermediate that forms N_2H_4 when decomposed by acid and forms NH_3 when the system is allowed to react further.

The second-order term in vanadium for N_2 reduction, and the corresponding 0·5 order for dihydrogen formation, are indicative of a polynuclear structure for the transition state during the reduction of dinitrogen. H_2 formation from the solvent protons requires two electrons. A comparison of the kinetic orders shows that eight electrons are available for the reduction of dinitrogen in the transition state, and it is reasonable to assume that the stoichiometry of the equation above reflects the reaction mechanism. It is possible, for example, to assume that at pH 8·5–13·5, a particle containing four vanadium atoms is present in the solution. This particle forms a complex with N_2 and then undergoes an equilibrium transformation into a hydrazine derivative of the type:

$$V{=}N{-}N{=}V$$

Irreversible reduction to ammonia becomes possible during collision with another tetranuclear particle. The N–N bond is disrupted in this case, and an ammonia derivative is formed which is then hydrolysed to NH_3. At the same time the two "extra" electrons are utilized for the formation of the H_2 molecule. An alternative possibility is the reaction of two binuclear particles with a $V^{2+} \xrightarrow{-2e} V^{4+}$ transfer occurring when dinitrogen is reduced to ammonia and H_2 is formed. A subsequent reaction of V^{4+} with excess V^{2+} to give $2V^{3+}$ will result in the formation of the final V(III) complex observed. Kinetic data, therefore, undoubtedly indicate that at least a four-nuclear, and taking into account the energy requirements more probably eight-nuclear transition state is responsible for N_2 reduction.

The 0·5 order for dihydrogen formation can be explained by an equilibrium dissociation of the complex into two reactive particles (each of them probably containing two V^{2+} ions). An undetectable reaction rate for the undissociated state indicates that the active centre is inaccessible for the solvent molecules because of the organic ligands surrounding the V^{2+} ions. This appears to be an important factor in determining why dinitrogen is a more reactive substrate than the proton-containing molecules of the solvent.

The catalytic role of CO in H_2 formation, and its inhibition of N_2 reduction, can be explained by assuming that reaction of the tetramer with CO gives a complex which can readily react with H_2O (possibly via dissociation into dimers).

The complex nature of the transition state is confirmed by data obtained with ligands other than catechol.[61] Phenols containing two *ortho* OH groups for complexing to the V(II) are active with respect to dinitrogen reduction. When one or two OH groups are substituted with SH, V(II) complexes are still formed, but they are inactive toward N_2, although they are able to form H_2 from the solvent and to reduce acetylene.

V(II) complexes with pyrogallol and gallic acid are capable of reducing dinitrogen at much higher pH values ($\geqslant 13$) than are catechol complexes, and the complex with protocatechuic acid (similar to gallic acid) is totally inactive with respect to N_2. These data, as well as the narrow pH range within which the V(II)–catechol complexes form ammonia from dinitrogen, are indicative of the necessity for the presence of only certain definite types of complexes, evidently in an associated form.

G. Other systems reducing dinitrogen in protic media; Schrauzer's model of nitrogenase

The systems described above, discovered in the author's laboratory and studied for the last nine years, are presently the most effective in terms of the nonbiological reduction of dinitrogen in protic media. At the same time, as already mentioned, a number of systems have been described in the literature which are thought capable of reducing dinitrogen with formation of small quantities of ammonia. A review[2] includes both the reports on these results and later results usually obtained with labelled nitrogen $^{15}N_2$, from which it appears that in these systems the ammonia is formed not from dinitrogen but from nitrogen-containing impurities in the reagents. On the other hand, there is no doubt that other substrates of nitrogenase: N_2O, CN^-, CH_3CN, CH_3NC and C_2H_2, are effectively reduced in aqueous solutions under the action of a large number of reducing systems that are inactive or only slightly active towards dinitrogen.[2] Systems of this kind have already been partially considered above.

Systems discovered and studied by Schrauzer and co-workers[62]† should be noted specifically. These systems involve complexes which include molybdenum and compounds containing the SH group (cysteine, glutathione,

† The first information on molybdenum–thiol complexes appeared in 1970,[63] only a few months after the first report of hydroxide systems with molybdenum participation.[22]

thioglycerol); in the presence of $NaBH_4$, they are capable of reducing acetylene, N_2O, CN^-, RCN, and N_3^-. Acetylene is reduced to ethylene and partially to ethane. As distinct from nitrogenase, ethylene can also be reduced to ethane and butadiene is produced in high yields from acetylene with ethylene and ethane.[64] Reaction with $NaBD_4$ shows that the reduction proceeds with the participation of hydrides.[65] In Schrauzer's opinion mononuclear Mo(IV) complexes display catalytic activity, and these systems are considered a close model of nitrogenase.

A distinctive feature of these systems is the activating role of ATP. However, the mechanism of its action differs from that in the reaction of enzymatic reduction. Thus it was shown[66] that ATP hydrolysis essentially does not take place, and so the energy of the macroergic bond is not utilized; this differs in principle from ATP action in biological systems. It is significant that, if the pH value of the added ATP solution is brought to that of the reaction medium, ATP does not display any activating effect. Moreover, other protic acids, such as ADP and even sulphuric acid, also have a strong activating effect,[65] and the acid effect is possibly associated with protonation of the ligands in the molybdenum complex, which can facilitate H^- transfer from $NaBH_4$.[65] In later work Schrauzer *et al.*[67] also assume a similar mechanism but suggest that ATP, ADP and phosphoric acid anions have an assisting effect.†

The possibility of dinitrogen being reduced in Schrauzer's systems, remains unclear. In one of Schrauzer's early works[68] data are presented showing small NH_3 yields (0·06–0·1% with respect to Mo, in five days with a N_2 pressure of 140 atm.). Elsewhere, a yield of NH_3 is reported corresponding to 4% of the Mo in 116 hr at 1 atm. N_2.[69] It was reported[70] that the yield of NH_3 in systems similar to Schrauzer's can even reach 1·5 moles of ammonia per mole of Mo compound, but later studies did not confirm this result.[20] Negative results have been reported in attempts to reduce N_2 with these systems,[71,72] labelled dinitrogen also being used.[72]

In his reports Schrauzer claims that the primary product of dinitrogen reduction is N_2H_2 "being accumulated during the first 40 min".[73] Taking into consideration the well known reactivity of N_2H_2, this assertion cannot but raise doubt. It is also difficult to understand why this reactive compound by itself is incapable of further reduction and it is assumed to be disproportionate with the formation of hydrazine, which is thought to undergo further reduction to ammonia. In any case, it is obvious that the activity of Schrauzer's systems towards dinitrogen (if it does exist) is very small and not comparable with the activity displayed by the systems described in the preceding sections.

---

† This assumption is difficult to reconcile with the fact that anions of ATP and ADP do not effect the rate, if the pH of the added solution is equalized to the pH of the reaction mixture. Differences in behaviour of different acids may be due to differences in pK of the anions, such as $H_2PO_4^-$ (pK_A = 7·2) and HSO_4^{2-} (pK_a = 2).

IV. Conclusions

The analysis of quite a large amount of experimental data obtained by studying dinitrogen reduction in protic media makes it possible to conclude that in all the cases presently known the reduction proceeds in the coordination sphere of a catalytic complex without intermediate formation of free N_2H and N_2H_2, and in some cases, including that of nitrogenase, without the intermediate formation of free hydrazine. For some substrates, there are undoubtedly no intermediate one- and two-electron reduction products (e.g., ethane is formed from acetylene without the intermediate formation of ethylene).

For dinitrogen it was shown in two cases that N_2 can be directly reduced to ammonia without the intermediate formation of free hydrazine. It seems clear that under these circumstances the intermediate formation of diazene, especially unfavourable from the thermodynamic point of view, appears to be virtually impossible, especially for such comparatively mild reducing agents (including nitrogenase) as are unable to reduce ethylene. Moreover, there seems to be no reason to suppose, as some authors do, that a diazene moiety is formed, bonded and stabilized by the active centre of the catalyst without ever leaving the coordination sphere. Indeed, if the elementary process of N_2 reduction consists in a rearrangement of the type:

$$\bar{M}{-}N{\equiv}N{-}\bar{M} \xrightarrow{H^+} M{=}N{-}\overset{+}{N}(H){=}M$$

(1) **(2)**

it follows that certainly neither the first nor the second complex is a diazene derivative. During the process of N_2 reduction in the coordination sphere of a catalytic complex, the N—N bond order is changing from 3 to 1 (and then to 0) and the N—N bond length naturally cannot miss that corresponding to bond order 2. The potential energy of this state, however, does not necessarily correspond to the energy of an N=N bond since some underbarrier movement of the N nuclei is in principle possible and the activation energy could be lower than the energy of N=N bond formation.

At present it is not clear in what sequence the protonation and disruption of the N—N and M—N bonds take place, but it seems likely that formation of the hydrazine derivative[2] will be stabilized as a result of the conjugation of double bonds; this can correspond to a lowering of the required reduction potential for the metal compounds forming the N_2 complex.

The results of studying dinitrogen reduction in protic media, as well as numerous negative results obtained with various reduction systems, make it possible to conclude that dinitrogen can only be reduced when several conditions are jointly met. These conditions can be formulated as follows:

(1) Dinitrogen must be activated in a complex with a metal of d^3 or d^2 electronic configuration, the most suitable being Mo(III) and V(II). It is becoming more and more probable that Mo(III), first found in model systems, is also an active Mo form in nitrogenase. These data are in agreement with the conclusion on the optimum properties of the d^3 configuration for the intermediate linear binuclear complexes which were discussed above.

(2) The reduction potential of the reducing agent present in the system must be sufficiently high, and in most cases presumably not lower than that required for the reaction $N_2 + 4e + 4H^+ \rightarrow N_2H_4$. Table 3 shows the reduction potentials for dinitrogen-fixing systems at certain pH values. It is seen that all of them satisfy the condition formulated except for nitrogenase, where the potential of the reducing agent (ferredoxin or dithionite) is lower. Taking into account, however, the possibility of increasing the potential at the active centre by hydrolysing ATP, this condition may also be satisfied in the enzymatic process. The stabilization of a hydrazine derivative as described seems to be small and can only ensure a small decrease in the potential as against that required for hydrazine production. In agreement with this assumption hydrazine-producing intermediates were found in systems forming NH_3 from N_2 in the coordination sphere of catalytic complexes and this phenomenon, first found in a model system (V(II)–catechol complex) was then also discovered for nitrogenase.

(3) The system must be multinuclear and have the possibility of transferring several electrons to the centre which activates dinitrogen. In the case of V(II) in hydroxide, functioning both as activator and reducer of dinitrogen to hydrazine, no less than four V(II) species participate in the active centre, changing their valency by one in the process of N_2 reduction and being oxidized to V(III). It is thought that the octavanadium transition state cluster, in the case of the V(II)–catechol complex, carries out the reduction of dinitrogen to ammonia with a simultaneous reduction of two protons to hydrogen. Systems with the participation of Mo(III) (mixed hydroxides and nitrogenase) include not less than two Mo(III) ions bonded with two or more atoms of a stronger reducing agent. Mo(III) by itself seems to be an insufficiently strong reducing agent for dinitrogen, and in principle may remain in the same oxidation state during the process of reduction if the electrons are supplied from the surrounding ions. The mechanisms of catalytic N_2 fixation in nitrogenase and Mo(III)-containing systems are similar from the point of view of electron transfer properties of ions bound to a molybdenum complex (Fig. 7). The functioning of the active centre also seems to be similar and some details of enzymatic N_2 reduction may be treated by model studies.

One can visualize the following mechanism for nitrogenase by analogy with the V(II)–catechol system where no intermediate hydrazine is formed and where the stoichiometry is similar to that of the nitrogenase reaction (25% of

Table 3. Reduction potentials of dinitrogen-fixing systems

Reaction	$E = E_{pHO} - a$ pH	Minimum pH for N_2 reduction	E(V) of reducing system	E(V)[a] N_2H_4
$S_2O_4^{2-} + 4OH^- - 2e \rightleftharpoons 2SO_3^{2+} + 2H_2O$ (4ATP → 4ADP + 4P)	+0·56 − 0·118 pH	7	−0·27[b] (−1·12)[c]	−0·73
V^{2+}(cat) + $2OH^- - e \rightleftharpoons VO^+$(cat) + $_HH_2O$	+0·10 − 0·118 pH	8·5	−0.903 [ref. *74*]	−0·82
$2Cr(OH)_2 + 2OH^- - 2e \xrightleftharpoons{Mo(III)} Cr_2O_3 + 3H_2O$	−0·56 − 0·059 pH	8·9	−1·06	−0·82
$2V(OH)_2 - 2OH^- - 2e \rightleftharpoons V_2O_3 + 3H_2O$	−0·55 − 0·059 pH	13	−1·32	−1·08
$Ti(OH)_3 + OH^- - e \xrightleftharpoons{Mo(III)} TiO_2 + 2H_2O$	−0·80 − 0·059 pH	10	−1·4	−0·91
$Na(Hg) - e \rightleftharpoons Na^+$	−1·85		−1·85	

[a] $E_{N_2H_4}$ is the reduction potential for the process $N_2 + 4H_2O + 4e \rightleftharpoons N_2H_4 + 4OH^-$.

[b] E_{max} = 1·12 V is the potential calculated on the assumption that a minimum of four ATP molecules are hydrolysed per $S_2O_4^{=1}$ ion, correspondingly increasing the reduction potential [ref. *75*].

[c] E_{min} = 0·27 V is the potential of $S_2O_4^{2-}$ without ATP.

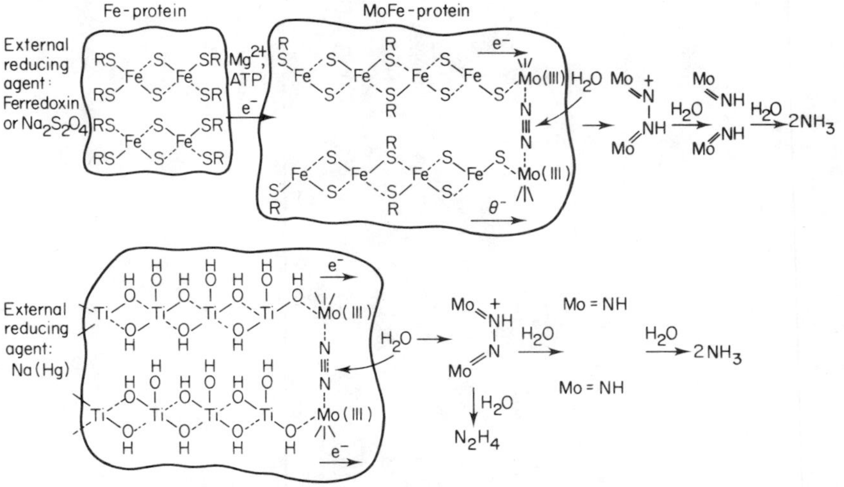

Fig. 7. Chemical mechanism of N_2 reduction in biological and model systems.

electrons are expended for H_2 formation). There are several Fe atoms together with two atoms of Mo in the active centre of nitrogenase capable of accepting electrons from a reducing agent. The number of electrons required to be transferred to the active centre is different for different substrates, and therefore the reduction potential of the acting enzyme is also different. A maximum number of electrons (presumably eight) is necessary for N_2 reduction and the potential is increased to its maximum value. The electrons are transferred by fours and twos, "extra" electrons being used for H_2 formation after N_2 reduction to two NH_3 molecules. Coupled H_2 formation itself proceeds possibly via hydride formed intermediately and this explains coupled H–D exchange in the presence of N_2 which was observed in the enzymatic system and so far has not received unequivocal explanation. Other substrates start their reduction with a smaller number of electrons in the active centre, and coupled H_2 formation does not proceed. Though the potential of the active centre for N_2 reduction is higher than that for other substrates, it is probably lower than for some known non-biological systems as indicated by inactivity of nitrogenase towards C_2H_4.

Thus with the three conditions mentioned above being satisfied optimally, i.e. when the system is well "organized" (with suitable interatomic distances to Mo(III), and the possibility of easy electron transfer among the ions of the cluster), dinitrogen can be easily reduced in protic media and can turn out to be a more reactive substrate than some other fairly active compounds, such as ethylene. When at least one of these conditions is not met, it seems generally that dinitrogen is essentially not reduced in protic media. These conditions are only applicable to dinitrogen and not to any other unsaturated molecules, e.g. other substrates of nitrogenase. When we consider cases like the reduction of acetylene to ethylene, or of nitrous oxide to N_2 and H_2O, none of the three conditions are essential, primarily because the processes are two-electron reactions. For acetylene to be reduced directly to ethane, or cyanide to methane and ammonia, it is essential that a polynuclear system participates, as these reactions require four and six electrons respectively. However, in these cases the participation of Mo(III) or V(II) in the activation of the substrate is not so essential, and the reduction properties of the system need not be so strongly manifested.

The poor (or zero) activity of Schrauzer's systems (Mo–thiol complexes + $NaBH_4$) with respect to dinitrogen is possibly associated with the insufficiently strong reducing properties for reduction of N_2 to hydrazine and/or the insufficiently good "organization" for the intermediate Mo—N≡N—Mo complex to be formed with direct reduction to ammonia. Thus it seems that the better the preliminary "organization" of the polynuclear system, the weaker is the reducing agent necessary for carrying out N_2 reduction. To justify such a generalization it will be necessary to study a greater number of systems than those presently known that are capable, like nitrogenase, of reducing dinitrogen directly to ammonia in protic media. It is hoped that such new catalytic systems will be found in the future.

References

1. A. E. Shilov, *In* "Biological Aspects of Inorganic Chemistry" (Ed. D. Dolphin), p. 197. Wiley, New York (1977).
2. A. E. Shilov, *Uspekhi Khim.* (1974) **43**, 863.
3. M. E. Vol'pin and V. B. Shur, *Doklady Akad. Nauk SSSR* (1964) **156**, 1102.
4. A. D. Allen and C. V. Senoff, *Chem. Commun.* (1965) 621.
5. A. E. Shilov, A. K. Shilova and Yu. G. Borod'ko, *Kinet. Katal.* (1966) **7**, 768.
6. Yu. G. Borod'ko and A. E. Shilov, *Uspekhi Khim.* (1969) **38**, 761.
7. J. Chatt and R. L. Richards, "The Chemistry and Biochemistry of Nitrogen Fixation", p. 57. Plenum Press, London (1971).
8. A. D. Allen, R. O. Harris, B. R. Loescher, J. R. Stevens and R. N. Whitely, *Chem. Rev.* (1973) **73**, 11.
9. J. Chatt, J. R. Dilworth and R. L. Richards, *Chem. Rev.*, (1978) **78**, 589.
10. A. E. Shilov, A. K. Shilova and E. F. Kvashina, *Kinet. Katal.* (1969) **10**, 1402.
11. A. E. Shilov and A. K. Shilova, *Zhur. Fiz. Khim.* (1970) **44**, 288.
12. J. H. Teuben and H. J. de Lief de Meijer, *Rec. Trav. Chim.* (1971) **90**, 360.
13. M. O. Broitman, N. T. Denisov, N. I. Shuvalova and A. E. Shilov, *Kinet. Katal.* (1972) **13**, 61.
14. Yu. G. Borod'ko, M. O. Broitman, L. M. Kachapina, A. E. Shilov and L. Yu. Ukhin, *Chem. Commun.* (1971) 1185.
15. J. M. Manriquez, R. D. Sanner, R. E. Marsh and J. E. Bercaw, *J. Amer. Chem. Soc.* (1976) **98**, 3042.

16. J. Chatt, *J. Organometallic Chem.* (1975) **100**, 17.
17. C. Krüger and Y-H. Tsay, *Angew. Chem.* (1973) **85**, 1051.
18. K. Jonas, D. J. Brauer, C. Krüger, P. J. Roberts and Y-H. Tsay, *J. Amer. Chem. Soc.* (1976) **98**, 74.
19. I. M. Treifel, M. T. Flood, R. E. Marsh and H. B. Gray, *J. Amer. Chem. Soc.* (1969) **91**, 6512.
20. J. Chatt and G. J. Leigh, *Chem. Soc. Rev.* (1972) **1**, 121.
21. R. W. F. Hardy, R. C. Burns and G. W. Parshall, *Adv. Chem. Ser.* (1971) **100**, 219.
22. G. I. Likhtenshtein, R. I. Gvozdev, L. A. Levchenko and L. A. Syrtsova, *Izvest. Akad. Nauk SSSR, Ser. Biol.* (1978) 165.
23. R. N. F. Thorneley, R. R. Eady and D. J. Lowe, *Nature (London)* (1978) **272**, 557.
24. R. I. Gvozdev, A. P. Sadkov, A. I. Kotelnikov and G. I. Likhtenshtein, *Izvest. Akad. Nauk SSSR, Ser. Biol.* (1973) 488.
25. S. P. Cramer, K. O. Hodgson, W. O. Gillum and L. E. Mortenson, *J. Amer. Chem. Soc.* (1978) **100**, 3398.
26. N. T. Denisov, V. F. Shuvalov, N. I. Shuvalova, A. K. Shilova and A. E. Shilov, *Kinet. Katal.* (1970) **11**, 813.
27. N. T. Denisov, O. N. Efimov, N. I. Shuvalova, A. K. Shilova and A. E. Shilov, *Zhur. Fiz. Khim.* (1970) **44**, 2694.
28. N. T. Denisov, V. F. Shuvalov, N. I. Shuvalova, A. K. Shilova and A. E. Shilov, *Doklady Akad. Nauk SSSR* (1970) **195**, 879.
29. A. E. Shilov, N. T. Denisov, O. N. Efimov, V. F. Shuvalov, N. I. Shuvalova and A. K. Shilova, *Nature (London)* (1971) **231**, 460.
30. Yu. G. Borod'ko, N. T. Denisov, M. S. Ioffe, A. I. Korosteleva and T. M. Moravskaja, *Doklady Akad. Nauk. SSSR* (1978) **238**, 596.
31. N. T. Denisov, A. E. Shilov, N. I. Shuvalova and T. P. Panova, *React. Kinet. Catal. Lett.* (1975) **2**, 237.
32. N. T. Denisov, A. I. Korosteleva, Yu. I. Morozov and S. I. Kobeleva, *Kinet. Katal.* (1977) **18**, 758.
33. N. T. Denisov and N. I. Shuvalova, *React. Kinet. Catal. Lett.* (1976) **5**, 431.
34. A. E. Shilov, A. K. Shilova and T. A. Vorontsova, *React. Kinet. Catal. Lett.* (1975) **3**, 143.
35. T. A. Bazhenova, M. S. Ioffe, L. M. Kachapina, R. M. Lobkovskaya, R. P. Shivaeva, A. E. Shilov and A. K. Shilova, *Zhur. Strukt. Khim.* (1978) **19**, 1047.
36. V. V. Abalyaeva, N. T. Denisov, M. L. Khidekel and A. E. Shilov, *Izvest. Akad. Nauk SSSR, Ser. Khim.* (1974) 196.
37. V. V. Abalyaeva, N. T. Denisov, M. L. Khidekel and A. E. Shilov, *Izvest. Akad. Naul SSSR, Ser. Khim.* (1975) 2638.
38. A. G. Ovcharenko, A. E. Shilov and L. A. Nikonova, *Izvest. Akad. Nauk SSSR, Ser. Khim.* (1975) 534.
39. G. V. Nikolaeva, O. N. Efimov, A. A. Brikenshtein and N. T. Denisov, *Zhur. Fiz. Khim.* (1976) **50**, 3030.
40. G. V. Nikolaeva, O. N. Efimov, A. A. Brikenshtein and A. E. Shilov, *React. Kinet. Catal. Lett.* (1977) **6**, 349.
41. L. P. Didenko, A. G. Ovcharenko, A. E. Shilov and A. K.Shilova, *Kinet. Katal.* (1977) **18**, 1078.
42. O. N. Efimov, V. N. Tsarev and A. A. Brikenstein, *Zhur. Fiz. Khim.* (1977) **51**, 1200.

43. A. F. Zueva, O. N. Efimov, A. D. Stirkas and A. E. Shilov, *Zhur. Fiz. Khim.* (1972) **46**, 760.
44. S. I. Kulakovskaya, A. E. Shilov and O. N. Efimov, *Kinet. Katal.* (1977) **18**, 1045.
45. M. Ichikawa and S. Mesitsuka, *J. Amer. Chem. Soc.* (1973) **95**, 3411.
46. D. A. Ledwith and F. A. Schultz, *J. Amer. Chem. Soc.* (1975) **97**, 6591.
47. O. N. Efimov, A. F. Zueva, G. N. Petrova and A. E. Shilov, *Koord. Khim.* (1976) **2**, 62.
48. G. N. Petrova, A. F. Zueva, O. N. Efimov and V. V. Strelets, *Zhur. Fiz. Khim.* (1978) **52**, 3155.
49. N. T. Denisov, E. I. Rudshtein, N. I. Shuvalova and A. E. Shilov, *Doklady Akad. Nauk SSSR* (1972) **202**, 623.
50. L. A. Nikonova, S. A. Isaeva, N. I. Pershikova and A. E. Shilov, *Kinet. Katal* (1977) **18**, 1606.
51. N. T. Denisov, N. I. Shuvalova and A. E. Shilov, *Kinet. Katal.* (1973) **14**, 1325.
52. N. T. Denisov, N. I. Shuvalova, I. N. Ivleva and A. E. Shilov, *Zhur. Fiz. Khim.* (1974) **48**, 2238.
53. Z. I. Zones, T. M. Vickrey, J. G. Palmer and G. N. Schrauzer, *J. Amer. Chem. Soc.* (1976) **98**, 7289.
54. Z. I. Zones, M. R. Palmer, J. G. Palmer, J. M. Doemeny and G. N. Schrauzer, *J. Amer. Chem. Soc.* (1978) **100**, 2113.
55. N. T. Denisov, A. G. Ovcharenko, V. G. Svirin, A. E. Shilov and N. I. Shuvalova, *Nouveau J. Chim.* (1979) **3**, 403.
56. L. A. Nikonova, A. G. Ovcharenko, O. N. Efimov, V. A. Avilov and A. E. Shilov, *Kinet. Katal.* (1972) **13**, 1602.
57. L. A. Nikonova, N. I. Pershikova, M. V. Bodeyko, G. L. Oleinik, D. N. Sokolov and A. E. Shilov, *Doklady Akad. Nauk SSSR* (1974) **216**, 140.
58. L. A. Nikonova, S. A. Isaeva, N. I. Pershikova and A. E. Shilov, *J. Mol. Catal.* (1975/76) **1**, 367.
59. N. P. Luneva, L. A. Nikonova and A. E. Shilov, *React. Kinet. Catal. Lett.* (1976) **5**, 149.
60. N. P. Luneva, L. A. Nikonova and A. E. Shilov, *Kinet. Katal.* (1977) **18**, 254.
61. S. A. Isaeva and L. A. Nikonova, *Izvest. Akad. Nauk SSSR, Ser. Khim.* (1977). 1968.
62. G. N. Schrauzer, *Angew. Chem. (Internat. Edn. Engl)* (1975) **14**, 514.
63. G. N. Schrauzer and G. Schlesinger, *J. Amer. Chem. Soc.* (1970) **92**, 1808.
64. J. L. Corbin, N. Pariyadath and E. J. Steifel, *J. Amer. Chem. Soc.* (1976) **98**, 7862.
65. A. P. Khrushch, T. A. Vorontsova and A. E. Shilov, *J. Amer. Chem. Soc.* (1974) **96**, 4987.
66. T. A. Vorontsova and A. E. Shilov, *Kinet. Katal.* (1973) **14**, 1326.
67. G. N. Schrauzer, G. W. Kiefer, K. Tano and P. R. Robinson, *J. Amer. Chem. Soc.* (1975) **97**, 6088.
68. G. N. Schrauzer, G. Schlesinger and P. A Doemeny, *J. Amer. Chem. Soc.* (1971) **93**, 1803.
69. G. N. Schrauzer, G. W. Kiefer, K. Tano and P. A. Doemeny, *J. Amer. Chem. Soc.* (1974) **96**, 641.
70. R. E. E. Hill and R. L. Richards, *Nature (London)* (1971) **233**, 144.
71. D. Werner, S. A. Rullel and H. J. Evans, *Proc. Natl. Acad. Sci., U.S.A.* (1973) **70**, 339.
72. T. A. Vorontsova, A. P. Khrushch and A. E. Shilov, *Kinet. Katal.* (1975) **16**, 1618.

73. G. N. Schrauzer, G. W. Kiefer, K. Tano and P. A. Doemeny, *J. Amer. Chem. Soc.* (1974) **96**, 641.
74. V. V. Streletz, O. N. Efimov, L. A. Nikonova and Ya. M. Zolotovitzkii, *Zhur. Fiz. Khim.* (1976) **50**, 1018.
75. G. D. Watt, W. A. Bulen, A. Burns and LaMont Hadfield, *Biochemistry* (1975) **14**, 4266.

Part 3

Nitrogen Fixation through Stable Dinitrogen Complexes

§3. NITROGEN FIXATION THROUGH STABLE DINITROGEN COMPLEXES

6

Preparation, Structure, Bonding and Reactivity of Dinitrogen Complexes

ARMANDO J. L. POMBEIRO

Complexo Interdisciplinar, Instituto Superior Técnico, Lisboa, Portugal

Abbreviations

acac = acetylacetone; Ar = aryl; Bu^n = n-butyl; Bu^t = t-butyl; CDT = 1,5,9-cyclododecatriene; COD = cyclo-octa-1,5-diene; Cp = η^5-C_5H_5; Cy = cyclohexyl; depe = $Et_2PCH_2CH_2PEt_2$; diars = 1,2-$(Me_2As)_2C_6H_4$; dmf = dimethylformamide; dmpe = $Me_2PCH_2CH_2PMe_2$; DMSO = dimethylsulfoxide; dmtpe = $(m\text{-Tol})_2PCH_2CH_2P(m\text{-Tol})_2$; dpae = $Ph_2AsCH_2CH_2AsPh_2$; dppae = $Ph_2AsCH_2CH_2PPh_2$; dppe = $Ph_2PCH_2CH_2PPh_2$; dppey = $Ph_2PCH{=}CHPPh_2$; dppm = $Ph_2PCH_2PPh_2$; dppp = $Ph_2P(CH_2)_3PPh_2$; dpte = $PhSCH_2CH_2SPh$; en = ethylenediamine; EPR = electron paramagnetic resonance; Et = ethyl; hmb = hexamethylbenzene; HOMO = highest occupied MO; IR = infrared; LUMO = lowest unoccupied MO; M = metal atom; Me = methyl; MO = molecular orbital; NMR = nuclear magnetic resonance; oep = octaethylporphyrin; Ph = phenyl; PR_3 = monotertiary phosphine; Pr^i = i-propyl; Pr^n = n-propyl; Py = pyridine; RT = room temperature; thf = tetrahydrofuran; Tol = tolyl; trien = triethylenetetraamine; ttp = tetra(*p*-tolyl)porphyrin; UV = ultraviolet; X = halide.

I. Introduction

FULLY characterized dinitrogen complexes have been prepared by conventional techniques with most of the main transition elements; the crossed elements below are the few which have not formed stable dinitrogen complexes as yet.

Ti	~~V~~	Cr	Mn	Fe	Co	Ni	~~Cu~~
Zr	Nb	Mo	~~Tc~~	Ru	Rh	~~Pd~~	~~Ag~~
~~Hf~~	~~Ta~~	W	Re	Os	Ir	~~Pt~~	~~Au~~

Some of the gaps may be filled by considering adducts of the type M–N≡N–Re (M = V, Ta) involving Re dinitrogen complexes, or complexes derived from low-temperature matrix isolation (e.g. for V, Pd, Pt, Cu), a technique which has been increasingly applied in the last few years. Also examples of bridging-dinitrogen complexes of Pd and Pt have been reported, although scantly, with attempted formulations which may be accepted with reserve. However, there is still a small number of dinitrogen complexes compared to the large number of well known carbonyl species.

Many of the dinitrogen complexes have metal atoms in low oxidation states and the co-ligands are generally good electron donors such as tertiary phosphines, ammonia, hydride and halides. Oxygen co-ligands are believed to play an important role during the reduction of N_2 to NH_3 or N_2H_4, due probably to their capacity to behave as π-donors particularly to lighter transition metals in higher oxidation states, but examples of dinitrogen complexes with such co-ligands are scarce.

A wide variety of preparative routes for dinitrogen complexes has been developed and examples of well defined dinitrogen complexes prepared in aqueous solution or from atmospheric dinitrogen have been reported. Also species such as carbonyl, isonitriles, dioxygen, hydrogen, water and ammonia may bind dinitrogen fixation sites, and displacements among these ligands are known to occur. These features observed in inorganic systems are of considerable importance to the understanding of biological dinitrogen fixation.

Different binding modes have been demonstrated by X-rays for N_2 and a still wider versatility of binding is expected. Although most of the reactions which the dinitrogen complexes undergo result in N_2 loss, activation of dinitrogen upon coordination has been observed and in a number of cases it leads to N—H (including NH_3) and N—C bond formation (Chapters 7 and 8). Hence dinitrogen can no longer be considered an inert molecule. Dinitrogen complexes have also been applied in catalysis. Interest in their study has bloomed and several reviews[1–14] have appeared.

II. Preparation of Dinitrogen Complexes

The preparation of any dinitrogen complex may be categorized under at least one of the following general preparative routes. (a) Direct reaction with dinitrogen: (i) reduction of a metal complex under N_2; (ii) addition of N_2 to a coordinatively unsaturated complex; (iii) displacement of a coordinated ligand by N_2; (iv) interaction of N_2 with metal atoms and surfaces. (b) Conversion of N—N or N species into the N_2 ligand. (c) Reaction of a N_2 complex with retention of N_2 (this route will be referred to in Section IV).

The boundaries between these routes are not clearly defined and, for instance, the preparation of a dinitrogen complex by reduction of a metal species appears to proceed also via displacement of a coordinated ligand by N_2 or by addition of N_2 to a coordinatively unsaturated species. Moreover, the same N_2 complex may be prepared in some cases by quite different routes. The most striking example is probably the first dinitrogen complex prepared

$$[Ru(NH_3)_5N_2]^{2+} \quad \textbf{(1)}$$

which was reported by Allen and Senoff[15a] in 1965. Since its preparation from aqueous $RuCl_3$ and hydrazine hydrate, other preparative reactions have been shown to generate the same species: reaction of other Ru(III) and (IV) chloro- and amine-complexes, (e.g., $[RuCl_6]^{2-}$, $[RuCl_5(H_2O)]^{2-}$, $[Ru(NH_3)_5(H_2O)]^{3+}$) with hydrazine hydrate;[15] $[Ru(NH_3)_5Cl]^{2+}$ + monoarylhydrazine;[16] $[Ru(NH_3)_5(H_2O)]^{3+}$ + NaN_3 or decomposition of $[Ru(NH_3)_5N_3]^{2+}$;[15b,17] $[Ru(NH_3)_5(N_2O)]^{2+}$ + Cr^{2+}/H^+ (reduction of the nitrous oxide);[18] $[Ru(NH_3)_5(H_2O)]^{2+}$ + $N_2CHCOOEt$ (reductive C—N bond cleavage of a diazoalkane);[19] $[Ru(NH_3)_6]^{2+}$ + NO/OH^- or $[Ru(NH_3)_5(NO)]^{3+}$ + OH^- (redox coupling);[20] $[Ru(NH_3)_5(H_2O)]^{2+}$ + N_2 (displacement of H_2O by N_2).[21]

A. Preparation of dinitrogen complexes by direct reaction with N_2

(i) *Reduction of a metal complex under dinitrogen*

A wide variety (see Table 1) of dinitrogen complexes may be prepared by reduction of a metal species to a low oxidation state followed by coordination of N_2. The reaction may be carried out in the presence of other ligands such as phosphines if those present in the starting complex are not suitable to stabilize the $M—N_2$ moiety.

Mild reductants such as amalgamated zinc in aqueous solution or stronger agents such as aluminium alkyls and Grignard reagents may be used, and in a few cases they appear in the final dinitrogen species; examples are $[\{(thf)_{1.5}MgCl_3Fe\}_2N_2]$[22] and $[Mg\{(N_2)Co(PMe_3)_3\}_2(thf)_4]$.[23]

Molybdenum(0) and tungsten(0) dinitrogen complexes of the type $[M(N_2)_2L_4]$ (M = Mo,[24] W;[25] L = phosphine or arsine) have been prepared by this route (e.g., reduction of $[MoCl_3(thf)_3]$, $[Mo(acac)_3]$ or $[MCl_4L_2]$) and the N_2 ligand is strongly activated towards electrophiles, undergoing, for example, adduct formation and yielding NH_3 upon protonation.

A dinitrogen–niobium species with bonded dioxygen has also been reported, $[\{Cp_2NbBu^n\}_2(O_2)(N_2)]$.[26a]

Dinitrogen metalloporphyrin complexes, $[Os(N_2)(thf)(L)]$ (L = octaethylporphyrin (oep),[26b] tetra(*p*-tolyl)porphyrin (ttp)[26c]) have been reported from reduction of the dioxoosmium(VI) species $[Os(O)_2(L)]$ by $AlHBu^i_2$ (also hydrazine hydrate in the former case) in thf and under N_2. If the reaction is carried out under argon, $[Os(thf)_2L)]$ is believed to be formed (L = oep). These dinitrogen complexes are interesting inasmuch as metalloporphyrins are known to catalyse the reduction of nitrogenase substrates such as acetylene.[26d]

Dinitrogen titanocene species, intermediates in the reduction of N_2 to NH_3 and N_2H_4 by Vol'pin-type systems, have been isolated at low temperatures from $[Cp_2TiCl_2]$(or$\{Cp_2TiCl\}_2$)—RMgX. They are thermally quite unstable,

Table 1. Preparation of dinitrogen complexes by reduction of metal species under N_2

Starting species	Reductant	Other reagents	N_2 complex	Reference
$[\{Cp_2TiCl\}_2]$[a]	RMgX[a']		$[\{Cp_2Ti\}_2N_2]$[a'']	*27, 32a*
	RMgX[b]		$[\{Cp_2TiR\}_2N_2]$[b']	*28, 33*
	RMgX[c]		$[\{Cp_2Ti\}_2N_2MgCl]$[c']	*29*
$[Cp_2ZrCl(R)]$[d]	Na–Hg		$[Cp_2Zr(N_2)R]$[e]	*95*
$[(\eta^5\text{-}C_5Me_5)_2ZrCl_2]$	Na–Hg		$[\{(\eta^5\text{-}C_5Me_5)_2Zr(N_2)\}_2\text{-}(N_2)]$	*31a, 31b*
$[Cp_2NbCl_2]$	Bu^nLi		$[\{Cp_2NbBu^n\}_2(O_2)(N_2)]$[f]	*26a*
$CrCl_2$	Mg	dppe	$[\{Cr(dppe)_2\}_2N_2]$	*96*
$[CrCl_2(PMe_3)_3]$	Mg	PMe_3	*cis*-$[Cr(N_2)_2(PMe_3)_4]$	*148a*
$[MoCl_3(thf)_3]$[g]	Mg[g']	LL[g'']	$[Mo(N_2)_2(LL)_2]$[g''']	*24d–24i*
$[Mo(acac)_3]$	AlR_3[h]	LL[h']	$[Mo(N_2)_2(LL)_2]$[h'']	*24a–24c*
	AlR_3	PPh_3 + Toluene	$[Mo(\eta^6\text{-}C_6H_5CH_3)(N_2)\text{-}(PPh_3)_2]$[i]	*24a–24c*
$[MoOCl_2(dppe)(thf)]$	Zn	dppe	*trans*-$[MoCl(N_2)\text{-}(dppe)_2]$[j]	*j'*
$[\{(\eta^6\text{-}C_6H_6)(\eta^3\text{-}C_3H_5)\text{-}Mo(\mu\text{-}Cl)\}_2]$	$LiBu^n$	PPh_3	$[\{(\eta^6\text{-}C_6H_6)(PPh_3)_2\text{-}Mo\}_2N_2]$	*42b*
$[(\eta^6\text{-}Me_3C_6H_3)(\eta^3\text{-}C_3H_5)\text{-}Mo(dmpe)]PF6$	Na–Hg		$[\{(\eta^6\text{-}Me_3C_6H_3)(dmpe)\text{-}Mo\}_2N_2]$	*k*
$[WCl_4(PPh_3)_2]$	Na–Hg	dppe	*trans*-$[W(N_2)_2(dppe)_2]$	*25*
$[WCl_4(PMe_2Ph)_2]$	Na–Hg	PMe_2Ph	*cis*-$[W(N_2)_2(PMe_2Ph)_4]$	*25*
$FeCl_3$	Pr^iMgCl	PPh_3	$[Pr^i(PPh_3)_2Fe(N_2)FeH\text{-}(PPh_3)_2Pr^i]$[l]	*l'*
$FeCl_3$	Mg	thf	$[\{(thf)_{1\cdot5}MgCl_3Fe\}_2N_2]$	*22*
$[M(acac)_2]$[m]	Bu^nMgBr	L[m]	$[ML_3(N_2)]$	*m'*
$RuCl_3$ aq.	Zn	thf	$[RuCl_2(N_2)(H_2O)_2(thf)]$[n]	*n'*
$[Ru(NH_3)_5Cl]^{2+}$	Zn–Hg		$[Ru(NH_3)_5N_2]^{2+}$[o]	*21a, 36*
cis-$[Ru(trien)Br_2]^+$	Zn–Hg		$[\{Ru(trien)X\}_2N_2]^{2+}$[p]	*p'*
mer-$[OsX_3L_3]$[q]	Zn–Hg		*mer*-$[OsX_2(N_2)L_3]$	*87*
$[Os(O)_2(L)]$[r]	$AlHBu^i_2$		$[Os(N_2)(thf)(L)]$	*26b, 26c*
$CoCl_2$	Mg	PMe_3 + thf	$[Mg\{(N_2)Co(PMe_3)_3\}_2\text{-}(thf)_4]$	*23*
$[CoCl(PR_3)_3]$[s]	Na		$[\{Co(PR_3)_3\}_2(N_2)]$[s']	*148c*
	Na		$Na[Co(N_2)(PR_3)_3]$[s'']	*148c*
$[CoX_2(PR_3)_2]$[t]	BH_4^-	PR_3[t]	$[CoH(N_2)(PR_3)_3]$[t']	*35*
$[Co(acac)_3]$	AlR_3[u]	PPh_3	$[CoH(N_2)(PPh_3)_3]$[u']	*u''*
$[Co(acac)_2]$	$AlEt_2(OEt)$	PPh_3	$[Co(N_2)(PPh_3)_3]$[v]	*46b*
$RhCl_3\cdot 3H_2O$	Na–Hg	PBu^t_2Ph	*trans*-$[RhH(N_2)\text{-}(PBu^t_2Ph)_2]$	*w*
$[Ni(acac)_2]$	$AlMe_3$	PCy_3	$[\{(PCy_3)_2Ni\}_2N_2]$[x]	*47d*
	$AlBu^i_3$	PR_3[y]	$[NiH(N_2)(PR_3)_2]$	*y'*

[a] Formed by reduction of $[Cp_2TiCl_2]$ (refs. *27b, 28b*). [a'] R = Me; X = Cl, I. Other strong reducing agents may be used, e.g., Na (ref. *32a*), Na–Hg or Li–Hg (ref. *32b*). [a''] Via $\{Cp_2Ti\}_n$ (n = 1, 2), see Table 2. [b] R = Et, Pr^i, $(CH_3)_3CCH_2$, Ph. [b'] Via $[Cp_2TiR]$, see Table 2. Two possibilities for the R group binding site have been suggested (ref. *28b*): Ti or Cp; however, IR and chemical data (ref. *33c*) corroborate the former formulation. [c] R = Pr^i. [c'] Via $[\{Cp_2TiR\}_2N_2]$. [d] R = $(Me_3Si)_2CH$. [e] Converts into $[\{Cp_2ZrR\}_2N_2]$ under vacuum or upon recrystallization in toluene. [f] Prepared under argon with 0·1% O_2 + 0·1% N_2. [g] Most convenient general preparative route to species $[Mo(N_2)_2(LL)_2]$. $[MoCl_4(LL)]$ may also be used as the starting complex. [g'] Na–Hg may also be used. [g''] LL = dppe, dppey, dppae, dpae, diars, $(PMe_2Ph)_2$, $(PMePh_2)_2$, $(PEt_2Ph)_2$. [g'''] All *trans*, except *cis*-$[Mo(N_2)_2(PMe_2Ph)_4]$.

undergoing decomposition with evolution of N_2 upon warming up, and their characterization has been a matter of controversy. However, established formulations appear to include the following: $[\{Cp_2Ti\}_2N_2]$,[27,32a] $[\{Cp_2TiR\}_2N_2]$[28,33] and $[\{Cp_2Ti\}_2(N_2)MgCl]$.[29] Since their instability is, at least in part, due to a facile hydrogen transfer from the Cp ligand to the metal, the difficulties encountered in their characterization have been substantially circumvented by studying analogous pentamethylcyclopentadienyl derivatives. These are much more stable, and their structures have been authenticated by X-rays in the cases $[\{(\eta^5\text{-}C_5Me_5)_2Ti\}_2N_2]$[30] and $[\{(\eta^5\text{-}C_5Me_5)_2Zr(N_2)\}_2N_2]$.[31] Their formation has been shown to proceed via N_2 addition to coordinatively unsaturated species of the types $\{Cp_2Ti\}_n$ ($n = 1,2$; Cp = C_5H_5, C_5Me_5)[30,32] and $[(\eta^5\text{-}C_5H_5)_2TiR]$[33,28c] (see also Table 2).

Only scant data on the mechanism of formation of other N_2 complexes have been reported. However, from kinetic studies,[11] it was concluded that the formation of $[RuCl_2(N_2)(H_2O)_2(thf)]$ from aqueous $RuCl_3$ and Zn amalgam also proceeds via N_2 addition to an intermediate coordinatively unsaturated

---

[h] R = Bu^i, Et. [h'] LL = dppe, dppp, dppm. [h''] All *trans*. [i] Not fully characterized. [j] Controversial formulation. It either readily disproportionates into $[MoCl_2(dppe)_2]$ + $[Mo(N_2)_2(dppe)_2]$ or is a mixture of these two species (J. Chatt, R. A. Head, G. J. Leigh, A. J. L. Pombeiro and R. L. Richards, unpublished observations). May also be prepared from $[MoOCl_3(thf)_3]$ + Zn + dppe or, by a different route, from $[Mo(N_2)_2(dppe)_2]$ + MeCl under irradiation [V. W. Day, T. A. George and S. D. A. Iske, *J. Amer. Chem. Soc.* (1975) **97**, 4127]. [j'] L. K. Atkinson, A. Mawby and D. C. Smith, *Chem. Commun.* (1971) 157. [k] M. L. H. Green and W. E. Silverthorn, *J.C.S. Dalton Trans.* (1974) 2164. [l] With Et_2O of crystallization. [l'] Yu. G. Borod'ko, M. O. Broitman, L. M. Kachapina, A. E. Shilov and L. Yu. Ukhin, *Chem. Commun.* (1971) 1185; L. M. Kachapina, Yu. G. Borod'ko, A. V. Sazhnikova, I. N. Ivleva, Yu. M. Shul'ga and M. O. Broitman, *Doklady Akad. Nauk S.S.S.R.* (1973) **208**, 135. [m] M = Fe; L = PPh_3. M = Co, Ni; L = $P\{p\text{-}(CH_3)_2C_6H_3\}_3$, $P(C_{10}H_7)_3$, $P(C_6H_5CH_2)_3$, $AsPh_3$, $As(C_{10}H_7)_3$, M = Co; L = $Sb\{p\text{-}(CH_3)_2C_6H_3\}_3$. M = Ni; L = PPh_3, $P(p\text{-}CH_3C_6H_4)_3$. [m'] D. Negoiu, C. Parlog and D. Sandulescu, *Rev. Roumaine Chim.* (1974) **19**, 387. [n] Increasing yield with N_2 pressure (almost quantitative at 40–60 atm.). [n'] Ref. *11*; Yu. G. Borod'ko, A. K. Shilova and A. E. Shilov, *Russ. J. Phys. Chem.* (1970) **44**, 349. [o] Air may be used as the source of N_2. Reaction proceeding through $[Ru(NH_3)_5(H_2O)]^{2+}$. The azide route is a better preparative one. [p] X = Br, I (from added KI). [p'] R. O. Harris and B. A. Wright, *Canad. J. Chem.* (1970) **48**, 1815. [q] X = Cl, Br; L = PMe_2Ph, PEt_2Ph. X = Cl; L = PPr^n_2Ph, PBu^n_2Ph, $PMePh_2$, $PEtPh_2$, PEt_3, $AsMe_2Ph$, $AsEt_2Ph$. [r] L = oep, ttp. [s] $R_3 = Ph_3$, Et_2Ph (the starting species is then $[CoCl_2(PEt_2Ph)_2]$ but reaction proceeds through $[CoCl(PEt_2Ph)_3]$, in the presence of this phosphine). [s'] Obtained from Co : Na = 1 : 1. [s''] From Co : Na = 1 : 2. [t] X = Cl, Br, I. $R_3 = Ph_3$, $EtPh_2$, Et_2Ph. [t'] Also prepared reversibly from $[CoH_3(PR_3)_3]$. [u] R = Bu^i, Et. $AlEt_2(OEt)$ may also be used. [u'] Intermediate formation of $[CoR(PPh_3)_3]$ is believed to occur. [u''] Ref. *34*; A. Misono, Y. Uchida and T. Saito, *Bull. Chem. Soc. Japan* (1967) **40**, 700; ref. *45*; B. R. Davis, N. C. Payne and J. A. Ibers, *J. Amer. Chem. Soc.* (1969) **91**, 1240. [v] Also prepared reversibly from $[CoH_2(PPh_3)_3]$ (ref. *46a*). [w] P. R. Hoffman, T. Yoshida, T. Okano, S. Otsuka and J. A. Ibers, *Inorg. Chem.* (1976) **15**, 2462. [x] May also be prepared from $[NiX_2(PCy_3)_2]$ (where X = Cl, Br) + Na (ref. *47e*) and is also formed by displacement reactions of PCy_3 (ref. *47e*) or olefin (ref. *47d*). [y] R = Et, Bu. [y'] S. C. Srivastava and M. Bigorgne, *J. Organometallic Chem.* (1969) **18**, P30.

Table 2. Preparation of dinitrogen complexes by N_2 addition to a coordinatively unsaturated species

Starting complex	Final N_2 complex	Reference
$[Cp_2TiR]^a$	$[\{Cp_2TiR\}_2N_2]$	*33, 28c*
$[\{Cp_2Ti\}_2]^b$	$[\{Cp_2Ti\}_2N_2]$	*32*
$[(\eta^5\text{-}C_5Me_5)_2Ti]^c$	$[\{(\eta^5\text{-}C_5Me_5)_2Ti\}_2N_2]^d$	*30, 32c*
	$[\{(\eta^5\text{-}C_5Me_5)_2Ti(N_2)\}_2N_2]^e$	*30, 32c*
$[Cp_2Ti\text{–}\mu\text{-}(\eta^1:\eta^5\text{-}C_5H_4)\text{–}TiCp]$	$[\{Cp_2Ti\text{–}\mu\text{-}(\eta^1:\eta^5\text{-}C_5H_4)\text{–}TiCp\}_2N_2]$	*40*
$[\{(\eta^5\text{-}C_5R_5)_2M\}_n]^f$	*g*	*37a, 37b*
$[Mo(CO)(dppe)_2]$	*trans*-$[Mo(CO)(N_2)(dppe)_2]$	*h*
$[ReR_4]^i$	$[\{R_4Re\}_2N_2]$	*j*
$[FeH(dppe)_2]BPh_4$	$[FeH(N_2)(dppe)_2]BPh_4$	*37c*
$[Fe(COD)_2]$	$[Fe(N_2)(dppe)_2]^k$	*148b*
$[MH_2(PR_3)_3]^l$	$[MH_2(N_2)(PR_3)_3]$	*37d, 39*
$[RhX(PCy_3)_2]^m$	*trans*-$[RhX(N_2)(PCy_3)_2]$	*38*

[a] R = Ph, *o*-, *m*-, *p*-$CH_3C_6H_4$,C_6H_5,$CH_2C_6F_5$. [b] Titanocene dimer prepared *in situ* from reduction of $[Cp_2TiCl_2]$ by Na (ref. *32a*) or Li amalgam (ref. *32b*), from $[Cp_2TiCl]$ and MeMgI (ref. *27a*), or from $[Cp_2TiH_2]$ (ref. *32c*). [c] Formed from a dihydride species (which is derived from $[(\eta^5\text{-}C_5Me_5)_2TiCl_2]$ + $LiCH_3$) by H_2 evolution. [d] From reaction in pentane at 0°C. [e] From reaction in toluene at −80°C. [f] R = H; M = Mo, W. R = Me; M = Mo. Unknown *n* ($\geqslant$2). Proposed transient intermediates in the N_2 fixation system $[(\eta^5\text{-}C_5R_5)_2MCl_2]$ + Na/Hg. [g] Uncharacterized species. [h] M. Sato, T. Tatsumi, T. Kodama, M. Hidai, T. Uchida and Y. Uchida, *J. Amer. Chem. Soc.* (1978) **100**, 4447. [i] R = CH_2SiMe_3,CH_2CMe_2Ph,CH_2CMe_3. Postulated intermediate in the system $[ReCl_4(thf)_2]$ + RMgCl (or MgR_2). Under argon, cyclotrimerization leads to $[Re_3R_{12}]$. [j] A. F. Masters, K. Mertis, J. F. Gibson and G. Wilkinson, *Nouveau J. Chimie* (1977) **1**, 389. [k] Prepared in the presence of dppe. [l] M = Fe; R_3 = $EtPh_2$, $BuPh_2$ (ref. *37d*). M = Ru; R = Ph (ref. *39*). [m] X = Cl, Br, I. The chloro species is in equilibrium with its dimer.

species $[RuCl_2L_3]$ formed in the reduction. In other cases (preparation of $[CoH(N_2)(PR_3)_3]$) it is believed that an intermediate alkyl complex is formed $[CoR(PPh_3)_3]$ from reaction of $[Co(acac)_3]$ with AlR_3 or $AlR_2(OR)$ (the alkylated species $[CoMe(PPh_3)_3]$ has been isolated after reaction under argon and also using $AlMe_2(OEt)$[34]) or a trihydride complex ($[CoH_3(PR_3)_3]$ has been isolated in the reaction of $[CoX_2(PR_3)_2]$ with BH_4^- in the presence of PR_3)[35], followed by displacement of the alkyl or two hydride ligands by N_2. The preparation of $[Ru(NH_3)_5(N_2)]^{2+}$ by reduction of $[Ru(NH_3)_5Cl]Cl_2$ with Zn–Hg[21a,36] (atmospheric dinitrogen may be used in the reaction) proceeds through the reduced intermediate species $[Ru(NH_3)_5(H_2O)]^{2+}$.

Hence the formation of a N_2 complex upon reduction of a metal centre appears to proceed via an initial reduction followed by a reaction in one of the categories described below (N_2 addition to a coordinatively unsaturated complex or N_2 displacement of a ligand).

(ii) *Addition of dinitrogen to a coordinatively unsaturated complex*

Well defined examples of this type of reaction are few when compared to the other methods used in the preparation of dinitrogen complexes, due to the paucity of isolated parent species. However the involvement of this route is much wider than shown by the examples of Table 2 since in some instances both the methods (i) and (iii) which follows are believed to occur via coordinatively unsaturated species which add N_2.

Dinitrogen addition to $[\{Cp_2Ti\}_2]$, $[Cp_2TiR]$, $[(\eta^5\text{-}C_5Me_5)_2Ti]$ and to related zirconium species has already been referred to. Similarly other metallocenes are believed to add N_2, and species $[\{(\eta^5\text{-}C_5R_5)_2M\}_n]$ (M = Mo, W; R = H and M = Mo; R = Me) are proposed transient intermediates in the dinitrogen fixation system $[(\eta^5\text{-}C_5R_5)_2MCl_2]$–Na/Hg. However, the instability of the dinitrogen derived species has precluded their characterization.[37a,37b]

Atmospheric N_2 may be used in the preparation of $[FeH(N_2)(dppe)_2]BPh_4$ from the five-coordinate $[FeH(dppe)_2]BPh_4$.[37c] $[FeH_2(N_2)(PR_3)_3]$ may also be prepared by N_2 addition to $[FeH_2(PR_3)_3]$.[37d]

The reactions are reversible in a few cases (e.g., with the Ti species) and various other molecules may add to these unsaturated systems. Hence, $[RhX(PCy_3)_2]$ for example undergoes addition reactions with N_2, CO, C_2H_4, PhC≡CPh, O_2 and H_2.[38] Also N_2 and CO add to $[RuH_2(PPh_3)_3]$;[39] and N_2 and H_2 to $[Cp_2Ti\text{–}\mu\text{-}(\eta^1:\eta^5\text{-}C_5H_4)\text{–}TiCp]$.[40]

The "unsaturated" parent species may be solvated: $[\{CpFe(dppe)\}_2N_2]^{2+}$ is probably derived from solvated $[CpFe(dppe)]^+$ which may be prepared *in situ* by irradiation of $[CpFe(CO)(dppe)]^+$ (see Table 3).[41]

(iii) *Displacement of a ligand by* N_2

N_2 may displace not only neutral ligands such as olefins, phosphines, oxygen donor ligands (H_2O, thf, acetone) and nitrogen donor ligands (NH_3) but also ionic ones such as halide (if assisted by a halide abstractor) and two hydride ligands from metal polyhydrides (see Table 3). In this case the metal is reduced and H_2 is evolved, the reaction often being reversible, such as the preparation of $[(\eta^6\text{-arene})Mo(N_2)(PR_3)_2]$ and $[\{(\eta^6\text{-}C_6H_6)Mo(PPh_3)_2\}_2N_2]$ from $[(\eta^6\text{-arene})MoH_2(PR_3)_2]$,[42] of $[FeH_2(N_2)(PRPh_2)_3]$ from $[FeH_4(PRPh_2)_3]$,[43,44] of $[CoH(N_2)(PPh_3)_3]$ from $[CoH_3(PPh_3)_3]$[35,45] and of $[Co(N_2)(PPh_3)_3]$ from $[CoH_2(PPh_3)_3]$.[46] Examples of replacement of two hydrides by other neutral ligands (CO, phosphine) are found in well established reactions.

Olefins may be displaced by N_2: in this category the bridging edge-on dinitrogen species $[\{(PhLi)_3Ni\}_2N_2\cdot 2Et_2O]_2$[47a,47b] and $[\{Ph(Na\cdot OEt_2)_2\text{-}(Ph_2Ni)_2N_2NaLi_6(OEt)_4\cdot OEt_2\}_2]$[47c] are prepared by reaction of [(CDT)Ni] in

Table 3. Preparation of dinitrogen complexes by displacement reactions

Starting complex	Displaced ligand	N_2 complex	Reference
$[(hmb)Cr(CO)_3]$	thf[a]	$[\{(hmb)Cr(CO)_2\}_2N_2]$	*51*
$[Mo(CO)(dmf)(dppe)_2]$	dmf	*trans*-$[Mo(CO)(N_2)(dppe)_2]$	*48*
$[(\eta^6\text{-arene})MoH_2(PR_3)_2]$	dihydrogen	$[(\eta^6\text{-arene})Mo(N_2)(PR_3)_2]$[b]	*42*
		$[\{(\eta^6\text{-}C_6H_6)Mo(PPh_3)_2\}_2N_2]$	*42*
$[CpRe(CO)_2(thf)]$	thf	$[CpRe(CO)_2(N_2)]$[c]	*76d*
$[CpFe(CO)(dppe)]PF_6$	acetone[d]	$[\{CpFe(dppe)\}_2N_2](PF_6)_2$	*41*
$[CpFe(dmpe)I]$	acetone[e]	$[\{CpFe(dmpe)\}_2N_2](BF_4)_2$	*50*
$[FeH_4(PRPh_2)_3]$[f]	dihydrogen	$[FeH_2(N_2)(PRPh_2)_3]$	*43, 44*
$[MHCl(depe)_2]$[g]	solvent[g]	*trans*-$[MH(N_2)(depe)_2]BPh_4$	*52*
$[Ru(NH_3)_{6-x}(H_2O)_x]^{2+}$	water	$[Ru(NH_3)_{6-x}(H_2O)_{x-1}(N_2)]^{2+}$	*h*
$[CoH(NH_3)(PPh_3)_3]$	NH_3	$[CoH(N_2)(PPh_3)_3]$	*34b, 34c*
$[CoH_2(PPh_3)_3]$	dihydrogen	$[Co(N_2)(PPh_3)_3]$	*46*
$[CoH_3(PPh_3)_3]$	dihydrogen	$[CoH(N_2)(PPh_3)_3]$	*35, 45*
$K[Co(olefin)(PMe_3)_3]$[i]	olefin	$K[Co(N_2)(PMe_3)_3]$	*110*
$[\{RhCl(C_8H_{14})_2\}_2]$	C_8H_{14}	$[RhCl(N_2)(PPr^i_3)_2]$[j]	*94*
$[Ni(CDT)]$	CDT	$[\{(PhLi)_3Ni\}_2N_2 \cdot 2Et_2O]_2$[k]	*47a, 47b*
	CDT	$[\{Ph(Na \cdot OEt_2)_2(Ph_2Ni)_2N_2NaLi_6(OEt)_4 \cdot OEt_2\}_2]$[l]	*47c*
$[(PCy_3)_2Ni(olefin)]$	olefin	$[Ni(N_2)(PCy_3)_3]$[m]	*47d*
		$[\{(PCy_3)_2Ni\}_2N_2]$[n]	*47d*
$[Ni(PCy_3)_3]$	PCy_3	$[\{(PCy_3)_2Ni\}_2N_2]$	*47e*
$[Ni(PR_3)_4]$	PR_3	$[Ni(N_2)(PR_3)_3]$[o]	*p*

[a] Reaction carried out in thf under irradiation and proceeding through the proposed intermediate $[(hmb)Cr(CO)_2(thf)]$. [b] Arene = MeC_6H_5; R_3 = Ph_3, Ph_2Me. Arene = *sym*-$Me_3C_6H_3$; R_3 = Ph_3. [c] Reaction under pN_2 = 100 bar. [d] Reaction carried out in acetone under irradiation and proceeding through a solvated intermediate. [e] Reaction in acetone in the presence of $TlBF_4$. [f] R = Me, Et, Bu^n, Ph. Complexes prepared *in situ* from reaction of $FeCl_2$ with BH_4^- in the presence of $PRPh_2$. Atmospheric N_2 may compete successfully with O_2. [g] M = Fe; solvent = acetone. M = Ru, Os; solvent = chloroform. Reactions believed to proceed through solvated intermediates. [h] x = 1 (ref. *21*), 2 (ref. *49a*), 6 (ref. *49b*). Dinuclear species $[\{RuL_5\}_2N_2]^{4+}$ (L = NH_3, H_2O) are formed (x = 1 and 6, respectively) from reaction of the mononuclear dinitrogen complexes with the parent aquocomplexes. [i] Olefin = propene, cyclopentene. [j] Prepared in the presence of PPr^i_3. Related complexes $[RhClL(PPr^i_3)_2]$ (L = C_2H_4, CO, O_2), have been prepared by the same route. [k] Prepared by reaction with LiPh, Et_2O. [l] Prepared by reaction with LiPh/NaPh, Et_2O. [m] Olefin = cyclohexene, *cis*-, *trans*-butene. Reversible reaction. [n] Olefin = ethylene. Reaction accelerated by PCy_3 addition and intermediate formation of $[Ni(PCy_3)_3]$. [o] R_3 = Et_3, Et_2Ph, Bu^n_3. Attempted formulation in solution. [p] Ref. *47e*; C. A. Tolman, D. H. Gerlach, J. P. Jesson and R. A. Schunn, *J. Organometallic Chem.* (1974) **65**, C23.

ether with LiPh and LiPh/NaPh, respectively. The bridging end-on dinitrogen complex $[\{(PCy_3)_2Ni\}_2N_2]$ may be formed by olefin displacement from $[(PCy_3)_2Ni(olefin)]$[47d] or by phosphine displacement from $[Ni(PCy_3)_3]$.[47e]

A mixed carbonyl–dinitrogen complex, *trans*-$[Mo(CO)(N_2)(dppe)_2]$, has been formed by displacement of dimethylformamide from $[Mo(CO)(dmf)(dppe)_2]$ in C_6H_6 (the latter being prepared by reflux of a dmf/C_6H_6 solution of *trans*-$[Mo(N_2)_2(dppe)_2]$).[48]

Dinitrogen may also displace coordinated water, for example in aqueous $[Ru(NH_3)_5(H_2O)]^{2+}$,[21] *cis*-$[Ru(NH_3)_4(H_2O)_2]^{2+}$,[49a] and $[Ru(H_2O)_6]^{2+}$,[47b] and also ammonia, e.g. from $[CoH(NH_3)(PPh_3)_3]$ reversibly.[34b, 34c]

The displacement of NH_3, H_2O and H_2 by N_2 is of biochemical significance. It shows that these species may coordinate the same metal site as dinitrogen, which may replace NH_3 (its final reduction product), thus closing the cycle of a hypothetical catalytic process. Also N_2 can compete favourably with H_2O for a metal site in aqueous solution (the preparation of the Allen and Senoff cation (**1**) from $[Ru(NH_3)_5(H_2O)]^{2+}$).[21] Atmospheric N_2 may be used in successful competition with O_2, as in the preparation of (**1**)[36] and of $[FeH_2(N_2)(PRPh_2)_3]$ from $[FeH_4(PRPh_2)_3]$[43,44] with dihydrogen evolution. All dinitrogen-fixing bacteria contain hydrogenase and can generate H_2, which is a competitive inhibitor of dinitrogen fixation—probably due to competition for the same active site.

The displacement reactions appear to proceed in some instances through solvated intermediates from which the solvent molecules are displaced by N_2. Hence $[\{CpFe(dmpe)\}_2(N_2)](BF_4)_2$ is formed by reaction of [CpFe(dmpe)I] with $TlBF_4$ (halide abstractor) in acetone via the solvated species $[CpFe(dmpe)(OCMe_2)]BF_4$.[50] Also the analogous complexes [CpFe(dmpe)I] $N_2](PF_6)_2$[41] and $[\{Cr(hmb)(CO)_2\}_2N_2]$[51] result from irradiation of $[CpFe(CO)(dppe)]PF_6$ (in acetone) and $[Cr(hmb)(CO)_3]$ (in thf), probably via displacement of coordinated solvent (reversibly in the former case), from the corresponding solvated intermediates, by N_2; dinitrogen does not replace CO directly. Similarly, the reactions of $[MHCl(depe)_2]$ with $NaBPh_4$ in acetone (M = Fe) or chloroform (M = Ru, Os) probably proceed through solvated intermediates $[MHS(N_2)(depe)_2]^+$.[52]

Kinetic studies have been undertaken, for instance on the reaction of N_2 with *cis*-$[Ru(LL)_2(H_2O)_2]^{2+}$ (LL = $(NH_3)_2$, en, $\frac{1}{2}$ triethylenetetramine),[53,49a] but no conclusive mechanistic results have been derived. The mechanism of these reactions remains unknown, the choice between S_N1 or S_N2 types being still undecided.

(iv) *Interaction of dinitrogen with metal surfaces and atoms. Matrix isolation studies*

Detection by IR spectroscopy of chemisorbed $N_2[\nu(N_2) = 2202\ cm^{-1}]$ on a metal surface (Ni) at room temperature was reported[54] at about the same time as the first isolation of a dinitrogen complex. However, interaction of N_2 with metal surfaces was recognized earlier, e.g. with W[55a] where the "surface" compounds formed are believed to contain both atomic (from dissociative chemisorption) nitrogen and molecular nitrogen as deduced from flash-filament mass spectrometry with $^{28}N_2$ and $^{30}N_2$.[55b] Dissociative sorption appears to be favoured by high temperatures[56] and accompanied by nitridation which is believed to result from a strong metal–dinitrogen interaction.[57] However understanding of the interaction of N_2 with metal surfaces has been a matter of controversy and when attempting to distinguish between physical adsorption and weak chemisorption, arguments favouring both have been advanced.[54,58,59] Large $\nu(N_2)$ shifts to lower values, high intensity of the band and a postulated negative dipole away from the metal surface (suggested by desorption spectra and field-emission spectroscopic studies) are better explained by chemisorption. Low heat of reaction in the right range for physical adsorption is ambiguous since this may also be explained by a small difference between the energy evolved during the formation of an M≡N bond and the energy absorbed on changing N≡N into N=N.

Study of the interaction between dinitrogen and metal surfaces is of great importance in attempting to establish a mechanism for the heterogeneous catalysis of dinitrogen reduction to ammonia (as in the Haber process) and is the subject of several reviews.[56,57,11,14] However, the dependence of the formation of some of the resulting dinitrogen species, on such factors as the crystallite size does not agree with the features expected of a true coordination compound and these species (which may be called "surface compounds")[11] will not be referred to further. The rest of this section will deal with the interaction of N_2 with metal atoms in frozen gas matrices, a method increasingly applied over the last few years.[60]

The preparation of binary $M(N_2)_m$ complexes (Table 4) has been reported from matrix co-condensation reactions of atomic metal vapours with dinitrogen gas at low temperature (6–25 K). The deposition temperature is generally critical since a balance between the thermal energy available and the activation energy for complex formation must be satisfied. The species are generally very unstable, decomposing upon evacuation or on raising the temperature, sometimes by only a few degrees. They are generally characterized by IR, Raman and electronic spectroscopy and are considered to be true coordination complexes due to the stronger character of the $M-N_2$ bond compared with N_2 chemisorbed on metal surfaces. This is substantiated by the lower values of $\nu(N_2)$ in the former species. The weaker $M-N_2$ interaction on chemisorption probably arises from the lower $M \rightarrow N_2$ π backbonding due to release of electrons from the surface metal atoms into the conduction band of the metal.[61a]

An unambiguously characterized side-on bonded N_2, $Co(\eta^2\text{-}N_2)$, has been reported, as being prepared by these techniques (co-condensation of Co atoms in N_2 matrices at 10 K). The dihapto structure has been confirmed by IR studies with labelled nitrogen (only one band at 2066·5 cm^{-1} has been observed for $\nu(^{29}N_2)$).[61b]

Octahedral $Ti(N_2)_6$ (a 16-electron species),[61c] $V(N_2)_6$ (17-electron),[61d]

Table 4. Dinitrogen species prepared by matrix isolation techniques

Species	Reference
LiN_2[a]	*67*
$Nb_n(N_2)_m$[b]	*c*
$M(N_2)_6$[d]	M = Ti,[61c] V,[61d] Cr[62a]
$Cr(N_2)_n$	(n = 1–6)[62]
$M(CO)_5(N_2)$	(M = Cr, Mo, W)[69]
$Mn(CO)_{4-n}(N_2)_n(NO)$	(n = 1, 2)[70a]
$Mn(N_2)(NO)_3$	*70b*
$Fe(CO)_{2-n}(N_2)_n(NO)_2$	(n = 1, 2)[70c]
$Co(CO)_{3-n}(N_2)_n(NO)$	(n = 1, 2)[70d]
$Co(\eta^2\text{-}N_2)$	*61b*
$M(N_2)_n$	M = Rh.[61a] Ni;[63,62b] n = 1–4 M = Pd,[63c,64] Pt;[65,62b] n = 1–3
$Ni(N_2)_m(CO)_{4-m}$	(m = 1–3)[66a]
$N(N_2)_n(O_2)$	(M = Ni, Pd, Pt; n = 1, 2)[66b,66c]
$Cu_n(N_2)_m$[b]	*62b*
$MX_2(N_2)$[e]	M = Ni;[71] Sn,[72a] Pb,[72a] Hg[72b]
CN_2	*68a, 68b*
SiN_2	*68c*

[a] Other lithium–dinitrogen species have been detected (see text). [b] Unassigned stoicheiometry. [c] D. W. Green, R. V. Hodges and D. M. Gruen, *Inorg. Chem.* (1976) **15**, 970. [d] $V_2(N_2)_n$, where n is probably 12, has also been reported (ref. *61d*). [e] X = halide.

$Cr(N_2)_6$ (18-electron),[62a] as well as other members of the series (e.g., $Cr(N_2)_n$ (n = 1–6))[62] and the CO analogues ($Ti(CO)_6$,[61c] $V(CO)_6$[61e] and $Cr(CO)_6$[61f]) have been similarly prepared. These $M(N_2)_6$ species constitute the sole examples of binary hexacoordinated N_2 complexes, although examples of CO analogues are well documented for VA and VIA group metals, rare earths and uranium.

Binary dinitrogen species of the Ni subgroup have been reported[63–65] and their stability follows the order Ni > Pt > Pd; the reversal of the Pd and Pt stabilities down the group follows the Allred–Rochow electronegativity values for the corresponding metals and is ascribed to the lanthanide contraction.[65b] An increase in stability with coordination number is observed for these species except for $Pt(N_2)_2$, which exhibits unusual spectroscopic properties and for which side-on bonded N_2 is postulated.[65b]

Mixed N_2/CO ($Ni(N_2)_m(CO)_{4-m}$ (m = 1–3)[66a]) and N_2/O_2 ($M(N_2)_n(O_2)$ (M = Ni, Pd, Pt; n = 1, 2)[66b,66c]) complexes have also been prepared using the same techniques. N_2 and CO are bonded in an end-on fashion, but O_2 is side-on.

Matrix isolation techniques have also been applied to the study of the interaction of N_2 with elements other than transition metals, such as alkali metals and non-metals. Hence, "supernitride" lithium species are formed by low-temperature matrix co-deposition of Li atoms and N_2.[67a] Two $\nu(N_2)$ absorptions in the IR spectra have been observed at very low frequencies, 1800 cm^{-1} and 1535 cm^{-1}, the former being ascribed to $Li^+N_2^-$ and the latter to $Li_2(N_2)_2$, probably derived from dimerization of LiN_2.[67a] However, theoretical calculations performed by other authors[67b] associate the lower frequency (1535 cm^{-1}) with a dinitrogen species such as Li_2N_2, in which the valence state of N_2 is formally −2.

Deposition of C or Si vapour in an inert gas matrix at low temperature yields the non-metal dinitrogen species diazocarbene, CN_2, and diazosilene, SiN_2, as well as their CO analogues (and also $C(CO)_2$ and $Si(CO)_2$). These species have been studied[68] by ESR, optical and IR spectroscopy, $\nu(N_2)$ exhibiting a high negative shift from the free value [~600 cm^{-1} for SiN_2 which has $\nu(N_2)$ at 1730 cm^{-1}].

Syntheses of N_2 complexes from other molecules (not simple metal atoms) have also been reported in frozen noble gas matrices at low temperature. $M(CO)_5(N_2)$ (M = Cr, Mo, W),[69] $Mn(CO)_{4-n}(N_2)_n(NO)$ (n = 1, 2),[70a] $Mn(N_2)(NO)_3$,[70b] $Fe(CO)_{2-n}(N_2)_n(NO)_2$ (n = 1, 2)[70c] and $Co(CO)_{3-n}(N_2)_n(NO)$ (n = 1, 2)[70d] have been prepared by photolysis of the carbonyl species $M(CO)_6$, $Mn(CO)_4(NO)$, $Mn(CO)(NO)_3$, $Fe(CO)_2(NO)_2$ and $Co(CO)_3(NO)$, respectively, in dinitrogen matrices at 20 K via CO replacement by N_2. MX_2L species (M = Ni;[71] Sn, Pb, Hg; [72] L = N_2, CO, NO. olefin; X = halide) have also been prepared in inert gas matrices from the coordinatively unsaturated metal dihalides MX_2 and L. The M—L bonding is believed to have a prominent σ character since the carbonyl compounds provide the only known examples, besides adsorbed CO species, of positive shifts of $\nu(CO)$ upon coordination. Nitrosyl analogues (M = Sn, Pb, Hg) also exhibit positive shifts of $\nu(NO)$ and dinitrogen species are also believed to be formed, although $\nu(N_2)$ has not been detected, probably due to the very weak $M—N_2$ σ-interaction.

B. Conversion of N—N or N species into the dinitrogen ligand

The formation of ligating N_2 may occur either by oxidation of N—N species (hydrazine and its derivatives; diazene or azide) or by reduction of such species (nitrous oxide N_2O or diazoalkanes). These reactions are believed to occur via the coordination of the N—N species (which is then activated) followed by its degradation to a N_2 ligand.

Redox reactions of two N (-containing) species (with one nitrogen in a positive oxidation state and the other negative, and at least one of them coordinated) may yield bonded N_2 as well.

(i) *Hydrazine and related species*

The first dinitrogen complex to be reported, $[Ru(NH_3)_5N_2]^{2+}$ (**1**), was prepared accidentally[15a] by this route by reaction of $[RuCl_3(H_2O)_3]$ with hydrazine hydrate in an attempt to synthesize $[Ru(NH_3)_6]^{2+}$ (hydrazine is known to be thermodynamically unstable to disproportionation forming $N_2 + NH_3$). The same dinitrogen species and other related Ru(II) and Os(II) dinitrogen complexes may also be prepared from aqueous hydrazine and chloro or amine complexes of Ru(III or IV)[15] and Os(IV).[73]

Oxidation of monoarylhydrazines leading to N_2 and the corresponding aromatic hydrocarbon (or phenol) is known to occur in reactions with some transition metal compounds. This has been used to prepared complex (**1**) from the reaction of phenyl-, benzyl- and β-naphthylhydrazine with $[Ru(NH_3)_5$-$Cl]Cl_2$ (benzene and traces of phenol and biphenyl have been detected as products of the phenylhydrazine reaction).[16] However, the azide route (*vide infra*) is a better preparative one for Ru.

Rhenium–dinitrogen complexes of the types $[ReCl(N_2)(LL)_2]$ (**2**) (LL = dppm, dppe, dppp, dppae, dppey, $(PMe_2Ph)_2$, $(PPh_2Me)_2$, $(PPh_2H)_2$, $\{P(CH_2OH)_3\}_2$, $(Me_2NPF_2)_2$) and $[ReCl(N_2)L_2(PR_3)_2]$ (**3**) (L = PF_3, PH_2Ph, $P(OCH_2)_3CMe$, $R_3 = Ph_3$; L = $P(OMe)_3$, $R_3 = Me_2Ph$) have been prepared[74] from a hydrazine derivative (benzoyl hydrazine), the latter according to the scheme:[74a–74c]

$$[ReOCl_3(PR_3)_2] + PhCONHNH_2 \longrightarrow \underset{\textbf{(4)}}{[ReCl_2(PR_3)_2(\overline{=N-N=C}OPh)]} \longrightarrow$$

$$\xrightarrow{L} \underset{\textbf{(5)}}{[ReCl_2(N{=}NCOPh)(PR_3)_2L]} \xrightarrow[MeOH]{L}$$

$$\underset{\textbf{(3)}}{[ReCl(N_2)L_2(PR_3)_2]} + PhCOOMe + Cl^-.$$

The intermediate derivatives rhenium(V)benzoylhydrazido(3-) (**4**) and benzoylazorhenium(III) (**5**) have both been isolated. However, X-ray photoelectron data[75] substantiate for (**4**) the Re(III) form:

$$[ReCl_2(PR_3)_2(\overline{-N{=}N-C(=O})Ph].$$

Displacement of PR_3 may also occur yielding type (**2**) complexes. When LL = dppae, the complex is less stable than the dppe analogue, whereas the analogous dpae species could not be obtained, showing the lower stability of the arsine–dinitrogen complexes compared to their phosphine analogues.[74] $[ReCl(N_2)(CO)_2(PR_3)_2]$ is formed when L = CO.[74b,74c]

Oxidation of the coordinated hydrazine in $[CpMn(CO)_2N_2H_4]$ by H_2O_2/Cu^{2+} yields $[CpMn(CO)_2(N_2)]$.[76a] The Cr[76b] and Re[76c] analogues may be prepared by the same route; [*trans*-N_2H_2 {$CpRe(CO)_2$}$_2$] is also formed.[76d] The Re—N_2 complex may also be obtained by disproportionation of $[CpRe(CO)_2N_2H_4]$ to the N_2 complex plus $[CpRe(Co)_2NH_3]$.[76d]

Dinitrogen complexes may also be formed from disproportionation of coordinated diazene. Hence, for example, [μ-N_2H_2{$Cr(CO)_5$}$_2$] disproportionates irreversibly to [{$Cr(CO)_5$}$_2N_2$] and the μ-hydrazine-*N,N'* species [μ-N_2H_4{$Cr(CO)_5$}$_2$] in the presence of catalytic amounts of base.[76e]

(ii) *Azides*

Azide may decompose according to either of the following ways, and both may generate a N_2 complex: $N_3^- \rightarrow N_2^- + \frac{1}{2}N_2$(**a**) or $N_3^- \rightarrow N_2 + N^-$(**b**).[8]

TYPE (a)

N_2 is evolved and the metal is reduced by one unit in type (**a**) reactions. They occur in the preparations of $[Ru(NH_3)_5(N_2)]^{2+}$ (**1**) and $[Ru(NH_3)_4(H_2O)(N_2)]^{2+}$ from reactions of azide with the Ru(III) species $[Ru(NH_3)_5(H_2O)]^{3+}$,[15b] and *cis*-$[Ru(NH_3)_4Cl_2]^+$,[77] respectively. An azide complex is formed—*e.g.* complex (**6**) of the scheme shown below in the preparation of (**1**)—from which a protonated nitrene (**8**) is believed to be derived upon protonation of the bonding N of the azide followed by evolution of N_2:

$$\underset{(\mathbf{6})}{[Ru(NH_3)_5(N_3)]^{2+}} + H^+ \rightarrow \underset{(\mathbf{7})}{[Ru(NH_3)_5\overset{\overset{\displaystyle H}{|}}{N}\text{—}N{\equiv}N]^{3+}} \rightarrow \underset{(\mathbf{8})}{[Ru(NH_3)_5NH]^{3+}} + N_2.$$

The dinitrogen complex (**1**) is then formed by reaction of the nitrene species (**8**) with (**6**), whereas the dinuclear complex $[\{Ru(NH_3)_5\}_2N_2]^{4+}$ (**9**) is derived from dimerization of the nitrene (**8**) (with loss of protons in both cases).[17]

TYPE (b)

Azide decomposition of type (**b**) may lead to a dinitrogen complex if N_2 becomes coordinated to the metal and the N^- fragment is trapped by another species. A few examples may be cited.

$$trans\text{-}[MCl(CO)(PPh_3)_2] + RN_3 \xrightarrow[\substack{CHCl_3,\\ EtOH}]{0°C} trans\text{-}[MCl(N_2)(PPh_3)_2] + RNCO$$

$$RNCO \xrightarrow{EtOH} RNHCOOEt$$

where M = Ir,[78] R = Ph, *p*-toluenesulphonyl, PhCO, furoyl; M = Rh,[79] R = $CH_3CH_2CH_2CO$. The isocyanate formed is prevented from displacing N_2 by reaction with EtOH to form a carbamate. Mechanistic studies of the formation of the Ir complexes have shown that the reaction is in the coordination sphere of Ir without evolution of N_2, the complex $[Cl(CO)(PPh_3)_2\overline{IrN{=}N{-}N}R]$ being formed as an intermediate.

The complexes *trans*-$[OsCl(N_2)(diars)_2]^+$ and *trans*-$[RuCl(N_2)(diars)_2]^+$ may be prepared by reaction of *trans*-$[OsCl(N_3)(diars)_2]$ with aqueous HCl[80] and of *trans*-$[RuCl(N_3)(diars)_2]$ with $NO^+PF_6^-$,[81] $NO_2^+SbF_6^-$ or HCl.[81c] Isotopic substitution experiments have shown that both Ru-$^{15}N^{14}N$(80%) and Ru-$^{14}N^{14}N$(20%) are formed from Ru-$^{15}NN_2$ and NO^+ and the formation of the former has been interpreted by considering the cyclic intermediate Ru-$\overline{^{15}N{-}^{14}N{-}^{14}N{-}O{-}^{14}N}$ derived from attack of NO^+ at the coordinated azide.[81c] The mechanisms for the exclusive formation of Ru-$^{15}N^{14}N$ from reaction of Ru-$^{15}NN_2$ with NO_2^+ and HCl have been suggested as occurring via the dinitrogen oxide complex $[RuCl(N_2O)(diars)_2]^+$ (which however has not been detected) and by protonation at the terminal azide nitrogen followed by NH liberation, respectively.[81c]

An unstable bis(dinitrogen) complex of ruthenium may be prepared by the same route, albeit azide displacement by water also occurs (with formation of $[Ru(N_2)(H_2O)(en)_2]^{2+}$)[82] as shown below:

$$cis\text{-}[Ru(N_2)(N_3)(en)_2]^+ + HNO_2(aq) + H^+ \rightarrow cis\text{-}[Ru(N_2)_2(en)_2]^{2+} + N_2O + H_2O.$$

Reactions of azides of organic acids with dinuclear carbonyl complexes of Rh(I) appear to yield bridged- or terminal-N_2 complexes (the former apparently containing the cyclic group

$$\overline{Rh{-}N_2{-}Rh}^{\,L},$$

where L is a chelated diphosphine ligand), depending on the azide structure.[83] However, their characterization must be accepted with reservation. Intermediate azide species are believed[84] to be formed in the preparation of some as yet ill-defined dinuclear Pd(II) species formulated with bridging N_2, obtained by treating a nitrito complex with hydrazinium sulphate ($NO_2^- + N_2H_5^+ \rightarrow N_3^- + H_3O^+ + H_2O$). The azide species may subsequently yield bridging N_2 following a mechanism analogous to that involving the formation of complex (**9**), although without a change in the metal oxidation state. Decomposition of azide according to scheme (**b**) may also lead to a nitride, rather than a dinitrogen, complex if the N^- fragment becomes coordinated to the metal, instead of N_2, and the available range of metal oxidation states is wide enough, e.g., the oxidation of Re(III) to Re(V) in the preparation of $[ReCl_2N(PR_3)_3]$ from $[ReCl_2(N_3)(PR_3)_3]$.[85]

(iii) *Nitrous oxide*

Various species which pick up N_2 can also yield the same dinitrogen complexes by reaction with N_2O. Hence, for example $[CpRe(CO)_2N_2]$,[76d] $[CoH(N_2)(PPh_3)_3]$,[86] $[OsX_2(N_2)(PR_3)_3]$[87] and Allen and Senoff's cation (**1**)[18] may be prepared from reaction of N_2 or N_2O with [CpRe(CO)$_2$(thf)] (a partial oxidation and decomposition of this species occurs by reaction with N_2O), $[CoH_3(PPh_3)_3]$ (in the presence of PPh_3 as a N_2O reducer leading to Ph_3PO), *mer*-$[OsX_3(PR_3)_3]$ (with Zn/Hg as a reducing agent) and $[Ru(NH_3)_5(H_2O)]^{2+}$ (in the presence of Zn/Hg, Cr^{2+} or V^{2+} as a N_2O reducing agent) respectively. The intermediate N_2O complex has been isolated in the Ru reaction and the coordinated N_2O is then reduced to N_2 by an external reducing agent:[18]

$$[Ru(NH_3)_5(H_2O)]^{2+} \underset{}{\overset{N_2O}{\rightleftharpoons}} [Ru(NH_3)_5(N_2O)]^{2+} \xrightarrow{2e,2H^+} [Ru(NH_3)_5N_2]^{2+} + H_2O.$$

The mode of N_2O coordination has not yet been ascertained. Bonding through a N atom is suggested by IR studies using ^{15}N-labelled N_2O[18b] and its linear character is inferred from the crystal symmetry obtained by X-ray powder diffraction patterns.[18f] However, coordination via the O atom is indicated by force-constant calculations.[18e]

(iv) *Diazo species*

Diazo species can undergo photolytic, thermal or catalytic cleavage of the C=N bond yielding carbenes and N_2. Carbene complexes have been prepared by this route, but only a few dinitrogen complexes (Ru species (**1**), (**9**)[19] and the dinuclear Mn complex (**10**)[88] where R = H, CH_3) have been reported from a diazo compound, as below.

$$[Ru(NH_3)_5(H_2O)]^{2+} \xrightarrow{N_2CHCOOEt(aq)} \underset{(\mathbf{1})}{[Ru(NH_3)_5(N_2)]^{2+}} + \underset{(\mathbf{9})}{[\{Ru(NH_3)_5\}_2N_2]^{4+}}$$

$$\underset{(\mathbf{11})}{[(\eta^5\text{-}RC_5H_4)Mn(CO)_2(thf)]} \xrightarrow{N_2CHCOOEt,Et_2O} \underset{(\mathbf{10})}{[\{(\eta^5\text{-}RC_5H_4)Mn(CO)_2\}_2N_2]}.$$

From kinetic data the formation of the dinitrogen complexes is believed to proceed via the coordination of the diazo species (which is thus activated by the metal centre) followed by reductive CN bond cleavage (CH_3COOEt and traces of $(CH_2COOEt)_2$ and $(CHCOOEt)_2$ have been detected in the Ru system[19]). Hence, the N_2 complex is not prepared from reaction of the N_2 evolved from the free diazo species by C=N cleavage. This is also substantiated by the fact that the mononuclear complex $[CpMn(CO)_2(N_2)]$ is the only species formed by direct reaction of $[CpMn(CO)_2(thf)]$ with N_2.[89]

(v) *Other N species*

As yet ill-defined bridging dinitrogen complexes of Pt[90a], $K_2[Pt_2(OH)_4(NO_2)_4(NH_3)_2N_2]\cdot 2H_2O$ and $K_2[Pt_2(ClO_4)_2(NO_2)_6(NH_3)_2N_2]\cdot 2KClO_4$ (or $0{\cdot}5K_2SO_4$), and of Rh[90b], $K_2[Rh_2(OH)_2(NO_2)_6(NH_3)_2N_2]$, have been reported as being prepared by reaction of the nitrito species $K_2[Pt(NO_2)_4]$ and $K[Rh(NO_2)_4(NH_3)_2]\cdot \frac{1}{2}H_2O$, respectively, with ammonium salts, following a route which bears resemblance to the one leading to the aforementioned Pd-bridging N_2 species.[84] Reactions have been carried out in aqueous solution and labelling experiments (using ammonium sulphate with ^{15}N) have shown that the nitrogen atoms of the bound N_2 originate partially from the ammonium ion,[90] which is suggestive of the reaction $NO_2^- + NH_4^+ \rightarrow N_2 + 2H_2O$.

The first bisdinitrogen complex was prepared by diazotization of an amine complex with nitrous acid:[91]

$$[Os(NH_3)_5(N_2)]^{2+} + HNO_2 \rightarrow cis\text{-}[Os(NH_3)_4(N_2)_2]^{2+} + 2H_2O$$

Reaction of $[Ru(NH_3)_6]^{3+}$ with NO in alkaline medium yields $[Ru(NH_3)_5(N_2)]^{2+}$ (**1**)[20a] probably through a coordinated amide (NH_2^-) formed by proton abstraction from NH_3 by OH^-. Labelling experiments[20b] using ^{15}NO substantiate an attack by NO (probably after previous coordination yielding the Ru(II) species $[Ru(NH_3)_5(NO)]^{3+}$) at the coordinated amide, but supporting testimony is also given[20c] for a nucleophilic attack of the amide at coordinated NO. The formation of the Allen and Senoff cation (**1**) from an alkaline solution of $[Ru(NH_3)_5(NO)]^{3+}$ is then represented by the following proposed[20c] equation:

$$[Ru(NH_3)_5(NO)]^{3+} + cis\text{-}[Ru(NH_2)(NH_3)_4(NO)]^{2+} \longrightarrow$$

$$[(H_3N)_5Ru\{N(O)NH_2\}Ru(NH_3)_4(NO)]^{5+} \xrightarrow{OH^-}$$

$$[Ru(NH_3)_5(N_2)]^{2+} + cis\text{-}[Ru(OH)(NH_3)_4(NO)]^{2+} + H_2O.$$

A related reaction is the formation of $[Ru(en)_2(N_2)(H_2O)]^{2+}$ from $[Ru(en)_3]^{3+}$ and NO in alkaline solution; no mechanistic studies have yet been reported, but the reaction is suggested to proceed via attack of NO on coordinated $NH_2CH_2CH_2NH^-$ followed by C–N(NO)H bond cleavage.[92]

Nucleophilic attack of phenylhydrazine at the coordinated nitrosyl of *trans*-$[OsCl(NO)(diars)_2]^{2+}$ yields *trans*-$[OsCl(NONNHPh)(diars)_2]$ and *trans*-$[OsCl(N_2)(diars)_2]^+$.[80]

III. Structure and Bonding

A. Modes of dinitrogen bonding

Among the possible modes of N_2 bonding which may be envisaged, only I, II, III and V have been reported from X-ray studies (Table 5):

M—N≡N (I, end-on), M···N≡N side-on (II, side-on or edge-on),

M—N≡N—M (III, bridging end-on),

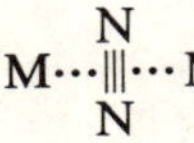

M···N≡N···M (IV) and M(N≡N)M (V) (bridging side-on),

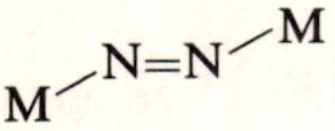

M—N=N—M (VI) (*trans*), M—N=N—M (VII) (*cis*), $M_2\overset{+}{N}=\overset{-}{N}$ (VIII).

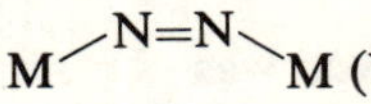

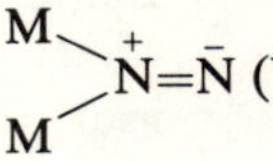

Dinitrogen in VI, VII, and VIII may also be visualized in the reduced form N_2^{2-}, i.e. as

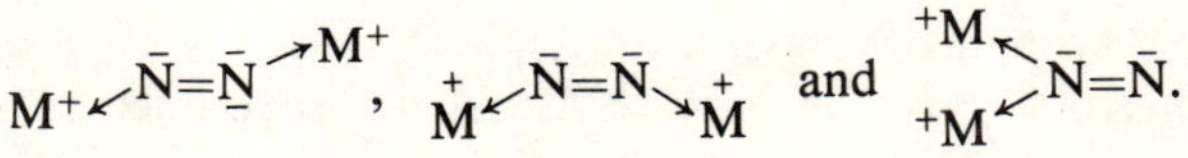

$M^+ \leftarrow \bar{N}=\bar{N} \rightarrow M^+$, $\overset{+}{M} \leftarrow \bar{N}=\bar{N} \rightarrow \overset{+}{M}$ and $(^+M)_2 \leftarrow \bar{N}=\bar{N}$.

End-on type bonding is the most common and the lesser stability of the edge-on type (which is analogous to a π-acetylene bond) relative to the former has been substantiated by theoretical calculations[93] (see also below). Side-on type II bonding has been reported by X-ray studies in only one case, $[RhCl(N_2)(PPr^i_3)_2]$,[94] but the high disorder appears to have prevented a clear structure determination. It has been unambiguously confirmed in another case ($Co(\eta^2-N_2)$ formed in the co-condensation reaction of atomic cobalt with dilute nitrogen–argon matrices at 10 K) by IR studies of $Co(^{28}N_2)$, $Co(^{29}N_2)$ and $Co(^{30}N_2)$.[61b] Contrary to some expectations, the di-hapto N_2 exhibits a high $\nu(N_2)$ which is comparable to the mono-hapto $\nu(N_2)$ in analogous species (in $Co(\eta^2-N_2)$, $\nu(^{28}N_2) = 2101\ cm^{-1}$ which is about $10\ cm^{-1}$ higher than in $Ni(\eta^1-N_2)$) and its intensity is high enough to be easily detected by IR spectroscopy. ESR data on $[Zr(\eta^5-C_5H_5)_2(N_2)R]$ (where $R = (Me_3Si)_2CH$) with $^{14}N_2$ and $^{15}N_2$ is also suggestive of a sideways-bound (type II) dinitrogen in this species, although a rapid zirconyl oscillation between the two nitrogen atoms of a corresponding end-on bonded N_2 complex cannot be ruled out.[95]

Bridging side-on structures (type V) have been authenticated by X-rays only in two species, (V) and (X) of Table 5. In these structures the NiN_2Ni moiety displays a distorted tetrahedral configuration, the side-on positions of N_2 being occupied by Ni atoms and the end-on positions by Na and/or Li atoms or clusters.[47a–c]

Bent structures such as VII have been postulated in a few cases, e.g., $[\{Cr(dppe)_2\}_2N_2]$,[96] $[\{Cp_2Ti\}_2N_2]$[27a] and $[\{Cp_2Ti\}_2N_2MgCl]$,[29] but the linear arrangement III for the Ti species is strongly suggested by the X-ray structure determination of the analogous species $[\{(\eta^5\text{-}C_5Me_5)_2Ti\}_2N_2]$[30c] and $[\{(\eta^5\text{-}C_5Me_5)_2ZrN_2\}_2N_2]$.[31b,31c]

Some of these possibilities are known in protonated forms, e.g., VIII in the μ-diazenido-N species[97] $[\{Pt(PPh_3)_2N_2H\}_2]^{2+}$, VI and VII in the μ-diazene-N,N′ complexes [*trans*-μ-$N_2H_2\{Cr(CO)_5\}_2$],[98,99a] [*trans*-μ-$N_2H_2\{CpM(CO)_2\}_2$] (M = Mn[99b] and Re[76d]), $[Cp(CO)_2Mn{-}N_2H_2{-}Cr(CO)_5]$ (with *trans*-μ-N_2H_2)[99c] and [*cis*-μ-$N_2H_2\{M(CO)_5\}_2$] (M = Cr[99a,99d,99e] and W[99f]). These species have been suggested as conceivable intermediates in the reduction of N_2 to NH_3, but they have been prepared by oxidation (e.g., by H_2O_2/Cu^{2+}) of hydrazine derivatives and not by protonation of bridging N_2.

B. Bonding parameters

Accurate determination of bond lengths is often prevented by disorder.

The reported N—N distances of coordinated N_2 are generally only slightly greater than in the free ligand, except for $[RhCl(N_2)(PPr^i_3)_2]$ whose accurate determination has been precluded by disorder,[94] and the difference may be within experimental error, although the decrease in $\nu(N_2)$ upon coordination is very pronounced. Only the asymmetrically bridged end-on type III $M{-}N_2{-}M'$ and mainly the bridging edge-on type V complexes[47b,47c] exhibit N—N distances with values considerably greater than that of the free ligand (e.g., 1·35 Å and 1·36 Å for species (V) and (X) respectively of Table 5 whereas 1·0976 Å is the value for free N_2). The asymmetric, bridged Re complexes (P)[100] and (Q)[101] have $\nu(N_2)$ values of 1660 cm^{-1} and 1800 cm^{-1}, respectively, which are well below the value for terminal N_2 in the parent Re complex (D)[102] (1920 cm^{-1}). This trend is in agreement with the observed lengthening of the N—N distance (and concomitant shortening of the Re—N distance) upon coordination of end-on N_2 to a second metal atom which is electron-deficient (Mo(V) and Mo(IV) in (P) and (Q), respectively). However, the second metal atom in the symmetrical, bridged end-on dinitrogen species (S)[103] is not electron-deficient and does not induce a greater N—N lengthening (it seems that the activation of

Table 5. X-Ray and IR data on dinitrogen complexes

Complex	M–N[a]	N–N[a]	∠M–N–N	BS[b]	ν(N$_2$)[c]	Reference
trans-[Mo(N$_2$)$_2$(dppe)$_2$] (A)	2·014(5)	1·118(8)	176·6(5)	I	2020 w 1970 s	*105*
trans-[Mo(CO)(N$_2$)(dppe)$_2$]·$\frac{1}{2}$C$_6$H$_6$ (B)	2·068(12)	1·087(18)	177·0(12)	I	2128 (Bz)	*d*
[WH(N$_2$)$_2$(dppe)$_2$]HCl$_2$·2thf (C)	2·05, 2·04	1·12(3), 1·07(3)	175, 177	I	1995	*e*
trans-[ReCl(N$_2$)(PMe$_2$Ph)$_4$] (D)	1·97(2)	1.06(3)	177(1)	I	1920	*102*
[Ru(NH$_3$)$_5$(N$_2$)]Cl$_2$ (E)	2·10(1)	1·12(9)	linear	I	2105	*f*
trans-[Ru(N$_3$)(N$_2$)(en)$_2$]PF$_6$ (F)	1·894(9)	1·11(1)	179·3(9)	I	2103	*g*
[Os(NH$_3$)$_5$(N$_2$)]Cl$_2$ (G)	1·84(1)	1·12(2)	178·3(13)	I	2025 2010	*h*
[CoH(N$_2$)(PPh$_3$)$_3$]·Et$_2$O (H)	1·80(4)	1·16(4)	175(4)	I	2088 (Tol)	*i*
[CoH(N$_2$)(PPh$_3$)$_3$][j] (I)	1·78(1) 1·83(1)	1.10(1) 1·12(1)	178(2) 178(1)	I I	2088	*k*
K[Co(N$_2$)(PMe$_3$)$_3$][l] (J)	1·70–1·71	1·16–1·18	linear	I	1795 s 1758 m	*110*
trans-[RhH(N$_2$)(PBu$^t_2$Ph)$_2$] (K)	1·970(4)	1·074(7)[m]	linear	I	2155	*n*
[RhCl(N$_2$)(PPr$^i_3$)$_2$] (L)	2·55(1), 2·51(1)	0·83(2)[o]	*p*	II[q]	2100	*94*
[{(η^5-C$_5$Me$_5$)$_2$ZrN$_2$}$_2$N$_2$][r] (M)	*s*	*t*	*u*	III + I	*v*	*31b, 31c*
[{(η^5-C$_5$Me$_5$)$_2$Ti}$_2$N$_2$][w] (N)	2·005(10), 2·016(10) 2·033(10), 2·013(10)	1·165(14) 1·155(14)	176·8(4), 178·1(4) 176·9(4), 177·8(4)	III		*30c*
[{(η^6-Me$_3$C$_6$H$_3$)(dmpe)Mo}$_2$N$_2$] (O)	2·042(4)	1·145(7)	175·6(4)	III	1989	*x*
[MoCl$_4$(OMe){(N$_2$)ReCl(PMe$_2$Ph)$_4$}] (P)	1·815 (15) (Re) 1·90(1) (Mo)	1·18(3)	179·6(14) (Re) 178·7(9) (Mo)	III	1660	*100*
[MoCl$_4${(N$_2$)ReCl(PMe$_2$Ph)$_4$}$_2$] (Q)	1·75(4) (Re) 1·99(4) (Mo)	1·28(5)	linear	III	1800	*101*
[{(η^5-CH$_3$C$_5$H$_4$)Mn(CO)$_2$}$_2$N$_2$] (R)	1·875(5)	1·118(7)	176·5(4)	III	1971	*88*
[{Ru(NH$_3$)$_5$}$_2$N$_2$](BF$_4$)$_4$·2H$_2$O (S)	1·928(6)	1·12(2)	178·3(5)	III	2100	*103*
[{Co(PMe$_3$)$_3$(N$_2$)}$_2$Mg(thf)$_4$] (T)	1·72 (Co) 2·04 (Mg)	1·18	linear (Co) 158 (Mg)	III	2068	*23*

$[\{(PCy_3)_2Ni\}_2N_2]$ (U)	1·77, 1·79	1·12	178·2, 178·3	III		*47d*
$[\{[(PhLi)_3Ni]_2N_2 \cdot 2Et_2O\}_2]$[y] (V)	1·92 (Ni1–N1), 1·91 (Ni1–N2) 1·93 (Ni2–N1), 1·94 (Ni2–N2)	1·35		V		*47b*
$[\{Ph(NaOEt_2)_2(Ph_2Ni)_2N_2NaLi_6$-$(OEt)_4 \cdot OEt_2\}_2]$ (X)	1·97(3) (Ni–N)[y']	1·359(18)		V		*47c*
N_2gas[z]		1·0976(1)[z']			2330[z'']	

[a] Values in Å. [b] Bonding structure (see text). [c] Values in cm^{-1} and in Nujol mull unless otherwise stated (Bz, Tol-benzene or toluene solution, respectively). [d] M. Sato, T. Tatsumi, T. Kodama, M. Hidai, T. Uchida and Y. Uchida, *J. Amer. Chem. Soc.* (1978) **100**, 4447. [e] Ref. *140b*; G. A. Heath, R. Mason and K. M. Thomas, unpublished results. [f] F. Bottomley and S. C. Nyburg, *Acta Cryst.* (1968) **24B**, 1289. [g] B. R. Davies and J. A. Ibers, *Inorg. Chem.* (1970) **9**, 2768. [h] J. E. Fergusson, J. L. Love and W. T. Robinson, *Inorg. Chem.* (1972) **11**, 1662. [i] H. Enemark, B. R. Davis, J. A. McGinnety and J. A. Ibers, *Chem. Commun.* (1968) 96. [j] Prepared in di-n-butylether. Exhibits two independent molecules in the asymmetric unit. [k] B. R. Davis, N. C. Payne and J. A. Ibers, *J. Amer. Chem. Soc.* (1969) **91**, 1240; *Inorg. Chem.* (1969) **8**, 2719. [l] Hexameric. Each N_2 also end-on linearly bonded to a K atom in the N–N direction (K–N bond lengths = 2·76–2·93 Å) and, in perpendicular directions, side-on bonded to 2 or 3 K atoms. [m] Shorter than in free N_2 but unreliable due to large anisotropic thermal motion of N_2. [n] P. R. Hoffman, T. Yoshida, T. Okano, S. Otsuka and J. A. Ibers, *Inorg. Chem.* (1976) **15**, 2462. [o] Unreliable due to disorder. [p] Not quoted. [q] However the PCy_3 analogue is believed to have end-on bonded N_2 *trans* to Cl (ref. *38b*). [r] Consider the notation N4–N3–Zr1–N1–N2=Zr2–N5–N6. [s] 2·087(3) (Zr1–N1), 2·075(3) (Zr2–N2), 2·188(4) (Zr1–N3), 2·188(4) (Zr2–N5). [t] 1·182(5) (N1–N2), 1·116(8) (N3–N4), 1·114(7) (N5–N6). [u] 176·7(3) (Zr1–N1–N2), 177·4(3) (Zr2–N2–N1), 177·9(5) (Zr1–N3–N4), 177·8(5) (Zr2–N5–N6). [v] 2040, 2003, 1578. Coupling of bridge and terminal N_2 stretching modes is believed to occur. [w] Two independent molecules. [x] R. A. Forder and K. Prout, *Acta Cryst.* (1974) **30B**, 2778. [y] N_1 end-on bonded to three Li atoms to form a tetrahedron; N_2 end-on bonded to one Li atom. The two N_2 moieties are linked by two Li atoms to form a six-membered ring. [y'] 2·05(5) (N–Li), 2·61(3) (N–Na). [z] Included for comparative purposes. [z'] B. P. Stoicheff, *Canad. J. Phys.* (1954) **82**, 630. [z''] (Raman). K. Nakamoto, "Infrared Spectra of Inorganic and Coordination Compounds". John Wiley, New York (1963).

each metal opposes that of the other one). These features may be rationalized by a simplified π-MO scheme which will be discussed later. The difference of the N—N bond lengths, between (Q) and (S), for example, may then be explained.

Although the M—N bond length is in some N_2 species considerably shorter than the expected value for a single bond, it is still not close to a formal M=N bond (compare, e.g., the Re—N distance in (P) and (Q) with 1·685(11) Å in $[ReCl_3(NMe)(PEtPh_2)_2]$).[104] Moreover, such behaviour is not general and in *trans*-$[Mo(N_2)_2(dppe)_2]$ (A) for example the Mo—N bond length is not all that much shorter than the usual M—N single bond.[105]

Comparison between terminal and bridging N_2 at the same metal site is possible in complex (M).[31b,31c] The distances N—N and Zr—N for the bridging N_2 are, respectively, longer and shorter than the corresponding ones involving terminal N_2.

C. Description of the dinitrogen bonding

(i) *General*

Bonding of N_2 (and its isoelectronic analogues such as CO, CNR, NCR, CN^-, acetylenes, NO^+, N_2R^+) to a transition metal is believed to bear close resemblances to the bonding of olefins and thus may be well accounted for by the MO model developed by Dewar, Chatt and Duncanson,[106] and its modifications.[107] The bonding is considered to have a σ and a π component. The former results from overlap of an occupied ligand orbital ($3\sigma_g$, $1\pi_u$, 5σ and $7a_1$ for end-on N_2,[93a] edge-on N_2,[47b] CO[108] and CNR,[108] respectively) with a vacant metal orbital, and the π component (π-backbonding) results from overlap of an occupied metal d orbital with a vacant antibonding π orbital of the ligand ($1\pi_g^*$, $2\pi^*$ and 3e for N_2,[93a] CO[108] and CNR,[108] respectively). Figure 1 represents such metal–dinitrogen bond types for end-on and edge-on coordination.

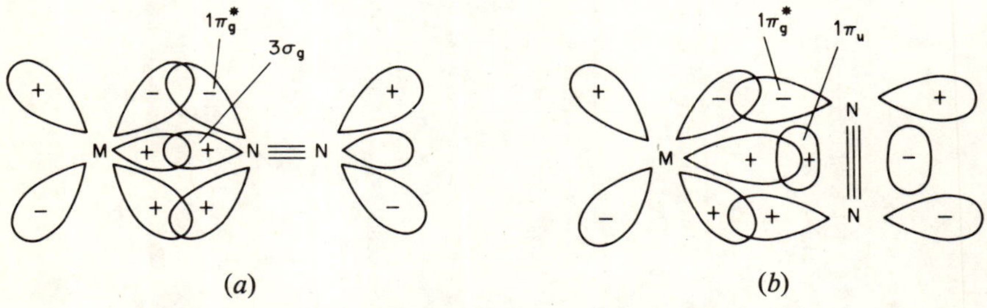

Fig. 1. Metal–dinitrogen bonding. (*a*) End-on. (*b*) Edge-on.

In end-on N_2, CO and CNR, the σ-donor ligand orbital appears to have a weakly antibonding character (not so weak for CO as for N_2) and the π-acceptor orbital displays a fairly antibonding character.

Dinitrogen has two π bonds at right angles and in the above-mentioned bridging edge-on type V structures, a metal atom is bound to each of them. A related bonding is observed in bridging acetylene species such as $[Co_2(CO)_2$-(μ-tolane)].[109]

N_2 coordinated to a transition metal may concomitantly bind electropositive main group elements (Li, Na or K) (within the species with type V structure[47a–c] and also in $K[Co(N_2)(PMe_3)_3]$[110] with type I structure), whose electron acceptor properties are considered to enhance the M $\rightarrow$ N_2 π-back-bonding and to stabilize the negative charge at N_2, the bonding being multi-centred and electron-deficient in nature.

The great stability of the N_2 molecule is believed to be mainly a consequence of the exceptionally low energy level of the ligand donor orbital (−15·6 eV for $3\sigma_g$ and −17·1 for $1\pi_u$[111] which are well below −14·0,[111] −11·24,[112] −13·14,[112] −13·53[113] and −13·22 eV[113] for CO, CNMe, NCMe, acetylene and ethylene, respectively). This generally precludes its overlap with an empty metal orbital. In order to compensate for the poor σ-donating character of N_2, an electon-rich metal site capable of strong π-backdonation is required. Moreover, there is a very large energy separation between the N_2 orbital and the $1\pi_g^*$ one, with which the metal π-donor orbital must overlap for bonding (8·6 eV and 10·1 eV for end-on and side-on bonded N_2, respectively, which is greater than the corresponding values of 8·00, 6·39 and 4·98 eV for CO, acetylene and ethylene, respectively). Hence high ligand field strength to promote the e_g–t_{2g} separation of the metal orbitals (in octahedral complexes) and the absence of empty π-accepting orbitals at an energy level which would compete with the N_2 molecule are favourable ligand features for N_2 coordination.[114]

These limitations based on energy level considerations are more pronounced if N_2 binds in a side-on fashion, which accounts for the rarity of this type of bonding. Moreover the π-backbonding is expected to be less extensive in this case since only the $1\pi_g^*$ N_2 orbital which lies in the M(η^2-N_2) plane can π-accept, whereas in the end-on coordination type both $1\pi_g^*$ N_2 orbitals can π-accept.

The relative atomic orbital percentages in molecular orbitals may also account for the weak N_2 end-on π ligand ability. The values of percentage of p (ligating atom) character in the π-antibonding orbital involved in bonding to a transition metal decrease in the order CO(66·8%) > CN^- (62·6%) > CNMe(59·3%) > N_2(50%).[108] Thus the worst overlap with metal d orbitals is expected to result for π-antibonding of N_2, ignoring energy level differences.

The poorer σ-donation and/or π-acceptance of N_2 compared with CO has also been substantiated by other data, namely from He(I) photoelectron spectroscopy,[115] SCF (self-consistent field) calculations[116] SCCC (self-consistent charge and configuration)[117] studies and electronic,[118] infrared,[119,66a,72a] and Mössbauer[51b,120] spectroscopy. In some instances N_2 appears to be an overall acceptor similar to CO in analogous complexes as substantiated by X-ray photoelectron data.[121] This may however result from N_2 being a weaker σ-donor *and* a weaker π-acceptor than CO.[116] Reports on the relative importance of the σ- and π-components of the $M{-}N_2$ bond have been published,[122,71,72] but often they are contradictory and the matter remains uncertain.

In contrast to CO, which exhibits a small dipole moment (0·2 D, the greater charge density being carried on the C atom), free N_2 is not polar. However, upon terminal coordination the N_2 carries a negative charge and exhibits a strong electronic polarization in mononuclear complexes with low $\nu(N_2)$ as revealed by ESCA and X-ray photoelectron studies,[121,123] by the extended Hückel method[122b] and LCAO–MO[122c] calculations and by the observed reactivity of coordinated N_2. Although it has not yet been unambiguously demonstrated, most authors believe that the higher negative charge is present at the terminal (*exo*) nitrogen atom which is also the most sensitive to the metal oxidation state.[123d]

(ii) *π-Bonding schemes for mononuclear terminal end-on dinitrogen complexes of fourfold symmetry*

(a) MONO N_2 COMPLEXES *TRANS*-[L—M—N_2], WHERE L IS NOT AN APPRECIABLE π-ACCEPTOR

A considerable number of dinitrogen complexes of this type have been characterized, such as *trans*-$[MoX(N_2)(dppe)_2]^-$ (X = SCN, N_3),[124] *trans*-$[ReX(N_2)(dppe)_2]$ (X = H, Cl, Br, I),[74,125] *trans*-$[ReCl(N_2)(LL)_2]$ (LL = dppm, dppe, dppp, dppae, dppey, $(PMe_2Ph)_2$, $(PPh_2Me)_2$, $(PPh_2H)_2$, $\{P(CH_2OH)_3\}_2$, $(Me_2NPF_2)_2$),[74] *trans*-$[ReCl(N_2)(dppe)_2]^+$,[74,126] *trans*-$[ReCl(N_2)(PMe_2Ph)_4)]^+$,[74] *trans*-$[MH(N_2)(depe)_2]^+$ (M = Fe, Ru, Os),[52] $[Ru(NH_3)_5N_2]^{2+}$,[15–21] $[Os(NH_3)_5N_2]^{2+}$,[73b] and *trans*-$[Os(N_2)(thf)(L)]$ (L = oep,[26b] ttp[26c]).

If the ligand L which is *trans* to N_2 is not appreciably involved in π-bonding, the simplified π-MO scheme of Fig. 2(*a*) is proposed to describe the π-bonding.

Apart from the Re(II) species, all the above-mentioned complexes are d^6 and the total number of π electrons involved is 10; for the Re(II) species this number is 9. In all cases, oxidation of the metal removes electrons from a N—N antibonding orbital (which has M—N bonding character) and the N—N bond strengthens (hence $\nu(N_2)$ increases by ~80 cm^{-1} on passing from the above-mentioned Re(I) to the corresponding Re(II) species) whereas the M—N

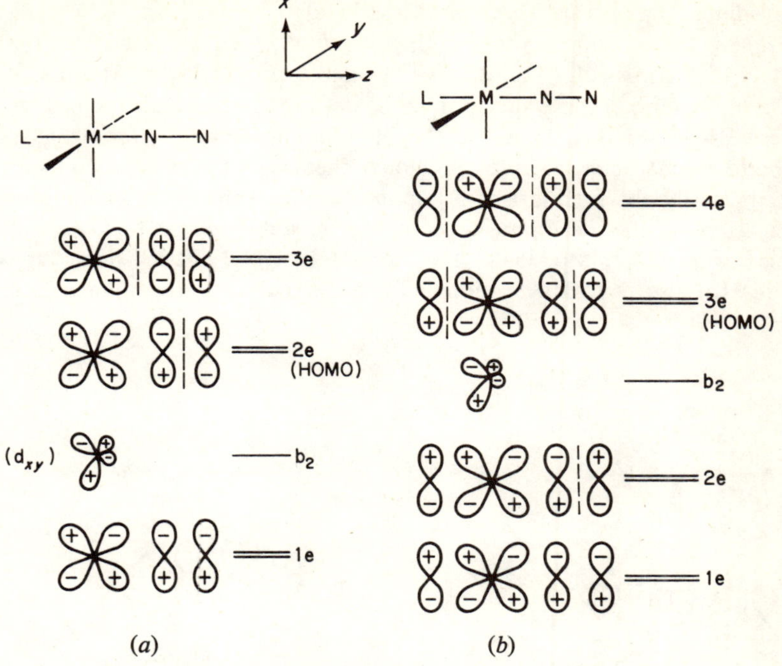

Fig. 2. Simplified π-MO schemes for *trans*-$[ML(N_2)L'_4]$.

bond weakens. Accordingly, the Re(II) complexes are less stable thermally than the Re(I) ones, and they may lose N_2 on standing. Upon reduction, electrons are added to a M–N antibonding orbital and N_2 is evolved (see Section IV.A).

If the ligand L behaves as a π-donor (this situation might occur, if at all, when L is a halide), the simplified π-MO scheme must include the L orbital involved in the π-donation. Hence, the scheme of Fig. 2(*b*) may be applied and the HOMO is now 3e (considering the π electronic contribution from the halide). Relative to the scheme of Fig. 2(*a*) the number of filled orbitals with N–N antibonding character has increased and thus a greater weakening of the N–N bond is expected.

(b) N_2 COMPLEXES, *TRANS*-$[L–M–N_2]$, WHERE L IS A π-ACCEPTOR

Aside from a few examples of monodinitrogen complexes of this type (*trans*-$[Mo(CO)(N_2)(dppe)_2]$,[48] *trans*-$[M(NCR)(N_2)(dppe)_2]$,[127] and $[Mo(CN)(N_2)(dppe)_2]^-$)[124] all the others which may be included in this section are bisdinitrogen species: *trans*-$[W(N_2)_2(dppe)_2]$,[25] *trans*-$[Mo(N_2)_2(LL)_2]$ (LL = dppe, dppp, dppm, dmtpe, dppey, dppae, dpae, diars, $(PMePh_2)_2$, $(PEt_2Ph)_2$),[24] *trans*-$[Mo(N_2)_2(dppe)_2]^+$,[128] and *trans*-$[W(N_2)_2(PMePh_2)_4]^+$.[129]

The π-bonding in these species may be described by the scheme of Fig. 3(*a*) (where B–A = N–N, O–C, C–N, N–C). The HOMO orbital has N–N antibonding character, as in the previous schemes, and it also accounts for the experimentally observed loss of N_2 upon oxidation (dinitrogen evolution is the initial step in the decomposition of the unstable *trans*-$[M(N_2)_2(dppe)_2]^+$ species, where M = Mo, W)[130] and increase of $\nu(N_2)$ of the oxidized species. Figure 3(*a*) also provides a rationale for N_2 evolution as the initial step in the

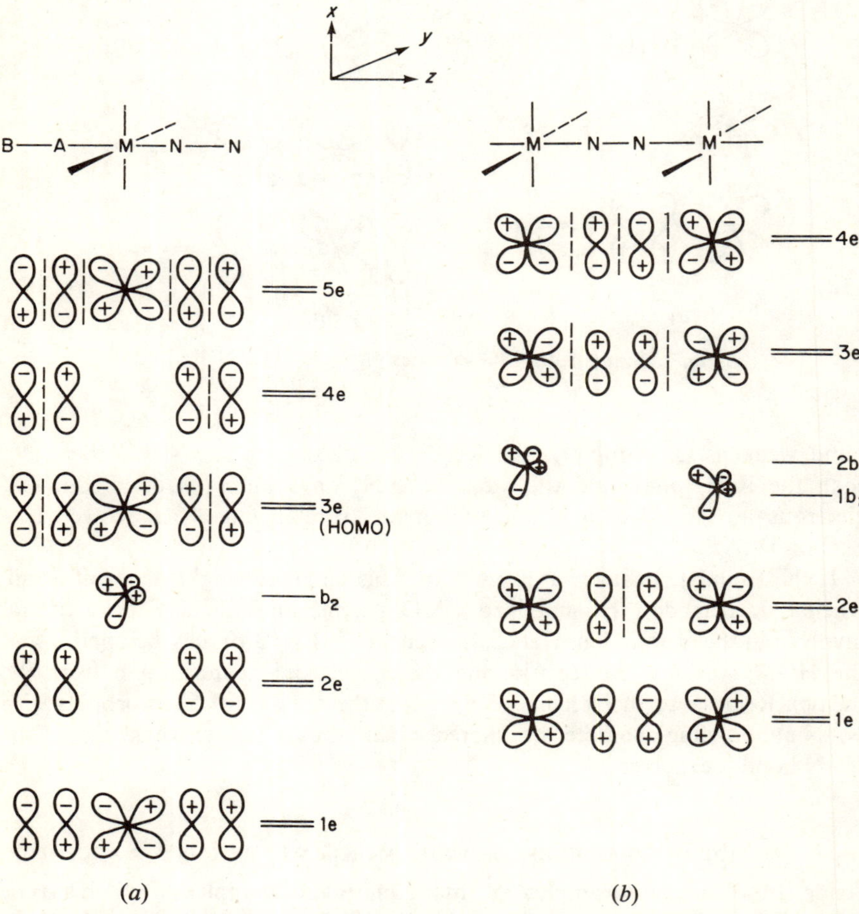

Fig. 3. Simplified π-MO schemes for *trans*-$[M(N_2)(AB)L'_4]$ (where AB is a π-acceptor) (*a*) and for bridging end-on dinitrogen complexes (*b*).

light-assisted alkylation of *trans*-$[M(N_2)_2(dppe)_2]$[131] since a weakening of the $M-N_2$ bond results upon photoexcitation of an electron from the HOMO orbital.

In the d^6 metal species, the numbers of filled π N–N antibonding and bonding orbitals are 2 and 4, whereas in the schemes of Fig. 2(*a*) and (*b*) they are 2, 2 and 4, 2, respectively; hence the N–N bond is not so much weakened in cases of Fig. 3(*a*) as in the other cases. This is expected, considering the competition of the π-acceptor ligand *trans* to N_2 for the available metal electrons in Fig. 3(*a*), as opposed to the situation in Fig. 2.

The scheme of Fig. 3(*a*) may be considered[132a] as an improvement which better fits the available experimental data of a related scheme with a different molecular orbital ordering, which was independently proposed.[132b]

If the dinitrogen ligand is considered to behave as a considerably stronger (or weaker) π-acceptor than the ligand in the *trans* position, it may be involved in π-acceptance from (or π-donation to) the latter ("push–pull" type effect, see below). Related π-MO schemes may be drawn which include

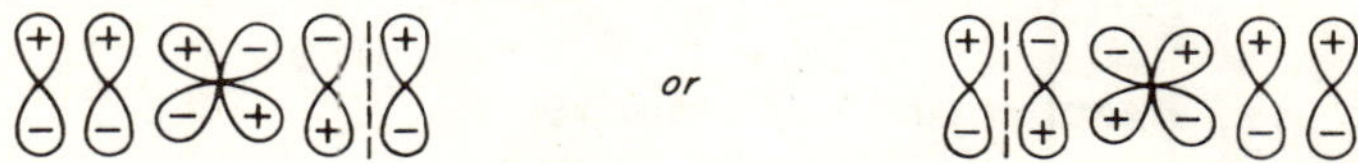

as 2e in the scheme 3(*a*) where the antibonding component derives from the stronger π-acceptor (N_2 in the nitrile- and cyanide-dinitrogen complexes; CO in the carbonyl-dinitrogen species).

(iii) π-*Bonding in linear bridging dinitrogen complexes*

Complexes of this type having fourfold symmetry whose X-ray structure determination has been reported are included in Table 5 (species (P), (Q) and (S)). Other examples of complexes which probably have related symmetry may be cited: $[TiCl_4\{(N_2)ReCl(PMe_2Ph)_4\}(thf)]$,[133] $[TiCl_4\{(N_2)ReCl(PMe_2Ph)_4\}_2]$,[133] $[CrCl_3\{(N_2)ReCl(PMe_2Ph)_4\}(thf)_2]$,[134] $[\{(H_2O)_5Ru\}_2N_2]^{4+}$,[49b] $[(NH_3)_5Ru(N_2)Os(NH_3)_5]^{n+}$ (n = 4, 5),[13] $[(H_2O)(NH_3)_4Ru(N_2)Os(NH_3)_5]^{4+}$,[135] $[\{OsCl(NH_3)_4\}_2(N_2)]^{3+}$,[136] and $[(L)(NH_3)_4Os(N_2)Os(NH_3)_5]^{n+}$ ($n = 5$, L = N_2, H_2O; $n = 4$; L = Cl).[136]

The π-bonding of these species has been described[134b,103] and the scheme of Fig. 3(*b*) has been proposed.[134b] If both metals M and M′ are electron-rich (e.g. d^6 as in complex (S)), the 3e orbitals are filled and the N–N bond is strong; i.e. the tendencies of both metals to backbond to N_2 oppose each other and the net effect does not lead to a N–N bond weakening. However, if one of the atoms is electronically poor, e.g., Mo(V) which is d^1, in complex (P), the 3e orbitals are empty (the number of π electrons in this species is 11) and lengthening of the N–N bond and strengthening of M–N and M′–N bonds occurs relative to the electron-rich case. The net effect may be visualized as a π-donation from the electron-rich $M-N_2$ group to the empty d orbital of the π-acceptor (electron-poor) M′. This effect has been described[31b] as a "push-pull" π-bonding in the lower symmetry, metallocene-type species $[\{(\eta^5\text{-}C_5Me_5)_2MN_2\}_2N_2]$ (M = Ti or Zr (complex (M) in Table 5))[31b,31c] and $[\{(\eta^5C_5Me_5)_2Ti\}_2N_2]$ (N)[30c] by considering the overlap of the $1\pi_g^*$ and $1\pi_u$ dinitrogen orbitals with the molecular orbitals of the $(\eta^5\text{-}C_5Me_5)_2M$ group. Consideration of the spatial distribution of the atoms is required in order to predict the possible overlaps. The π-backbonding is considered to be, as usual, from filled metal orbitals to $1\pi_g^*$ of N_2 and high charge build-up on the bridging N_2 is prevented by π-donation from this ligand to empty metal orbitals with convenient orientation. Hence, N_2 is considered to bridge a π-donor to a π-acceptor metal, resulting in a strong M-bridge N_2 bond and a high μ-N—N bond weakening. This provides a rationalization for ^{15}N NMR and ^{1}H NMR data which show no evidence for rotation about any M-bridge N_2 bond, or for exchange between terminal and bridging N_2, although a fluxional behaviour has been observed which is suggested to involve $(\eta^5\text{-}C_5Me_5)$ site exchange coupled with terminal N_2 dissociative exchange with free, dissolved N_2.[137]

IV. Reactivity of N_2 Complexes

The N_2 molecule exhibits a high ionization potential (360 kcal mol^{-1}) and a large negative electron affinity (estimated as −84 kcal mol^{-1}).[11] Hence it is difficult to oxidize or reduce. Moreover, the N—N bond is hard to weaken due to the high stabilization energy between the double and the triple bond (125 kcal mol^{-1} compared to 65 for CO and 53 for C_2H_2), which has been rationalized[138] by considering the lower repulsion energy of the electron lone pairs of N≡N. Generally the activation which the N_2 molecule attains upon coordination is not enough to induce further attack upon it and most of the reactions involving N_2 complexes lead to the rupture of the M—N bond with N_2 liberation.

Reactions which involve attack at the N_2 ligand (adduct, N—H and N—C bond formation) will be referred to in a later section and we deal here only with those reactions of dinitrogen complexes which do not proceed through attack at the N_2 ligand. They may be classified as follows: (a) oxidations and reductions; (b) displacements; (c) N_2 complexes in catalysis; (d) additions; (e) reactions of co-ligands. These types, except (c), may proceed either with retention or with evolution of N_2.

A. Oxidation and reduction reactions

Attempted oxidations (and also reductions) of N_2 complexes generally result in N_2 evolution (see Table 8) in agreement with expectations based upon the π-MO schemes discussed earlier. Hence, oxidations may be used for the

Table 6. Oxidation of dinitrogen complexes without complete loss of N_2

Starting complex	Oxidizing agent	Product	Reference
trans-$[Mo(N_2)_2(LL)_2]$[a]	I_2/MeOH	$[Mo(N_2)_2(LL)_2]I_3$	*128*
trans-$[Mo(N_2)_2(dppe)_2]$	PhSCl[b]	$[Mo_2(N_2)_3(SPh)_2(dppe)_4]Cl_2$	*140c*
	$HClO_4$[c]	$[Mo_2H_2(N_2)_3(dppe)_4](ClO_4)_2$	*140c*
	hν, MeX[6]	*trans*-$[MoX(N_2)(dppe)_2]$[e]	*f*
trans-$[W(N_2)_2(dppe)_2]$	HCl	$[WH(N_2)_2(dppe)_2]HCl_2$	*140b*
trans-$[W(N_2)_2(PMePh_2)_4]$	$FeCl_3$	*trans*-$[W(N_2)_2(PMePh_2)_4][FeCl_4]$	*129*
trans-$[ReH(N_2)(dppe)_2]$	HBF4	$[ReH_2(N_2)(dppe)_2]BF_4$	*125*
trans-$[ReCl(N_2)(dppe)_2]$	*g*	$[ReCl(N_2)(dppe)_2]^+$	*126*
$Na[Co(N_2)(PEt_2Ph)_3]$	H_2O	$[CoH(N_2)(PEt_2Ph)_3]$	*148c*

[a] LL = dppe, dppae. [b] In toluene at −20°C. [c] In toluene at −10°c. [d] X = Cl. Br. [e] See note *j* of Table 1. [f] V. W. Day, T. A. George and S. D. A. Iske, *J. Amer. Chem. Soc.* (1975) **97**, 4127. [g] Ag^+, $FeCl_3$, $CuCl_2$ or halogens.

quantitative assay of bonded N_2. However, in some cases where the M—N bond is strong enough, oxidation or reduction, generally by one unit, may give a stable new species with bonded N_2 (Table 6).

Electron-rich dinitrogen complexes are generally easily oxidized and reagents such as hydrogen or iodine may oxidize them readily. Halogenated species are also effective and, for example, the complexes $[MoX_2(dppe)_2]$ (X = I, Cl) may be formed from reaction of *trans*-$[Mo(N_2)_2(dppe)_2]$ with ICN and the imidoyl halide $[HC(Cl){=}NMe_2]Cl$, respectively.[139] In reactions with acids

Table 7. Electrochemical reversible oxidation of dinitrogen complexes[a]

Starting complex	Product	Reference
trans-$[Mo(N_2)_2(LL)_2]$	*trans*-$[Mo(N_2)_2(LL)_2]^+$[b]	*128b, 130*
trans-$[M(N_2)(NCR)(dppe)_2]$[c]	$[M(N_2)(NCR)(dppe)_2]^{1+}$	*144*
	$[M(N_2)(NCR)(dppe)_2]^{2+}$[d]	*144*
$[ReX(N_2)(PMe_2Ph)_4]$[e]	$[ReX(N_2)(PMe_2Ph)_4]^+$	*f*
$[\{Ru(NH_3)_5\}_2N_2]^{4+}$	$[\{Ru(NH_3)_5\}_2N_2]^{5+}$	*g*
$[(H_2O)(NH_3)_4Ru(N_2)Os(NH_3)_5]^{4+}$	$[(H_2O)(NH_3)_4Ru(N_2)Os(NH_3)_5]^{5+}$	*135*
$[Os(NH_3)_5(N_2)]^{2+}$	$[Os(NH_3)_5(N_2)]^{3+}$	*g*

[a] Reaction at a Pt electrode relative to a standard calomel electrode. [b] LL = dppe, dppae. The kinetics of its decomposition (LL = dppe) in thf–MeOH–0·1N LiX electrolyte (X = Cl, ClO_4) have been studied by electrochemical techniques, proceeding with loss of both N_2 via a first-order dissociative mechanism (ref. *130*). [c] M = Mo, W; R = Me, Ph, Pr^n, $Si(OMe)_3(CH_2CH_2CH_2)$. One-electron reversible oxidation has also been reported for $[M(N_2)$ (electrode-CN)$(dppe)_2]$ (M = Mo, W) attached to a tin oxide electrode via the organic nitrile moiety (ref. *144*). [d] Formed by one-electron reversible oxidation of $[M(N_2)(NCR)(dppe)_2]^+$. [e] X = Cl, Br. Other Re—N_2 related species have also been studied. [f] Ref. *121b*; C. M. Elson, *J.C.S* (*Dalton*) (1975) 2401. [g] C. Elson, J. Gulens, I. Itzkovitch and J. Page, *Chem. Commun.* (1970) 875.

or other hydrogen sources, facile protonation at the metal often occurs[140,125] thus substantiating the electron-rich metal character. Other chemical oxidizing agents include alkyl halides,[47d] acyl halides,[141] and thiols,[142] but oxidations may

Table 8. Oxidation of N_2 complexes with complete loss of N_2

Starting complex	Oxidizing agent	Product	Reference
$[\{Cp_2TiR\}_2N_2]$[a]	HCl	$[Cp_2TiCl_2]$[b]	*33c*
	Br_2	$[Cp_2TiBr_2]$	*33c*
$[\{(\eta^5\text{-}C_5Me_5)_2ZrN_2\}_2N_2]$	H_2	$[(\eta^5\text{-}C_5Me_5)_2ZrH_2]$	*c*
trans-$[Mo(N_2)_2(dppe)_2]$	ICN	$[MoI_2(dppe)_2]$	*139*
	$[HC(Cl){=}NMe_2]Cl$	$[MoCl_2(dppe)_2]$	*139*
	HCl	$[MoH_2Cl_2(dppe)_2]$	*140b*
	RSH[d]	$[Mo(SR)_2(dppe)_2]$	*142*
	"H"[e]	$[MoH_4(dppe)_2]$	*140d*
	H_2	$[MoH_2(dppe)_2]$[f]	*24c*
trans-$[Mo(N_2)_2(dmtpe)_2]$	H_2	$[MoH_4(dmtpe)_2]$	*g*
trans-$[Mo(N_2)_2(dppe)_2]I_3$	ICN	$[MoI_2(dppe)_2]$	*139*
cis-$[Mo(N_2)_2(PMe_2Ph)_4]$	H_2	$[MoH_4(PMe_2Ph)_4]$	*24i*
trans-$[ReH(N_2)(dppe)_2]$	H_2	$[ReH_3(dppe)_2]$	*125*
$[RuH_2(N_2)(PR_3)_3]$[h]	H_2	$[RuH_4(PR_3)_3]$	*i*
$[Ru(NH_3)_5(N_2)]^{2+}$[j]	*hν*, aq. HCl	$[RuCl(NH_3)_5]^{2+}$[k]	*143*
	hν, neutral H_2O	$[Ru(NH_3)_5(OH)]^{2+}$	*143a*
	hν, aq. NH_3	$[Ru(NH_3)_6]^{3+}$	*143a*
	HX[l]	$[RuX(NH_3)_5]^{2+}$	*15a, 15b*
cis-$[Os(NH_3)_4(N_2)_2]^{2+}$	HX[l]	*cis*-$[Os(NH_3)_4X_2]^+$	*m*
$[Os(NH_3)_5(N_2)]^{2+}$	I_2	$[Os(NH_3)_5I]^{2+}$	*73b*
$[CoH(N_2)(PPh_3)_3]$	CCl_4	$[CoCl_2(PPh_3)_2]$	*35a*
	$SiH(OEt)_3$	$[CoH_2\{Si(OEt)_3\}(PPh_3)_3]$	*166*
trans-$[IrCl(N_2)(PPh_3)_2]$	RCOCl[n]	$[IrCl_2(R)(CO)(PPh_3)_2]$	*141*
	CS_2	$[IrCl(C_2S_5)(PPh_3)_2]$	*o*
	HCl	$[IrHCl_2(PPh_3)_2]$	*140a*
$[\{(PCy_3)_2Ni\}_2N_2]$	RX[p]	$[NiXR(PCy_3)_2]$[q]	*47d*

[a] R = Ph, *o*-, *m*-, *p*-$CH_3C_6H_4$, C_6F_5, $CH_2C_6H_5$. [b] The intermediate Cp_2TiCl is formed at −78°C. [c] J. M. Manriquez, D. R. McAlister, R. D. Sanner and J. E. Bercaw, *J. Amer. Chem. Soc.* (1978) **100**, 2716. [d] R = Et, Pr^n, Bu^n, Ph. [e] Hydrogen source: 2-propanol or a hydroaromatic such as thf, pyrrolidine, indoline. [f] Using benzene or thf as a solvent; $[\{MoH_2(dppe)\}_2\mu\text{-}(dppe)]$ is formed in toluene (ref. *24c*). Formation of $[MoH_4(dppe)_2]$ is also suggested (ref. *g*). [g] L. J. Archer and T. A. George, *J. Organometalic Chem.* (1973) **54**, C25. [h] R = Ph, *p*-Tol. [i] M. H. Knoth, *J. Amer. Chem. Soc.* (1972) **94**, 104. [j] The photo-oxidized species are reversibly converted to this Ru(II) complex (and to the dinuclear$[\{Ru(NH_3)_5\}_2N_2]^{4+}$) by reduction with amalgamated Zn in dilute H_2SO_4 under N_2 (ref. *143a*). $[\{Ru(NH_3)_5\}_2N_2]^{4+}$ (ref. *143a*) and $[Ru(NH_3)_5L]^{2+}$ (L = Py, NCMe) (ref. *k'*) undergo analogous photo-oxidation reactions. [k] The first step appears to be photo-oxidation to $[Ru(NH_3)_5(N_2)]^{3+}$ followed by N_2 displacement by H_2O (ref. *143b*). H_2 is also formed (ref. *143b*). [k'] P. C. Ford, D. H. Stuermer and D. P. McDonald, *J. Amer. Chem. Soc.* (1969) **91**, 6209. [l] X = Cl, Br, I. [m] A. D. Allen and J. R. Stevens, *Canad. J. Chem.* (1973) **51**, 92. [n] R = Me, Et, $C_6H_5CH_2$, C_6H_5, C_6F_5. [o] M. Kubota and C. R. Carey, *J. Organometallic Chem.* (1970) **24**, 491. [p] R = H; X = Cl, OC_6H_5. R = Me; X = I. R = Et; X = Br. [q] The reaction proceeds by oxidative-addition to the coordinatively unsaturated species $[(PCy_3)_2Ni]$ derived in solution from the parent complex.

also be accomplished by photochemical excitation[143] or by electrochemical methods. One-electron, reversible, electrochemical oxidations may be carried out in some cases (Table 7), but generally on attempting to oxidize further, rapid secondary irreversible reactions occur with loss of N_2. Moreover the species derived from one-electron reversible oxidation are often transient and decompose with loss of N_2. However, the M—N_2 bond in *trans*-[M(N_2)(NCR)(dppe)$_2$] (M = Mo, W) appears to be stabilized by the nitrile (which is probably a weaker π-acceptor than N_2) in the *trans* position and the analogous M(I) and M(II) species may be reversibly formed.[144]

The mechanisms of these reactions have not yet been fully ascertained. Dinitrogen evolution may result from oxidation or reduction, or photochemical excitation of the metal (e.g., in *trans*-[Mo(N_2)$_2$(dppe)$_2$]),[132a] but in some cases N_2 displacement is believed to precede the metal oxidation (see notes *b* and *q* in Table 8).

B. Displacement reactions

The M—N bond of a dinitrogen complex may be sufficiently stable for displacement of co-ligands to occur without splitting of that bond, thus allowing the preparation of new dinitrogen complexes from others by simple metathesis (Table 9). The preparation of some dinitrogen complexes with sulphur-bonded ligands, such as *trans*-[Mo(N_2)$_2$(PMe$_2$Ph)$_2$(dpte)][24i] and *mer*-[Re(S$_2$CNR$_2$)(N_2)(PMe$_2$Ph)$_3$],[145] has been carried out by this route.

Table 9. Displacement reactions with retention of the M—N_2 bond

Starting complex	Displacing agent	Product	Reference
cis-[Mo(N_2)$_2$(PMe$_2$Ph)$_4$]	dpte	*trans*-[Mo(N_2)$_2$(PMe$_2$Ph)$_2$(dpte)]	*24i*
trans-[M(N_2)$_2$(PMePh$_2$)$_4$]	dppe	[M(N_2)$_2$(PMePh$_2$)$_2$(dppe)][a]	*129*
	L[b]	[W(N_2)$_2$(L)(PMePh$_2$)$_3$]	*129*
trans-[ReH(N_2)(dppe)$_2$]	MeX[c]	*trans*-[ReX(N_2)(dppe)$_2$]	*125*
trans-[ReCl(N_2)(PMe$_2$Ph)$_4$]	L[d]	*mer*-[Re(L)(N_2)(PMe$_2$Ph)$_3$]	*145*
[FeH$_2$(N_2)(PEtPh$_2$)$_3$]	AlEt$_3$	[FeH$_2$(N_2)(PEtPh$_2$)$_2$][e]	*43b*
[RuCl$_2$(N_2)(H$_2$O)$_2$(thf)]	L[f]	*g*	*h*
mer-[OsX$_2$(N_2)(PR$_2$Ph)$_3$][i]	NaBH$_4$	*mer*-[OsHX(N_2)(PR$_2$Ph)$_3$]	*j*

[a] M = Mo, W. [b] L = NC$_5$H$_5$, NC$_5$H$_4$Me-3, NC$_5$H$_4$Me-4, NC$_5$H$_4$COOH-4. [c] X = Br, I. [d] L = S$_2$CNMe$_2^-$, S$_2$CNEt$_2^-$, S$_2$COEt$^-$, S$_2$PPh$_2^-$. [e] Not fully characterized. Et$_3$AlPEtPh$_2$ is also formed. Reversible reaction. [f] NC$_5$H$_5$, NH$_3$, ethylenediamine, dmf. [g] Not fully characterized ruthenium dinitrogen species derived from H$_2$O and thf replacement by the stronger bases L. [h] Yu. G. Borod'ko, A. K. Shilova and A. E. Shilov, *Russ. J. Phys. Chem.* (1970) **44**, 349. [i] X = Cl, Br, R = Me; X = Cl, R = Et. [j] J. Chatt, D. P. Melville and R. L. Richards, *J. Chem. Soc.* (A) (1971) 895.

Table 10. Displacement reactions with N_2 evolution

Parent complex	Displacing agent	Product	Reference
	(i) Replacement of N_2 by another ligand		
$[\{(\eta^5\text{-}C_5Me_5)_2ZrN_2\}_2N_2]$	CO	$[\{(\eta^5\text{-}C_5Me_5)_2Zr(CO)\}_2N_2]$	*137*
cis-$[Cr(N_2)_2(PMe_3)_4]$	CO	$[Cr(CO)_2(PMe_3)_4]$	*148a*
trans-$[Mo(N_2)_2(dppe)_2]$	CO	$[Mo(CO)_2(dppe)_2]$[a]	*24c, 24i, 128b*
	"CO"[b]	*trans*-$[Mo(CO)(N_2)(dppe)_2]$	*48, 140d*
	C_2H_4	*trans*-$[Mo(C_2H_4)_2(dppe)_2]$	*149b*
		$[Mo(C_2H_4)(dppe)_2]$[c,o]	*149a*
	NCR[d]	*trans*-$[Mo(N_2)(NCR)(dppe)_2]$	*127*
	NBu^n_4X[e]	*trans*-$[MoX(N_2)(dppe)_2]NBu^n_4$	*124*
trans-$[M(N_2)_2(dppe)_2]$[f]	CH_2N_2, *hν*	*trans*-$[M(CH_2)_2(dppe)_2]$	*152b*
	$N_2CHCOOEt$	$[M(N_2CHCOOEt)_2(dppe)_2]$	*152b*
	CNR[g]	*trans*-$[M(CNR)_2(dppe)_2]$	*146*
cis-$[M(N_2)_2(PMe_2Ph)_4]$	CNR, *hν*	*trans*-$[Mo(CNR)_2(PMe_2Ph)_4]$[h]	*152*
		cis-$[M(CNR)_2(PMe_2Ph)_4]$[i]	*152*
trans-$[Mo(N_2)_2(PMePh_2)_4]$	CNR, *hν*	*mer*-$[Mo(CNR)_3(PMePh_2)_3]$[j]	*152*
trans-$[Mo(CO)(N_2)(dppe)_2]$	L[k]	$[Mo(CO)L(dppe)_2]$	*48*
$[(\eta^6\text{-arene})Mo(PR_3)_2(N_2)]$[l]	CO	$[(\eta^6\text{-arene})Mo(PR_3)_2(CO)]$[m]	*42b*
trans-$[ReH(N_2)(dppe)_2]$	L[n]	*trans*-$[ReHL(dppe)_2]$	*125*
$[CpMn(CO)_2N_2]$	thf[o]	$[CpMn(CO)_2(thf)]$	*151*
$[Fe(N_2)(dppe)_2]$	CO	$[Fe(CO)(dppe)_2]$	*148b*
$[FeH(N_2)L]BPh_4$[p]	L′[q]	$[FeHL'L]BPh_4$	*147a*
$[FeH_2(N_2)(PEtPh_2)_3]$	CO	$[FeH_2(CO)(PEtPh_2)_3]$	*37d, 43b*
$[\{CpFe(dmpe)\}_2N_2]^{2+}$	CO	$[CpFe(dmpe)(CO)]^+$	*50*
	$LiAlH_4$	$[CpFeH(dmpe)]$	*50*
$[Ru(NH_3)_5(N_2)]^{2+}$	NH_3	$[Ru(NH_3)_6]^{2+}$	*15b*
	Py	$[Ru(NH_3)_4(Py)_2]^{2+}$	*15b*
		$[Ru(NH_3)_2(Py)_4]^{2+}$	*15b*
	DMSO	$[Ru(NH_3)_5(DMSO)]^{2+}$	*15b*
$[Os(oep)(N_2)(thf)]$	CO	$[Os(oep)(CO)(thf)]$	*26b*
	Py	$[Os(oep)(Py)_2]$	*26b*
$[Os(N_2)_2(NH_3)_4]^{2+}$	L[r]	$[Os(N_2)(NH_3)_4(L)]^{n+}$	*150*
$[CoH(N_2)(PPh_3)_3]$	CO	$[CoH(CO)_n(PPh_3)_{4-n}]$[s]	*45*
	C_2H_4	$[CoH(C_2H_4)(PPh_3)_3]$	*34b, 34c*
	NCR[o,t]	$[CoH(NCR)(PPh_3)_3]$	*147b*
	NO	$[Co(NO)(PPh_3)_3]$	*35a*
	dppe	$[CoH(dppe)_2]$	*35a*
$Na[Co(N_2)(PEt_2Ph)_3]$	CO	$Na[Co(CO)(PEt_2Ph)_3]$	*148c*
$[RhX(N_2)(PCy_3)_2]$[u]	L[u′]	$[RhX(L)(PCy_3)_2]$	*38*
$[IrCl(N_2)(PPh_3)_2]$	CO	$[IrCl(CO)(PPh_3)_2]$	*78a, 78c*
	PPh_3	$[IrCl(PPh_3)_3]$	*78c*
	ArN_2^+	$[IrCl(N_2Ar)(PPh_3)_2]^+$[v]	*153*
$[\{[(PhLi)_3Ni]_2N_2\cdot 2Et_2O\}_2]$	C_2H_4	$[(PhLi)_2NiC_2H_4\cdot Et_2O]$	*47a*
$[\{(PCy_3)_2Ni\}_2N_2]$	CDT	$[(PCy_3)_2Ni(CDT)]$	*47d*
	(ii) N_2 evolution without replacement by the displacing agent		
$[\{(\eta^5\text{-}C_5Me_5)_2Ti\}_2N_2]$	Vacuum[o]	$[(\eta^5\text{-}C_5Me_5)_2Ti]$	*30a*
trans-$[Mo(CO)(N_2)(dppe)_2]$	Vacuum (argon)	$[Mo(CO)(dppe)_2]_n$	*48*
$[FeH_2N_2(PEtPh_2)_3]$	HCl	$[FeCl_2(PEtPh_2)_2]$	*37d*
$[RuH_2N_2(PPh_3)_3]$	HCl	$[RuCl_2(PPh_3)_3]$	*168c*
$[CoHN_2(PPh_3)_3]$	HX[w]	$[CoX(PPh_3)_3]$[w′]	*x*
	I_2	$[CoI_n(PPh_3)_{4-n}]$[y]	*35a, 34c*
$[\{(PCy_3)_2Ni\}_2N_2]$	[z]	$[(PCy_3)_2Ni]$[z′]	*47d*

However in most of the reactions the M—N_2 grouping is not retained and N_2 is evolved (Table 10). These reactions may occur with simultaneous displacement of other neutral or ionic co-ligands and may be subclassified as follows: (i) displacement with N_2 replacement (N_2 is replaced by another ligand without change in the oxidation state of the metal; the reactions with concomitant oxidation have been included previously); (ii) displacement without N_2 replacement.

Various neutral ligands such as isonitriles,[146] nitriles.[147,127,48] carbon monoxide,[78,24i,148,45] olefins,[149,47a,47d] ammonia,[14b] hydrogen (these reactions have been categorized as oxidations), phosphines,[35a,78c] pyridine[15b,26b] and other N heterocycles,[150] dimethylsulfoxide[15b] and thf[151] may replace N_2; diazo compounds may also cause dinitrogen displacement to give compounds tentatively formulated as $[M(CH_2)_2(dppe)_2]$ (M = Mo or W) and $[M(N_2CHCO_2Et)_2(dppe)_2]$, by reaction, under irradiation, of *trans*-$[M(N_2)_2$-$(dppe)_2]$ with diazomethane and ethyl diazoacetate, respectively.[152b]

Ionic species may also replace N_2, for example, arydiazonium,[153] thiocyanate, azide and cyanide.[124]

Some of these displacing agents (mainly isonitriles) which are isoelectronic with N_2, are activated upon coordination to the dinitrogen fixation site in such a way that they undergo reactions[152] which may be related to those observed for N_2 when ligating the same metal centres. This subject will be dealt with in the next chapter.

Coordinatively unsaturated species (such as $[(PCy_3)_2Ni]$ which may undergo addition and oxidative addition reactions) are derived by N_2 evolution

---

[a] The *trans* isomer is formed in toluene at 0°C and the *cis* one at RT; the former converts irreversibly into the latter. Dinitrogen replacement by CO also occurs in the related species $[Mo(N_2)(LL)_2][LL$ = dppae, $(PMe_2Ph)_2$, $(PMePh_2)_2$, $(PEt_2Ph)_2]$. [b] CO source such as thf, MeOH, EtOH, DMF, HCOOMe. [c] Mixed (i) + (ii) type reaction. [d] R = Me, Et, Ph, 2-$CH_3C_6H_4$, *p*-XC_6H_4 (X = Me, MeO, COMe, Cl, NH_2). Reversible displacement for R = alkyl. [e] X = SCN, N_3, CN. [f] M = Mo, W. [g] R = Me, Bu^t, Ph, *p*-$CH_3C_6H_4$, *p*-$MeOC_6H_4$, *p*-ClC_6H_4, 2,6-$Cl_2C_6H_3$. [h] R = Me. *p*-Tol, Bu^t. Upon irradiation complete *trans* to *cis* isomerization occurs except when R = Bu^t for which the *trans* isomer always predominates. [i] M = Mo, W; R = Me, *p*-Tol. M = Mo; R = Bu^t. [j] R = Me, *p*-Tol, Bu^t. Displacement of phosphine also occurs. [k] L = C_2H_4, PhCN, 4-ClC_6H_4CN, Py, DMF, *N*-methylimidazole. [l] Arene = MeC_6H_5, *sym*-$Me_3C_6H_3$. [m] The same product is obtained from $[\{(\eta^6\text{-}C_6H_6)\text{-}Mo(PR_3)_2\}_2N_2]$. [n] L = CO, C_2H_4. [o] Reversible displacement. [p] L = $N(CH_2CH_2PPh_2)_3$, $P(CH_2CH_2PPh_2)_3$. [q] L′ = CO, NCMe, NCPh. [r] Nitrogen heterocycle, e.g., pyrazine, *N*-methylpyrazinium, isonicotinamide, isonicotinic acid. [s] n = 1, 2. $[\{Co(CO)_3(PPh_3)\}_2]$ is also formed (reduction of the metal has occurred). [t] R = Me, Et. [u] X = Cl, Br, I. [u′] L = CO, O_2, C_2H_4. The rate of displacement decreases in the order 1 > Br > Cl. C_2H_4 does not displace N_2 from the chlorocomplex. [v] Parent dinitrogen species prepared *in situ* from reaction of $[IrCl(CO)(PPh_3)_2]$ with azide (see Section II.B(ii)). [w] X = Cl, Br. [w′] $[CoX_2(PPh_3)_2]$ is formed upon further reaction. [x] Refs. *35a*, *140a*, *34c*; J. Lorberth, H. Nöth and P. V. Rinze, *J. Organometallic Chem* (1969) **16**, P1. [y] n = 1 (for Co : I_2 = 1 : 0·5) (ref. *35a*), 2 (ref. *34c*). [z] Benzene or toluene solution. [z′] $[(PCy_3)_2NiN_2]$ is the other product formed in benzene solution.

without replacement in the precursor complex. These reactions may occur with concomitant hydride replacement by halide and may be involved in the partial reduction of species with $-CX_3$ groups (X = halide) by $[CoH(N_2)(PPh_3)_3]$ and $[CoH_3(PPh_3)_3]$ in diglyme with N_2 evolution. Displacement of a halogen atom of the $-CX_3$ (or $-CHX_2$) moiety by hydrogen has been reported[154] to occur in a few cases, e.g., CCl_4 (or CCl_3Br) → $CHCl_3$, benzotrichloride → benzylidene chloride, benzylidene bromide → benzyl bromide, hexachloroethane → pentachloroethane.

C. Dinitrogen complexes in catalysis

Dinitrogen fixation sites have been shown to catalyse various reaction types. However this section deals only with well defined dinitrogen complexes.

(i) *Olefin hydrogenation, isomerization and polymerization*

Due to facile N_2 displacement, various N_2 species behave as catalyst precursors for olefin hydrogenations, isomerizations and polymerizations.

Thus, the coordinatively unsaturated complex

$$[Cp_2\overline{Ti-\mu\text{-}(\eta^1:\eta^5\text{-}C_5H_4)-Ti}Cp] \qquad \textbf{(12)}$$

which reversibly adds N_2 or H_2, and which reduces N_2 to NH_3 if treated with potassium naphthalene followed by acid hydrolysis, has been shown to display catalytic activity in the hydrogenation and the double-bond migration of olefins. Ethylene, internal olefins and diolefins are rapidly reduced to alkanes, terminal olefins are rapidly isomerized into internal olefins (e.g., 1-heptene into *trans* (~95%) and *cis* (~5%) 2-heptenes) and cycloocta-1,5-diene is isomerized into cycloocta-1,3-diene.[40a] The species (**12**) also has the striking ability to dissociate NH_3 into H_2, homogeneously, leading to the formation of $[Cp_2TiNH]_2H$ (**13**). The iron catalysts used in the reduction of N_2 by H_2 are also able to dehydrogenate ammonia to yield H_2. Species (**13**), which is capable of causing dinitrogen reduction to NH_3 in the presence of potassium naphthalene followed by HCl addition, is a homogeneous catalyst for the hydrogenation of olefins, but not for their isomerization.[155]

It has been shown[156] by mechanistic studies that the coordinatively unsaturated complex $[RhCl(PR_3)_2]$ (**14**), which can bind dinitrogen for certain phosphines, is an important reactive intermediate in the hydrogenation of olefins by Wilkinson's catalyst $[RhCl(PR_3)_3]$ from which it is derived by loss of one phosphine ligand in solution. The next step of the reaction is not olefin coordination to (**14**) but the addition of hydrogen to this species with formation of the five-coordinated $[RhH_2Cl(PR_3)_2]$ (**15**) which may also be formed by phosphine loss from the species derived from direct hydrogenation of

$[RhCl(PR_3)_3]$. Species (**15**) adds the olefin which then inserts into a Rh—H bond yielding an alkyl complex from which a rapid elimination of alkane occurs and the intermediate catalytic species (**14**) is regenerated. If the initial Rh complex is hydridic, e.g., $[RhH(CO)(PR_3)_3]$ (which however has no dinitrogen analogue), olefin coordination, which follows the loss of phosphine, precedes the metal hydrogenation step.[157]

Benzene solutions of $[FeH_2(N_2)(PEtPh_2)_3]$ (**16**) with $AlEt_3$ catalyse the hydrogenation of propylene and 1-heptene to propane and heptane. The unsaturated species $[FeH_2(N_2)(PEtPh_2)_2]$, formed by phosphine displacement by $AlEt_3$, was suggested as a possible intermediate.[43b] Complex (**16**) is also a catalyst of hydrogenation of ethylene to ethane, $[FeH(C_2H_5)(PEtPh_2)_2(L)]$ (L = solvent or $PEtPh_2$) being suggested as the active intermediate.[158a] This suggestion has some support from the reported insertions of cyclohexene and CO_2 into the Fe—H bond of $[FeH_2L(PEtPh_2)_3]$ (L = N_2, H_2) yielding the cyclohexyl species $[FeH(C_6H_{11})(PEtPh_2)_2(L)]$ (L = solvent),[158b] which has only been identified in solution, and the diformate $[Fe(OCHO)_2(PEtPh_2)_2]$.[158c]

$[RuH_2(N_2)(PPh_3)_3]$ is known to catalyse double bond migration in n-pentenes and spectroscopic and kinetic studies suggest that $[RuH_2(PPh_3)_3]$ and $[Ru(PPh_3)_3]$ are the active species.[159] $[CoH(N_2)(PPh_3)_3]$ has been used as a catalyst precursor for olefin hydrogenations (e.g., of vinyl compounds),[160] isomerizations (e.g., *cis* to *trans* or double bond migration which may also be catalysed by $[Co(N_2)(PPh_3)_3]$[161a])[34c,161b,161c] and polymerizations (e.g., dimerization of ethylene and propylene, polymerization of vinyl monomers).[162,34c] Kinetic studies have been performed in a number of cases but the active species has not yet been identified.

The presence of N_2 may modify the reaction pathways and the product distribution. This must be taken into account particularly when the reactions are carried out under N_2. The catalytic activity of $[CoH(N_2)(PPh_3)_3]$ (and $[CoH_3(PPh_3)_3]$)[159b,161c] and of $[Co(N_2)(PPh_3)_3]$[163] in olefin isomerization is promoted by the presence of N_2, whereas the double bond migration catalysed by $[RuH_2(N_2)(PPh_3)_3]$, and by $[RuH_4(PPh_3)_3]$,[159] is inhibited by the presence of N_2. It is believed that in the Co species the rate determining step is the displacement of product olefin by the reactant olefin which is promoted by N_2.[159b] The inhibiting effect of N_2 in the Ru system is explained by the competition of N_2 for the active metal site ($[RuH_2(PPh_3)_3]$ and $[Ru(PPh_3)_3]$).[159b,164]

(ii) *Other reactions*

Trans-$[W(N_2)_2(dppe)_2]$ has been shown to catalyse the reduction of aryldiazonium species *p*-$XC_6H_4N_2Cl$ (X = H, CH_3, CH_3O, Cl, NO_2) by alcohols to hydrocarbons p-XC_6H_5. There occurs quantitative loss of N_2 and the α-hydrogens of the alkyl group of the alcohol are shown to be the hydrogen source by deuteration studies. The true catalyst appears to be a product, yet uncharacterized, resulting from the reaction of the dinitrogen complex with the diazonium species.[165]

$[CoH(N_2)(PPh_3)_3]$, which undergoes oxidative addition reactions with the silane $SiH(OEt)_3$ yielding $[CoH_2\{Si(OEt)_3\}(PPh_3)_3]$, is an efficient catalyst for O-silylation and other related reactions. Thus $Si(OEt)_4$ and n-$C_6H_{11}Si(OEt)_3$ are catalytically formed from reaction of $SiH(OEt_3)_3$ with EtOH and with hex-1-ene, respectively, in the presence of the dinitrogen complex.[166] Reduction of nitrous oxide to N_2 with concomitant oxidation of PPh_3 to Ph_3PO has also been shown[86,34c] to be catalysed by this cobalt dinitrogen complex. Displacement of N_2 by N_2O is believed to occur and the parent N_2 species is regenerated by cleavage of the N—O bond of N_2O.

D. Addition reactions

Olefins may add to unsaturated dinitrogen complexes. For example the four-coordinate Ir(I) complex *trans*-$[IrCl(N_2)(PPh_3)_2]$ undergoes addition reactions with diethylmaleate (DEM)[78a,78c] and with fumaronitrile (FMN)[167] yielding the higher-coordinate $[IrCl(N_2)(L)(PPh_3)_2]$ (L = DEM, FMN). Displacement of N_2 by the olefin may also occur and $[IrCl(FMN)(PPh_3)_2]$ is also formed in the latter reaction.

E. Reactions of co-ligands

These reactions may proceed with either retention or loss of N_2. Examples of displacing reactions of co-ligands have already been given (see Section IV.B) and now a few examples of other types of reactions may be cited.

HD formation during enzymatic dinitrogen reduction under N_2/D_2 is known to occur, and the exchange of D_2 with some hydridodinitrogen complexes has been studied. Hydrogen–deuterium exchange of both the hydride ligands and the *ortho*-protons of the phenyl rings of the phosphine ligands has been reported for $[CoH(N_2)(PPh_3)_3]$[168a] and $[RuH_2(N_2)(PPh_3)_3]$.[168b,168c] The mechanism has been postulated[168a] to involve a reversible oxidative addition of the *ortho*-phenyl C—H bond to a coordinatively unsaturated species, formed by N_2 evolution, yielding, for example, in the former case, the proposed intermediate (**17**). Photochemically induced reversible formation, with H_2 evolution, of the related species (**18**) has been reported[37d] from $[FeH_2(N_2)$-$(PEtPh_2)_3]$ without loss of N_2.

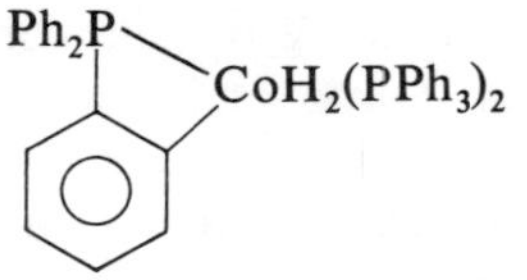

Ph_2P — $CoH_2(PPh_3)_2$ **(17)**

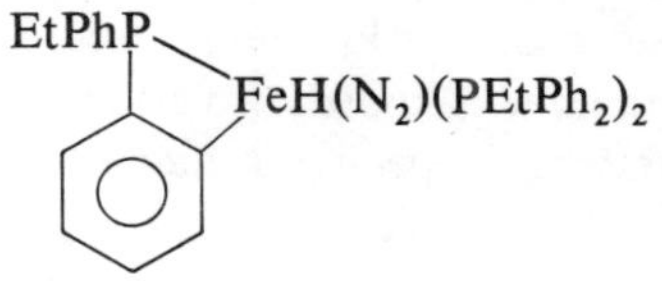

$EtPhP$ — $FeH(N_2)(PEtPh_2)_2$ **(18)**

Insertions of ethylene,[158a] CO_2[158c] and cyclohexene[158b] into the Fe—H bond of $[FeH_2L(PEtPh_2)_3]$ ($L = N_2$, H_2) are related to the olefin reaction catalysis and have been previously referred to. Other examples of insertion reactions involving dinitrogen binding sites will be cited later (Chapter 10).

P—C bond cleavage by a strong reducing agent (n-BuLi) in $[Co(N_2)(PR_3)_3]$ appears to yield the anionic species $[Co(N_2)(PR_3)_2(PR_2)]^-$ (where R = Et, Ph) and $[Co(N_2)(PPh_3)(PPh_2)_2]^{2-}$ which have only been detected in solution.[169]

Diazotization of a co-ligand in a dinitrogen complex may lead to a bis-dinitrogen species and *cis*-$[Os(NH_3)_4(N_2)_2]^{2+}$ has been prepared[91] by this route from reaction of $[Os(NH_3)_5(N_2)]^{2+}$ with HNO_2 as previously referred to (see Section II.B(V)).

V. Final Remarks

Although a great deal of effort has been focused on the attempt to induce the reactivity of what is the most abundant gas in the atmosphere, mainly since the middle of the sixties, and a few relevant achievements have been reported, important difficulties have not yet been overcome which preclude catalytic nitrogen fixation under mild conditions with industrial and agricultural applications and the derivation of an effective model for enzymatic nitrogen fixation.

Coordination of dinitrogen to a metal site is believed to be a key step in its activation, and so further developments in this field are desirable. Hence, for example, further research on species with molybdenum, iron and sulphur ligands is suggested by the presence of these elements in the natural systems. Polynuclear complexes and clusters appear particularly attractive since the activation of the N_2 molecule may then be more effectively done by various metal atoms which may share the supply of the six electrons required for reduction to ammonia without too large a variation of their oxidation states. As in olefin catalysis, the intermediates in the catalytic dinitrogen reduction systems are fairly unstable and further efforts are required to procure the isolation and characterization of the dinitrogen complexes involved. The use of low temperatures may be helpful, although the systems may lose their catalytic features, since unstable dinitrogen complexes have been prepared and characterized at low temperature (e.g., within the group IV transition metals) and the use of matrix isolation techniques is leading to a considerable number of dinitrogen species. Aqueous photolytic and electrolytic processes may also become valuable techniques in the future. Mechanistic studies must be undertaken more frequently since only few examples have been reported so far and attempts to classify preparative reactions into general types are often not fully supported.

References

1. J. Chatt, J. R. Dilworth and R. L. Richards, *Chem. Rev.* (1978) **78**, 589; J. Chatt, "Biological Aspects of Inorganic Chemistry" (Ed. A. W. Addison, W. R. Cullen, D. Dolphin and B. R. James), p. 197. Wiley, New York (1976); J. Chatt and G. J. Leigh, *Chem. Soc. Rev.* (1972) **1**, 121: J. Chatt, *Proc. Roy. Soc. B* (1969) **172**, 337.
2. D. Sellmann, *Angew. Chem.* (*Internat. Edn.*) (1974) **13**, 639.
3. A. D. Allen, R. O. Harris, B. R. Loeschor, J. R. Stevens and R. N. Whiteley, *Chem. Rev.* (1973) **73**, 11; A. D. Allen and F. Bottomley, *Accounts Chem. Res.* (1968) **1**, 360.
4. G. Henrici-Olivé and S. Olivé, "Coordination and Catalysis" Monographs in Modern Chemistry, Vol. 9, p. 289. Verlag Chemie (1977); *Angew. Chem.* (*Internat. Edn.*) (1969) 650.
5. (a) "Recent Developments in Nitrogen Fixation" (Ed. W. Newton, J. R. Postgate and C. Rodriguez-Barrueco). Academic Press, New York and London (1977); (b) "The Chemistry and Biochemistry of Nitrogen Fixation" (Ed. J. R. Postgate). Plenum Press, New York (1971).
6. R. W. F. Hardy, R. C. Burns and G. W. Parshall, "Inorganic Biochemistry" Vol. II, p. 745. Elsevier, Amsterdam (1973).
7. S. W. Schneller, *J. Chem. Educ.* (1972) **49**, 786.
8. G. J. Leigh, In "Preparative Inorganic Reactions" (Ed. W. L. Jolly), Vol. 6, p. 165. Wiley (1971).
9. M. E. Vol'pin and V. B. Shur *In* "Organometallic Reactions" (Ed. E. I. Becker and M. Tsutsui), Vol. I, p. 55. Wiley–Interscience, New York (1970).
10. J. E. Fergusson and J. L. Love, *Rev. Pure Appl. Chem.* (1970) **20**, 33.
11. Yu. G. Borod'ko and A. E. Shilov, *Russ. Chem. Rev.* (1969) **38**, 355.
12. K. Kuchynka, *Catalysis Rev.* (1969) **3**, 111.
13. C. V. Senoff, *Chem. Canada* (1969) 31.
14. R. Murray and D. C. Smith, *Coord. Chem. Rev.* (1968) **3**, 429.
15. (a) A. D. Allen and C. V. Senoff, *Chem. Commun.* (1965) 621; (b) A. D. Allen, F. Bottomley, R. O. Harris, V. P. Reinsalu and C. V. Senoff, *J. Amer. Chem. Soc.* (1967) **89**, 5595; (c) J. Chatt, R. L. Richards, J. E. Fergusson and J. L. Love, *Chem. Commun.* (1968) 1522; (d) J. E. Fergusson and J. L. Love, *Chem. Commun.* (1969) 399; (e) F. Bottomley, *Canad. J. Chem.* (1970) **48**, 351.
16. V. B. Shur, I. A. Tikhonova, E. F. Isaeva and M. E. Vol'pin, *Izvest. Akad. Nauk. S.S.S.R. Ser. Khim.* (1971) 2357 (English translation, p. 2247).
17. (a) L. A. P. Kane-Maguire, F. Basolo and R. G. Pearson, *J. Amer. Chem. Soc.* (1969) **91**, 4609; (b) L. A. P. Kane-Maguire, P. S. Sheridan, F. Basolo and R. C. Pearson, *J. Amer. Chem. Soc.* (1970) **92**, 5865.

18. (a) J. N. Armor and H. Taube, *J. Amer. Chem. Soc.* (1969) **91**, 6874; (b) (1970) **92**, 2560; (c) (1971) **93**, 6476; (d) A. A. Diamantis and G. J. Sparrow, *Chem. Commun.* (1969) 469; (e) A. A. Diamantis, G. J. Sparrow, M. R. Snow and T. R. Norman, *Austral. J. Chem.* (1975) **28**, 1231; (f) F. Bottomley and J. R. Crawford, *J. Amer. Chem. Soc.* (1972) **94**, 9092.
19. I. A. Tikhonova, V. B. Shur and M. E. Vol'pin, *Izvest. Akad. Nauk, S.S.S.R. Ser. Khim* (1976) 229.
20. (a) S. Pell and J. Armor, *J. Amer. Chem. Soc.* (1972) **94**, 686; (b) (1973) **95**, 7625; (c) F. Bottomley, E. M. R. Kiremire and S. G. Clarkson, *J. C. S.* (*Dalton*) (1975) 1909.
21. (a) D. F. Harrison and H. Taube, *J. Amer. Chem. Soc.* (1967) **89**, 5706; (b) D. F. Harrison, E. Weissberger and H. Taube, *Science* (1968) **159**, 320.
22. B. Jezowska-Trzebiatowska and P. Sobota, *J. Organometallic Chem.* (1972) **46**, 339.
23. R. Hammer, H. F. Klein, U. Schubert, A. Frank and G. Huttner, *Angew. Chem.* (*Internat. Edn.*) (1976) **15**, 612.
24. (a) M. Hidai, K. Tominari, Y. Uchida and A. Misono, *Chem. Commun.* (1969) 814; (b) (1969) 1392; (c) M. Hidai, K. Tominari and Y. Uchida, *J. Amer. Chem.* Soc. (1972) **94**, 110; (d) J. Chatt and A. G. Wedd, *J. Organometallic Chem.* (1971) **27**, C15; (e) M. W. Anker, J. Chatt, G. J. Leigh and A. G. Wedd, *J. C. S.* (*Dalton*) (1975) 2639; (f) T. A. George and C. D. Seibold, *J. Organometallic Chem.* (1971) **30**, C13; (g) *Inorg. Nuclear Chem. Letters* (1972) **8**, 465; (h) *Inorg. Chem.* (1973) **12**, 2544; (i) M. Aresta and A. Sacco, *Gazz. Chim. Ital.* (1972) **102**, 755.
25. B. Bell, J. Chatt and G. J. Leigh, *Chem. Commun.* (1970) 842.
26. (a) D. A. Lemenovskii, T. V. Baukova, E. G. Pereralova and A. N. Nesmeyanov, *Izvest. Akad. Nauk S.S.S.R., Ser. Khim* (1976) 2404; (b) J. W. Buchler and P. D. Smith, *Angew Chem.* (*Internat. Edn.*) (1974) **13**, 745; (c) J. W. Buchler and M. Folz, *Z. Naturforsch* (1977) **32b**, 1439; (d) E. B. Fleischer and M. Krishnamurthy, *Ann. N.Y. Acad. Sci.* (1973) **206**, 32.
27. (a) Yu. G. Borod'Ko, I. N. Ivleva, L. M. Kachapina, S. I. Salienko, A. K. Shilova and A. E. Shilov, *Chem. Commun.* (1972) 1178; (b) *Zhur. Struk. Khim.* (1973) **14**, 1112 (English translation, p. 1041).
28. (a) A. E. Shilov, A. K. Shilova and E. F. Kvashina, *Kinet. Katal.*, (1969) **10**, 1402; (b) A. E. Shilov, A. K. Shilova, E. F. Kvashina, T. A. Vorontsova, *Chem. Commun.* (1971) 1590; (c) F. W. Van Der Weij, H. Scholtens and J. H. Tauben, *J. Organometallic Chem.* (1977) **127**, 299.
29. Yu. G. Borod'Ko, I. N. Ivleva, L. M. Kachapina, E. F. Kvashina, A. K. Shilova and A. E. Shilov, *J. C. S. Chem. Commun.* (1973) 169.
30. (a) J. E. Bercaw, *J. Amer. Chem. Soc.* (1974) **96**, 5087; (b) J. E. Bercaw, E. Rosenberg and J. D. Roberts, *J. Amer. Chem. Soc.* (1974) **96**, 612; (c) R. D. Sanner, D. M. Duggan, T. C. McKenzie, R. E. Marsh and J. E. Bercaw, *J. Amer. Chem. Soc.* (1976) **98**, 8358.
31. (a) J. M. Manriquez and J. E. Bercaw, *J. Amer. Chem. Soc.* (1974) **96**, 6229; (b) R. D. Sanner, J. M. Manriquez, R. E. Marsh and J. E. Bercaw, *J. Amer. Chem. Soc.* (1976) **98**, 8351; (c) J. M. Manriquez, R. D. Sanner, R. E. Marsh and J. E. Bercaw, *J. Amer. Chem. Soc.* (1976) **98**, 3042.
32. (a) E. E. van Tamelen, W. Cretney, N. Klaentschi and J. S. Miller, *J.C.S. Chem. Commun.* (1972) 481; (b) C. Ungurenasu and E. Streba, *J. Inorg. Nuclear Chem.* (1972) **34**, 3753; (c) J. E. Bercaw, R. H. Marvich, L. G. Bell and H. H. Brintzinger, *J. Amer. Chem. Soc.* (1972) **94**, 1219.

33. (a) J. H. Teuben and H. J. L. Meijer, *Recl. Trav. Chim. Pays-Bas* (1971) **90**, 360; (b) *J. Organometallic Chem.* (1972) **46**, 313; (c) J. H. Teuben, *J. Organometallic Chem.* (1973) **57**, 159.
34. (a) A. Yamamoto, S. Kitazume, L. S. Pu and S. Ikeda, *Chem. Commun.* (1967) 79; (b) *J. Amer. Chem. Soc.* (1967) **89**, 3071; (c) *J. Amer. Chem. Soc.* (1971) **93**, 371.
35. (a) A. Sacco and M. Rossi, *Inorg. Chim. Acta* (1968) **2**, 127; (b) *Chem. Commun.* (1967) 316.
36. A. D. Allen and F. Bottomley, *Canad. J. Chem.* (1968) **46**, 469.
37. (a) J. L. Thomas and H. H. Brintzinger, *J. Amer. Chem. Soc.* (1972) 1386; (b) J. L. Thomas, *J. Amer. Chem. Soc.* (1973) **95**, 1838; (c) P. Giannoccaro, M. Rossi and A. Sacco, *Coord. Chem. Rev.* (1972) **8**, 77; (d) A. Sacco and M. Aresta, *Chem. Commun.* (1968) 1223.
38. (a) H. L. M. Van Gaal, F. G. Moers and J. J. Steggerda, *J. Organometallic Chem.* (1974) **65**, C43; (b) H. L. M. Van Gaal and F. L. A. Van Den Bekeron, *J. Organometallic Chem.* (1977) **134**, 237.
39. T. I. Eliades, R. O. Harris and M. C. Zia, *Chem. Commun.* (1970) 1709.
40. (a) G. P. Pez and S. C. Kwan, *J. Amer. Chem. Soc.* (1976) **98**, 8079; (b) G. P. Pez, *J. Amer. Chem. Soc.* (1976) **98**, 8072.
41. (a) D. Sellmann and E. Kleinschmidt, *Angew. Chem.* (*Internat. Edn.*) (1975) **14**, 571; (b) *J. Organometallic Chem.* (1977) **140**, 211.
42. (a) M. L. H. Green and W. E. Silverthorn, *Chem. Commun.* (1971) 557; (b) *J. C. S.* (*Dalton*) (1973) 301.
43. (a) M. Aresta, P. Giannocaro, M. Rossi and A. Sacco, *Inorg. Chim. Acta* (1971) **5**, 115: (b) 203.
44. Yu. G. Borod'ko, M. O. Broitman, L. M. Kachapina, A. K. Shilova and A. E. Shilov, *Zhur. Strukt. Khim.* (1971) **12**, 545 (English translation, p. 498).
45. A. Misono, Y. Uchida, M. Hidai and T. Kuse, *Chem. Commun.* (1968) 981.
46. (a) A. Misono, Y. Uchida, T. Saito and K. M. Song, *Chem. Commun.* (1967) 419; (b) G. Speier and J. Markó, *Inorg. Chim. Acta* (1969) **3**, 126.
47. (a) K. Jonas, *Angew. Chem.* (*Internat. Edn.*) (1973) **12**, 997; (b) C. Krüger and Y.-H. Tsay, *Angew. Chem.* (*Internat. Edn.*) (1973) **12**, 998; (c) K. Jonas, D. J. Brauer, C. Krüger, P. J. Roberts and Y.-H. Tsay, *J. Amer. Chem. Soc.* (1976). 74; (d) P. W. Jolly, K. Jonas, C. Krüger, Y.-H. Tsay, *J. Organometallic Chem.* (1971) **33**, 109; (e) M. Aresta, C. F. Nobile and A. Sacco, *Inorg. Chim. Acta* (1975) **12**, 167.
48. T. Tatsumi, H. Tominaga, M. Hidai and Y. Uchida, *J. Organometallic Chem.* (1976) **114**, C27.
49. (a) C. M. Elson, I. J. Itzkovitch and J. A. Page, *Canad. J. Chem.* (1970) **48**, 1639; (b) C. Greutz and H. Taube, *Inorg. Chem.* (1971) **10**, 2664.
50. W. E. Silverthorn, *Chem. Commun.* (1971) 1310.
51. D. Sellmann and G. Maisel, *Z. Naturforsch.* (1972) **27b**, 718.
52. (a) G. M. Bancroft, M. J. Mays and B. E. Prater, *Chem. Commun.* (1969) 585; (b) G. M. Bancroft, M. J. Mays, B. E. Prater and F. P. Stefanini, *J. Chem. Soc.* (*A*) (1970) 2146.
53. L. A. P. Kane-Maguire, *J. Inorg. Nuclear Chem.* (1971) **33**, 3964.
54. R. P. Eischens and J. Jacknow, "Proc. 3rd Internat. Congress on Catalysis" p. 627. North Holland, Amsterdam (1965).
55. (a) G. Ehrlich, *J. Phys. Chem.* (1956) **60**, 1388; (b) J. T. Yates, Jr., T. E. Madey, *J. Chem. Phys.* (1965) **43**, 1055 and references therein.
56. A. Jones and B. D. McNicol, *J. Catalysis* (1977) **47**, 384.

57. G. J. Leigh in ref. (5b) p. 44.
58. (a) R. van Hardeveld and A. V. Montfoort, *Surface Sci.* (1966) **4**, 396; (b) (1969) **17**, 90.
59. (a) D. A. King, *Surface Sci.* (1968) **9**, 375; (b) A. Ravi, D. A. King and N. Sheppard, *Trans. Faraday Soc.* (1968) **64**, 3358; (c) T. A. Egerton and N. Sheppard, *J. Chem. Soc. Faraday I* (1974) **70**, 1357.
60. G. A. Ozin and A. V. Voet *In* "Progr. Inorg. Chem." (Ed. S. J. Lippard) **19**, p. 105. Interscience, New York (1975).
61. (a) G. A. Ozin and A. V. Voet, *Canad. J. Chem.* (1973) **51**, 3332; (b) 637; (c) R. Busby, W. Klotzbücher and G. A. Ozin, *Inorg. Chem.* (1977) **16**, 822; (d) H. Huber, T. A. Ford, W. Kotzbücher and G. A. Ozin, *J. Amer. Chem. Soc.* (1976) **98**, 3176; (e) H. Huber, T. A. Ford, M. Moskovits, G. A. Ozin and W. Klotzbücher, *Inorg. Chem.* (1976) **15**, 1666; (f) E. P. Kündig and G. A. Ozin, *J. Amer. Chem. Soc.* (1974) **96**, 3820.
62. (a) T. C. DeVore, *Inorg. Chem.* (1976) **15**, 1315; (b) J. K. Burdett, M. A. Graham and J. J. Turner, *J. C. S.* (*Dalton*) (1972) 1620.
63. (a) J. K. Burdett and J. J. Turner, *J. C. S. Chem. Commun.* (1971) 885; (b) M. Moskovits and G. A. Ozin, *J. Chem. Phys.* (1973) **58**, 1251; (c) H. Huber, E. P. Kündig, M. Moskovits and G. A. Ozin, *J. Amer. Chem. Soc.* (1973) **95**, 332.
64. G. A. Ozin, M. Moskovits, P. Kündig and H. Huber, *Canad. J. Chem.* (1972) **50**, 2385.
65. (a) D. W. Green, J. Thomas and D. M. Gruen, *J. Chem. Phys.* (1973) **58**, 5453; (b) E. P. Kündig, M. Moskovits and G. A. Ozin, *Canad. J. Chem.* (1973) **51**, 2710.
66. (a) E. P. Kündig, M. Moskovits and G. A. Ozin, *Canad. J. Chem.* (1973) **51**, 2737; (b) W. E. Klotzbücher and G. A. Ozin, *J. Amer. Chem. Soc.* (1973) **95**, 3790; (c) G. A. Ozin and W. E. Klotzbücher, *J. Amer. Chem. Soc.* (1975) **97**, 3965.
67. (a) R. C. Spiker, Jr., L. Andrews and C. Trindle, *J. Amer. Chem. Soc.* (1972) **94**, 2401; (b) V. E. Avdeev, I. I. Zakharov, Yu. A. Borisov and N. N. Bulgakov, *Izvest. Akad, Nauk S.S.S.R., Ser. Khim.* (1976) 239 (English translation, p. 227).
68. (a) D. E. Milligan and M. E. Jacox, *J. Chem. Phys.* (1966) **44**, 2850 and references therein; (b) R. L. DeKock and W. Weltner, Jr., *J. Amer. Chem. Soc.* (1971) **93**, 7106; (c) R. R. Lembke, R. F. Ferrante and W. Weltner, Jr., *J. Amer. Chem. Soc.* (1977) **99**, 416.
69. (a) J. K. Burdett, J. M. Grzybowski, R. N. Perutz, M. Poliakoff, J. J. Turner and R. F. Turner, *Inorg. Chem.* (1978) **17**, 147; (b) J. K. Burdett, A. J. Downs, G. P. Gaskill, M. A. Graham, J. J. Turner and R. F. Turner, *Inorg. Chem.* (1978) **17**, 523.
70. (a) O. Crichton and A. J. Rest, *J. C. S.* (*Dalton*) (1978) 208; (b) (1978) 202; (c) (1977) 656; (d) (1977) 536.
71. (a) C. W. DeKock and D. A. Van Leirsburg, *J. Amer. Chem. Soc.* (1972) **94**, 3235; (b) D. A. Van Leirsburg and C. W. DeKock, *J. Phys. Chem.* (1974) **78**, 134.
72. (a) D. Tevault and K. Nakamoto, *Inorg. Chem.* (1976) **15**, 1282; (b) D. Tevault, D. P. Strommen and K. Nakamoto, *J. Amer. Chem. Soc.* (1977) 2997.
73. (a) Yu.G. Borod'Ko, V. S. Bukreev, G. I. Kozub, M. L. Khidekel' and A. E. Shilov, *Zhur. strukt. Khim.* (1967) **8**, 542 (English translation, p. 480); (b) A. D. Allen and J. R. Stevens, *Chem. Commun.* (1967) 1147.

74. (a) J. Chatt, J. R. Dilworth and G. J. Leigh, *Chem. Commun.* (1969) 687; (b) *J. Organometallic Chem.* (1970) **21**, P49; (c) *J. C. S.* (*Dalton*) (1973) 612; (d) D. J. Darensbourg and D. Madrid, *Inorg. Chem.* (1974) **13**, 1532.
75. V. I. Nefedov, M. A. Porai-Koshits, I. A. Zakharova and M. E. Dyatkina, *Doklady Akad. Nauk S.S.S.R.* (1972) **202**, 605 (English translation, p. 78).
76. (a) D. Sellmann, *Angew. Chem.* (*Internat. Edn.*) (1971) **10**, 919; (b) D. Sellmann, G. Maisel, *Z. Naturforsch* (1972) 465; (c) D. Sellmann, *J. Organometallic Chem.* (1972) **36**, C27; (d) D. Sellmann and E. Kleinschmidt, *Z. Naturforsch* (1977) **32b**, 795; (e) D. Sellmann, A. Brandl and R. Endell, *J. Organometallic Chem.* (1975) **90**, 309.
77. A. D Allen, T. Eliades, R. O. Harris and P. Reinsalu, *Canad. J. Chem.* (1969) **47**, 1605.
78. (a) J. P. Collman and J. W. Kang, *J. Amer. Chem. Soc.* (1966) **88**, 3459; (b) P. Collman, M. Kubota, J.-Y. Sun and F. D. Vastine, *J. Amer. Chem. Soc.* (1967) **89**, 169; (c) J. P. Collman, M. Kubota, F. D. Vastine, J.-Y. Sun and J. W. Kang, *J. Amer. Chem. Soc.* (1968) **90**, 5430.
79. L. Yu. Ukhin, Yu. A. Shvetsov and M. L. Khidekel', *Izvest. Akad. Nauk. S.S.S.R., Ser. Khim.* (1967) 957 (English translation, p. 934).
80. F. Bottomley and E. M. R. Kiremire, *J. C. S.* (*Dalton*) (1977) 1125.
81. (a) P. G. Douglas, R. D. Feltham and H. G. Metzger, *Chem. Commun.* (1970) 889; (b) *J. Amer. Chem. Soc.* (1971) **93**, 84; (c) P. G. Douglas and R. D. Feltham, *J. Amer. Chem. Soc.* (1972) **94**, 5254.
82. L. A. P. Kane-Maguire, P. S. Sheridan, F. Basolo and R. G. Pearson, *J. Amer. Chem. Soc.* (1968) **90**, 5295.
83. G. I. Kozub and V. V. Karpov, *Izvest. Akad. Nauk. S.S.S.R., Ser. Khim.* (1972) 671 (English translation, p. 639).
84. V. M. Volkov, A. I. Korosteleva and S. S. Cherniko, *Russ. J. Inorg. Chem.* (1972) **17**, 1061.
85. J. Chatt, C. D. Falk, G. J. Leigh and R. J. Paske, *J. Chem. Soc.* (*A*) (1969) 2288.
86. L. S. Pu, A. Yamamoto and S. Ikeda, *Chem. Commun.* (1969) 189.
87. (a) J. Chatt, G. J. Leigh and R. L. Richards, *Chem. Commun.* (1969) 515; (b) *J. Chem. Soc.* (*A*) (1970) 2243.
88. M. L. Ziegler, K. Weidenhammer, H. Zeiner, P. S. Skell and W. A. Herrmann, *Angew. Chem.* (*Internat. Edn.*) (1976) **15**, 695.
89. B. Bayerl, K. Schmidt and M. Wahren, *Z. Chem.* (1975) **15**, 277.
90. (a) V. M. Volkov and L. S. Volkova, *Russ. J. Inorg. Chem.* (1971) **16**, 1382; (b) L. S. Volkova, V. M. Volkov and S. S. Chernikov, *Russ. J. Inorg. Chem.* (1971) **16**, 1383.
91. H. A. Scheidegger, J. N. Armor and H. Taube, *J. Amer. Chem. Soc.* (1968) **90**, 3263.
92. S. Pell and J. N. Armor, *J. C. S. Chem. Commun.* (1974) 259.
93. (a) L. E. Orgel, "An Introduction to Transition Metal Chemistry", Methuen, London (1966); (b) K. B. Yatsimirskii and Yu. A. Kraglyak, *Doklady Akad. Nauk S.S.S.R.* (1969) **186**, 885.
94. C. Busetto, A. D'Alfonso, F. Maspero, G. Perero and A. Zazzetta, *J. C. S.* (*Dalton*) (1977) 1828.
95. M. J. S. Gynane, J. Jeffery and M. L. Lappert, *J. C. S. Chem. Commun.* (1978) 34.
96. P. Sobota and B. J. Trzebiatowska, *J. Organometallic Chem.* (1977) **131**, 341.
97. G. C. Dobinson, R. Mason, G. B. Robertson, R. Ugo, F. Conti, D. Morelli, S. Cenini and F. Bonati, *Chem. Commun.* (1967) 739.

98. (a) G. Huttner, W. Gartzke and K. Allinger *J. Organometallic Chem.* (1975) **91**, 47; (b) *Angew Chem.* (*Internat. Edn.*) (1974) **13**, 822.
99. (a) D. Sellmann, A. Brandl and R. Endell, *J. Organometallic Chem.* (1976) **111**, 303; (b) D. Sellmann, *J. Organometallic Chem.* (1972) **44**, C46; (c) D. Sellmann and K. Jodden, *Angew. Chem.* (*Internat. Edn.*) (1977) **16**, 464; (d) D. Sellmann, A. Brandl and R. Endell, *J. Organometallic Chem.* (1973) **49**, C22; (e) *Angew. Chem. Internat. Edn.* (1973) **12**, 1019 (German, p. 1122); (f) *Angew. Chem. Internat. Edn.* 1019 (German, p. 1121).
100. (a) M. Mercer, R. H. Crabtree and R. L. Richards, *J. C. S. Chem. Commun.* (1973) 808; (b) M. Mercer, *J. C. S.* (*Dalton*) (1974) 1637.
101. P. D. Cradwick, J. Chatt, R. H. Crabtree and R. L. Richards, *J. C. S. Chem. Commun.* (1975) 351.
102. B. R. Davis and J. A. Ibers, *Inorg. Chem.* (1971) **10**, 578.
103. I. M. Treitel, M. T. Flood, R. E. Marsh and H. B. Gray, *J. Amer. Chem. Soc.* (1969) **91**, 6512.
104. D. Bright and J. A. Ibers, *Inorg. Chem.* (1969) **8**, 703.
105. (a) T. Uchida, Y. Uchida, M. Hidai and T. Kodama, *Bull. Chem. Soc. Japan* (1971) **44**, 2883; (b) *Acta Cryst.* (1975) **B31**, 1197.
106. M. J. S. Dewar, *Bull. Soc. Chim. France* (1951) C.71; (b) J. Chatt and L. A. Duncanson, *J. Chem. Soc.* (1953) 2939.
107. J. H. Nelson, K. S. Wheelock, L. C. Cusachs and H. B. Jonassen, *J. Amer. Chem. Soc.* (1969) **91**, 7005.
108. A. C. Sarapu and R. F. Fenske, *Inorg. Chem.* (1975) **14**, 247.
109. W. G. Sly, *J. Amer. Chem. Soc.* (1959) **81**, 18.
110. R. Hammer, H.-F. Klein, P. Friedrich and G. Huttner, *Angew. Chem.* (*Internat. Edn*) (1977) 485.
111. R. S. Mulliken, *Canad. J. Chem.* (1958) **36**, 10.
112. R. F. Lake and H. W. Thompson, *Spectrochim. Acta* (1971) **27A**, 783.
113. R. Hoffmann, *J. Chem. Phys.* (1963) **39**, 1397.
114. J. Chatt, *Pure Appl. Chem.* (1970) **24**, 425.
115. D. L. Lichtenberger and R. F. Fenske, *J. Organometallic Chem.* (1976) **117**, 253.
116. K. F. Purcell, *Inorg. Chim. Acta.* (1969) 540.
117. W. J. Chambers and N. J. Fitzpatrick, *Proc. Roy. Irish Acad.* (1971) **71B**, 97.
118. A. B. P. Lever and G. A. Ozin, *Inorg. Chem.* (1977) **16**, 2012.
119. (a) J. Chatt, D. P. Melville and R. L. Richards, *J. Chem. Soc.* (*A*) (1969) 2841; (b) D. J. Darensbourg, *Inorg. Chem.* (1971) **10**, 2399; (c) Yu. G. Borod'Ko, S. M. Vinogradova, Yu. P. Myagkov and D. D. Mozzhukhin, *Zhur. Strukt. Khim.* (English translation) (1970) **11**, 269; (d) M. S. Quinby and R. D. Feltham, *Inorg. Chem.* (1972) **11**, 2468.
120. G. M. Bancroft, R. E. B. Garrod, A. G. Maddock, M. J. Mays and B. E. Prater, *J. Amer. Chem. Soc.* (1972) **94**, 647.
121. (a) V. I. Nefedov, V. S. Lenenko, V. B. Shur, M. E. Vol'pin, J. E. Salyn and M. A. Porai-Koshits, *Inorg. Chim. Acta* (1973) 499; (b) J. Chatt, C. M. Elson, N. E. Hooper and G. J. Leigh, *J. C. S.* (*Dalton*) (1975) 2392.
122. (a) Yu. A. Kruglyak and K. B. Yatsimirskii, *Teor. Eks. Khim.* (1969) **5**, 308; (b) S. M. Vinogradova, M. G. Kaplunov and Yu. G. Borod'Ko, *Zhur. Strukt. Khim.* (1972) **13**, 67; (c) S. M. Vinogradova and Yu. G. Borod'Ko, *Russ. J. Phys. Chem.* (1973) **47**, 449; (d) V. I. Baranovskii, N. V. Ivanova and A. B. Nikol'skii, *Zhur. Strukt. Khim.* (1973) **14**, 133.

123. (a) G. J. Leigh, J. N. Murrell, W. Bremser and W. G. Proctor, *Chem. Commun.* (1970) 1661; (b) B. Folkesson, *Acta Chem. Scand.* (1973) **27**, 287; (c) H. Binder and D. Sellmann, *Angew. Chem.* (*Internat. Edn.*) (1973) **12**, 1017; (d) P. Brant and R. D. Feltham, *J. Less Common Metals* (1977) **54**, 81.
124. J. Chatt, G. J. Leigh, C. J. Pickett, personal communication.
125. M. E. Tully and A. P. Ginsberg, *J. Amer. Chem. Soc.* (1973) **95**, 2042.
126. J. Chatt, J. R. Dilworth, H. P. Gunz, G. P. Leigh and J. R. Sanders, *Chem. Commun.* (1970) 90.
127. T. Tasumi, M. Hidai and Y. Uchida, *Inorg. Chem.* (1975) 2530.
128. (a) T. A. George and C. D. Seidbold, *J. Amer. Chem. Soc.* (1972) **94**, 6859; (b) *Inorg. Chem.* (1973) **12**, 2548.
129. J. Chatt, A. J. Pearman and R. L. Richards, *J. C. S.* (*Dalton*) (1977) 2139.
130. C. M. Elson, *Inorg. Chim. Acta* (1976) **18**, 209.
131. J. Chatt, R. A. Head, G. J. Leigh, C. J. Pickett, *J. C. S. Chem. Commun.* (1977) 299.
132. (a) J. Chatt, G. J. Leigh, C. J. Pickett, A. J. L. Pombeiro and R. L. Richards, *Nouveau J. Chimie*, (1978) **2**, 541; (b) D. L. DuBois and R. Hoffmann, *Nouveau J. Chimie* (1977) **1**, 479.
133. R. Robson, *Inorg. Chem.* (1974) **13**, 475.
134. (a) J. Chatt, J. R. Dilworth, G. J. Leigh and R. L. Richards, *Chem. Commun.* (1970) 955; (b) J. Chatt, R. C. Fay and R. L. Richards, *J. Chem. Soc.* (*A*) (1971) 702.
135. C. M. Elson, J. Gulens and J. A. Page, *Canad. J. Chem.* (1971) **49**, 207.
136. R. H. Magnuson and H. Taube, *J. Amer. Chem. Soc.* (1972) **94**, 7213.
137. J. M. Manriquez, D. R. McAlister, E. Rosenberg, A. M. Shiller, K. L. Williamson, S. I. Chan and J. E. Bercaw, *J. Amer. Chem. Soc.* (1978) **100**, 3078.
138. L. Pauling, *Tetrahedron* (1962) **17**, 229.
139. A. J. L. Pombeiro, R. L. Richards and J. Chatt, presented at the *First National Meeting of Chemistry, Lisbon, January, 1978.*
140. (a) J. Chatt, R. L. Richards, J. R. Sanders and J. E. Fergusson, *Nature* (*London*) (1969) **221**, 551; (b) J. Chatt, G. A. Heath and R. L. Richards, *J. C. S.* (*Dalton*) (1974) 2074; (c) M. Aresta and C. F. Nobile, *Inorg. Chim. Acta* (1976) **17**, L17; (d) T. Tatsumi, H. Tominaga, M. Hidai and Y. Uchida, *Chem. Letters* (*Chem. Soc. Japan*) (1977) 37.
141. M. Kubota and D. M. Blake, *J. Amer. Chem. Soc.* (1971) **93**, 1368.
142. J. Chatt, J. P. Lloyd and R. L. Richards, *J. C. S.* (*Dalton*) (1976) 565.
143. (a) C. Sigwart, J. T. Spence, *J. Amer. Chem. Soc.* (1969) **91**, 3991; (b) R. E. Hintze and P. C. Ford, *J. Amer. Chem. Soc.* (1975) **97**, 2664.
144. G. J. Leigh and C. J. Pickett, *J. C. S.* (*Dalton*) (1977) 1797.
145. J. Chatt, R. H. Crabtree, J. R. Dilworth and R. L. Richards, *J. C. S.* (*Dalton*) (1974) 2358.
146. (a) J. Chatt, K. W. Muir, A. J. L. Pombeiro, R. L. Richards, G. H. D. Royston and R. Walker, *J. C. S. Chem. Commun.* (1975) 708; (b) J. Chatt, C. M. Elson, A. J. L. Pombeiro, R. L. Richards and G. H. D. Royston, *J. C. S.* (*Dalton*) (1978) 165.
147. (a) P. Stoppioni, F. Mani and L. Sacconi, *Inorg. Chim. Acta* (1974) **11**, 227; (b) A. Misono, Y. Uchida, M. Hidai and T. Kuse. *Chem. Commun.* (1969) 208.
148. (a) H. H. Karsch, *Angew. Chem. Internat. Edn.* (1977) **16**, 56; (b) R. A. Cable, M. Green, R. E. MacKenzie, P. L. Timms and T. W. Turney, *J. C. S. Chem. Commun.* (1976) 270; (c) M. Aresta, C. F. Nobile, M. Rossi and A. Sacco, *Chem. Commun.* (1971) 781.

149. (a) T. Ito, T. Kokubo, T. Yamamoto, A. Yamamoto and S. Ikeda, *J. C. S.* (*Dalton*) (1974) 1783; (b) J. W. Byrne, H. U. Blaser and J. A. Osborn, *J. Amer. Chem. Soc.* (1975) **97**, 3871.
150. R. H. Magnuson and H. Taube, *J. Amer. Chem. Soc.* (1975) **97**, 5129.
151. D. Sellmann, *Angew. Chem. Internat. Edn.* (1972) **11**, 534.
152. (a) A. J. L. Pombeiro, D. Phil. Thesis, University of Sussex (1976); (b) A. J. L. Pombeiro and R. L. Richards, unpublished results; (c) *Transition Metal Chemistry* (1980) **5**, 55; (d) J. Chatt, A. J. L. Pombeiro and R. L. Richards, *J. C. S.* (*Dalton*), in press.
153. B. L. Haymore and J. A. Ibers, *J. Amer. Chem. Soc.* (1973) **95**, 3052.
154. J. B. Lee and B. Cubberly, *Tetrahedron Letters* (1969), 1061.
155. (a) J. N. Armor, *Inorg. Chem.* (1978) **17**, 203; (b) 213.
156. (a) J. A. Osborn, F. H. Jardine, J. F. Young and G. Wilkinson, *J. Chem. Soc.* (*A*) (1966) 1711; (b) R. W. Mitchell, J. D. Ruddick and G. Wilkinson, *J. Chem. Soc.* (*A*) (1971) 3224; (c) H. Arai and J. Halpern, *Chem. Commun.* (1971) 1571; (d) J. Halpern and C. S. Wong, *J. C. S. Chem. Commun.* (1973) 629; (e) C. A. Tolman, P. Z. Meakin, D. L. Lindner and J. P. Jesson, *J. Amer. Chem. Soc.* (1974) **96**, 2762; (f) J. Halpern *In* "Organotransition-Metal Chemistry" (Ed. Y. Ishii and M. Tsutsui). Plenum, New York (1975).
157. C. O'Connor and G. W. Wilkinson, *J. Chem. Soc.* (*A*) (1968) 2665.
158. (a) V. D. Bianco, S. Doronzo and M. Aresta, *J. Organometallic Chem.* (1972) **42**, C63; (b) V. B. Bianco, S. Doronzo and N. Gallo, *J. Organometallic Chem.* (1977) **124**, C43; (c) V. B. Bianco, S. Doronzo and M. Rossi, *J. Organometallic Chem.* (1972) **35**, 337.
159. (a) F. Pennella and R. L. Banks, *J. Catalysis* (1974) **35**, 73; (b) F. Pennella, *Co-ordination Chem. Rev.* (1975) **16**, 51.
160. (a) S. Tyrlik, K. Falkowski and K. Leibler, *Inorg. Chim. Acta* (1972) 291; (b) S. Tyrlik, *J. Organometallic Chem.* (1973) **50**, C46; (c) **59**, 365.
161. (a) J. Kovács, G. Speier and L. Markó, *Inorg. Chim. Acta* (1970) 412; (b) S. Tyrlik, *J. Organometallic Chem.* (1972) **39**, 371; (c) F. Pennella, *J. Organometallic Chem.* (1974) **78**, C10.
162. L. S. Pu, A. Yamamoto and S. Ikeda, *J. Amer. Chem. Soc.* (1968) **90**, 7170.
163. F. K. Schmidt, L. O. Nindakova, S. M. Krasnopolskaya, T. V. Dmitrieva and G. V. Ratovskii, *Reaction Kinetics and Catalysis Letters* (1977) **7**, 247.
164. F. Pennella, *J. Organometallic Chem.* (1974) **65**, C17.
165. V. S. Lenenko, A. G. Knizhnik, E. I. Mysov, V. B. Shur and M. E. Vol'pin, *Izvest. Akad. Nauk S.S.S.R., Ser. Khim.* (1975) 2380 (English translation, p. 2271).
166. N. J. Archer, R. N. Haszeldine and R. V. Parish, *Chem. Commun.* (1971) 524.
167. P. Uguagliati, G. Deganello, L. Busetto and U. Belluco, *Inorg. Chem.* (1969) **8**, 1625.
168. (a) G. W. Parshall, *J. Amer. Chem. Soc.* (1968) **90**, 1669; (b) G. W. Parshall, W. H. Knoth and R. A. Schunn, *J. Amer. Chem. Soc.* (1969) **91**, 4990; (c) W. H. Knoth, *J. Amer. Chem. Soc.* (1968) **90**, 7172.
169. G. Speier and J. Markó, *J. Organometallic Chem.* (1970) **21**, P46.

7

The Reduction of Ligating Dinitrogen

R. L. RICHARDS

A.R.C. Unit of Nitrogen Fixation, The University of Sussex, Brighton, England

I. The Basicity of Ligating Dinitrogen

THE accepted bonding scheme for dinitrogen (Chapter 6) involves its donating σ-electron density to the metal and receiving π-density from the metal into its π^*-orbitals. Such a synergic process raises the question of charge distribution within the metal–dinitrogen system, which will clearly relate to the susceptibility of ligating dinitrogen to attack by various reagents. There is no direct method of measuring this property, but it has been inferred as a result of a number of spectroscopic studies of mononuclear complexes, in particular infrared (IR) absorption intensities and X-ray photoelectron (XPE) spectral measurements.

Examples of binding energies and IR intensity data for some complexes are shown in Table 1. Generally the complexes chosen are stable, with low $\nu(N_2)$ values which imply strong back-bonding. On the basis of binding energy correlations and a comparison with dipole moment and electrochemical data, it has been concluded that in these particular complexes the dinitrogen is overall negatively charged.[1] The two nitrogen atoms also often carry sufficiently different charges for their clear resolution in the XPE spectrum (see Table 1). This seems to be a feature confined to stable complexes of low $\nu(N_2)$, as evidenced by the spectrum of the very stable complex *trans*-$[ReCl(N_2)(PMe_2Ph)_4]$ ($\nu(N_2) = 1925\ cm^{-1}$),[2] as compared to that of its less stable cation in the salt $[ReCl(N_2)(PMe_2Ph)_4][FeCl_4]$ ($\nu(N_2) = 2055\ cm^{-1}$),[2] where the resolution is lost and the average N(1s) binding energy is higher.[1]

On the basis of an apparent effective oxidation state of rhenium (determined by XPE spectroscopy) in a closely related series of complexes of N_2 and the isostructural ligands CO and NO, it was concluded that N_2 and CO are about equal in their ability to withdraw negative charge from the metal.[1] Both are

Table 1. XPS and IR intensity (I) data for dinitrogen complexes

Complex	n(1s) binding energies (eV)		$I(N_2) \times 10^{-4}$ (M^{-1} cm^{-2})	Refs (*a*)	Refs (*b*)
$[Mo(N_2)_2(dppe)_2]$	399·6	398·6	19·4	(*c*)	(*d*)
$[W(N_2)_2(dppe)_2]$			18.9		(*d*)
$[Mn(\eta^5\text{-}C_5H_5)(CO)_2(N_2)]$	403·0	401·8		(*e*)	
$[ReCl(N_2)(PMe_2Ph)_4]$	400·1	398·4	10·85	(*1*)	(*d*)
	400·0	398·6		(*f*)	
	400·3	398·2		(*q*)	
$[ReCl(N_2)(PMe_2Ph)_4][FeCl_4]$	400·1			(*1*)	
$[ReCl(N_2)(dppe)_2]$	400·9	398·8	8·95	(*1*)	(*d*)
	400·4	398·5		(*g*)	
$[ReCl(N_2)(dppe)_2][FeCl_4]$	400·3			(*1*)	
$[ReCl(N_2)(Py)(PMe_2Ph)_3]$	399·8	398·3		(*1*)	
$[FeH_2(N_2)(PPh_3)_3]$	400·1	399·0		(*h*)	
$[Ru(NH_3)_5(N_2)]Cl_2$	399·6(2·0)[i]		5·4	(*h*)	(*j*)
$[RuCl(N_2)(diars)_2]SbF_6$	402·3	400·7		(*k*)	
$[Os(NH_3)_5(N_2)]Cl_2$	399·5(2·2)[i]		6·5	(*h*)	(*j*)
$[IrCl(N_2)(PPh_3)_2]$	400·4	399·1		(*h*)	

[a] References to XPE. [b] References to IR intensities. [c] P. Brant and R. D. Feltham, *Proceedings 2nd Climax Internat. Conf. on Chemistry of Molybdenum* (Ed. P. C. H. Mitchell). Oxford University Press, Oxford (1976). [d] D. J. Darensbourg, *Inorg. Chem.* (1972) **11**, 1436; with C. L. Hyde, *Inorg. Chem.* (1971) **10**. 431. [e] H. Binder and D. Sellman, *Angew. Chem. Internat. Edn,* (1973) **12**, 1017. [f] V. I. Majedov, M. A. Porai-Koshits, I. A. Zakjarova and M. E. Dyatkina, *Doklady Akad. Nauk. USSR.* (1972) **202**. 605. [g] B. Folkesson, *Acta Chem. Scand.* (1973) **27**, 287. [h] B. Folkesson, *Acta Chem. Scand.* (1973) **27**, 1441. [i] Broad unresolved line, half-width in parentheses where quoted. [j] B. Folkesson, *Acta Chem. Scand.* (1972) **26**, 4008; (1973) **27**, 276. [k] M. S. Quinby and R. D. Feltham, *Inorg. Chem.* (1972) **11**, 2468. py = C_5H_5N, dppe = $Ph_2PCH_2CH_2PPh_2$, diars = 1,2-$(AsMe)_2C_6H_4$.

rather less electron-withdrawing than NO, which is about equal to the chlorine atom.[1]

The conclusion to be derived from this work therefore, is that upon ligation dinitrogen develops a negative charge which is asymmetrically distributed between the nitrogen atoms, the extent of asymmetry varying according to the complex. It has not been possible to determine unequivocally by physical methods whether the nitrogen atom adjacent to the metal (hereafter N_α) or the terminal nitrogen atom (hereafter N_β) is the more negatively charged. However, chemical studies, such as the reactions which will now be described, lead to the conclusion that N_β is the more negative atom.

The first demonstration of the basic character of ligating dinitrogen was provided by the displacement of ligating water from ruthenium(II) to give a binuclear, μ-dinitrogen complex, equation (**1**).[3]

$$[Ru(NH_3)_5(H_2O)]^{2+} + [Ru(NH_3)_5(N_2)]^{2+} \rightleftharpoons [\{Ru(NH_3)_5\}_2(N_2)]^{4+} + H_2O \qquad (1)$$

This type of reaction is the basis of the synthesis of a variety of binuclear, μ-dinitrogen complexes which are described in Chapter 6. In these complexes, ligating dinitrogen may be regarded as exhibiting both σ-donor and π-acceptor properties towards the electron-rich metal centre in the original complex, together with σ-donor or $(\sigma + \pi)$-donor properties towards an electron-deficient centre;[4] for example $(\sigma + \pi)$-donor behaviour is seen in the complex $[Cl(PMe_2Ph)_4ReN_2MoCl_4(OMe)]$.[5] The $(\sigma + \pi)$-donor behaviour has been discussed in more detail elsewhere[4] and here we confine our discussion to the σ-donor ability alone.

The σ-donor strengths, relative to Et_2O, of the terminal nitrogen atom in a few robust dinitrogen complexes have been measured in benzene solution by determining the equilibrium constants K of a series of equilibria of the type (**2**).[6] These determinations, obtained by an NMR method, and similar data for analogous carbonyl complexes are shown in Table 2, which lists the complexes examined in order of σ-donor strength towards trimethyl aluminium, relative to diethylether.

$$M{-}N_2 + Me_3Al\cdot OEt_2 \underset{}{\overset{K}{\rightleftharpoons}} M{-}N_2{-}Al\cdot Me_3 + Et_2O \qquad \textbf{(2)}$$

$$(M = \text{metal} + \text{co-ligands})$$

It can be seen that the basicity of ligating dinitrogen can approach that of tetrahydrofuran in certain complexes and that the order of basicity follows the order of $\nu(N_2)$ values qualitatively, but not exactly. The co-ligands of dinitrogen have a marked effect upon its basicity, for example if two PMe_2Ph ligands of *trans*-$[ReCl(N_2)(PMe_2Ph)_4]$ are replaced by the less basic $P(OMe)_3$,

Table 2. The $\nu(N_2)$ and relative σ-donor strengths [K, reaction type (2)] of the substances (A) against $AlMe_3$ ($Et_2O = 1$) in benzene[a]

(A)	$\nu(N_2)$ (cm^{-1})	K
Tetrahydrofuran	—	70
trans-$[ReCl(N_2)(PMe_2Ph)_4]$	1923	20·6
trans-$[Mo(N_2)_2(dppe)_2]$	1979	33
	2020 w	(16·5)[b]
trans-$[W(N_2)_2(dppe)_2]$[c]	1948	15
	2015 w	(7·5)[b]
trans-$[ReCl(N_2)(PMe_2Ph)_2\{P(OMe)_3\}_2]$	2000	5·5
trans-$[ReCl(CO)(PMe_2Ph)_4]$	—	3·3
Et_2O	—	1
mer-$[OsCl_2(N_2)(PEt_2Ph)_3]$[c]	2063	0·3

[a] At $34 \pm 2°$, data ref. *6*. [b] Halved for comparative purposes to account for two basic sites on the molecule. [c] Adduct slowly decomposes in solution.
dppe = $Ph_2PCH_2CH_2PPh_2$.

$\nu(N_2)$ is raised and K is lowered.[6] Thus the greater the transfer of electron density from the metal into the π^*-orbitals of the dinitrogen ligand, the more basic is the terminal nitrogen atom. It is also to be noted that in the equivalent complexes *trans*-$[ReCl(XY)(PMe_2Ph)_4]$ (XY = CO or N_2), dinitrogen is more basic than is carbon monoxide.[6]

II. Protonation Reactions Leading to Formation of Hydrazine, Ammonia and Ligating Nitrogen Hydrides

An obvious extension to the study of basicity described above is an investigation of the attack of electrophiles in general upon ligating dinitrogen. Such attack does indeed occur, with the formation of nitrogen–hydrogen and nitrogen–carbon bonds (the latter is discussed in Chapter 8). First we discuss the attack of protic reagents upon dinitrogen complexes and second discuss the relevance of these reactions to the possible mode of action of nitrogenase at the atomic level.

From the earliest days of their discovery, attempts were made to protonate dinitrogen complexes at nitrogen. Generally such a reaction does not occur, even in complexes such as *trans*-$[ReCl(N_2)(PMe_2Ph)_4]$ which has a basic N_β atom. Rather, protonation at the electron-rich metal site occurs which generally leads to loss of dinitrogen as the gas, but a few examples of dinitrogen retention are known, e.g. (**3**).[7]

$$trans\text{-}[W(N_2)_2(dppe)_2] + 2HCl \rightarrow [WH(N_2)_2(dppe)_2]HCl_2 \qquad \textbf{(3)}$$

$$(dppe = Ph_2PCH_2CH_2PPh_2)$$

The pentagonal bipyramid structure of $[WH(N_2)_2(dppe)_2][HCl_2]$ has been established by X-rays.[7] Under the appropriate conditions, however, protonation of dinitrogen ligating molybdenum and tungsten does occur and in fact is limited in terminal complexes to these metals, with the possible exception of cobalt. Very recently cobalt complexes of the formula $[Co(N_2)(PPh_3)_3Mg(thf)_2]$, which give ~0·1–0·3 moles of hydrazine on treatment with acids, have been described. It is not clear, however, whether the dinitrogen is terminal, or bridged between cobalt and magnesium.[8]

When treated with an excess of halogen acid at 20°C the complexes *trans*-$[M(N_2)_2(dppe)_2]$ (M = Mo or W) give complexes of the N_2H_2 ligand; see equation (**4**).[7]

$$trans\text{-}[M(N_2)_2(dppe)_2] + 2HX \rightleftharpoons [MX_2(N_2H_2)(dppe)_2] + N_2 \qquad \textbf{(4)}$$

$$(M = W, X = Cl, Br \text{ or } I; M = Mo, X = Br \text{ or } I)$$

On the basis of the wide splitting of the N–H stretching IR bands and a single, fairly low field N–H resonance in these complexes, apparently seven-coordinate for X = Br in dichloromethane solution, the N_2H_2 group was originally assigned the diazene structure. More recent evidence, however, suggests that the IR splitting and large chemical shift are caused by asymmetric hydrogen bonding of the N_2H_2 ligand in the hydrazido (2—) form (:N—NH_2) to halogens, as is discussed in detail later in this section.

One halide in the above complexes is labile and in the salt produced by replacing it with such non-coordinating anions as BPh_4, the N_2H_2 group certainly has the hydrazido(2—) form in which it acts as a 4-electron donor ligand, so that the metal retains an 18-electron configuration (equation (**5**)).[7]

$$[MX_2(N_2H_2)(dppe)_2] + NaY \rightleftharpoons [MX(NNH_2)(dppe)_2]Y + NaX \qquad \textbf{(5)}$$

$(M = W, X = Cl \text{ or } Br, Y = BPh_4, ClO_4, PF_6 \text{ or } BF_4; M = Mo.$
$X = Br, Y = BF_4)$

The structure of *trans*-$[WCl(NNH_2)(dppe)_2]BPh_4$ (Fig. 1) shows that the M≡N═N unit is essentially linear and the N—N bond distance indicates a bond order of greater than unity, the short W—N distance confirming the expected essentially triple bond character of that bond (see also Table 4).[9]

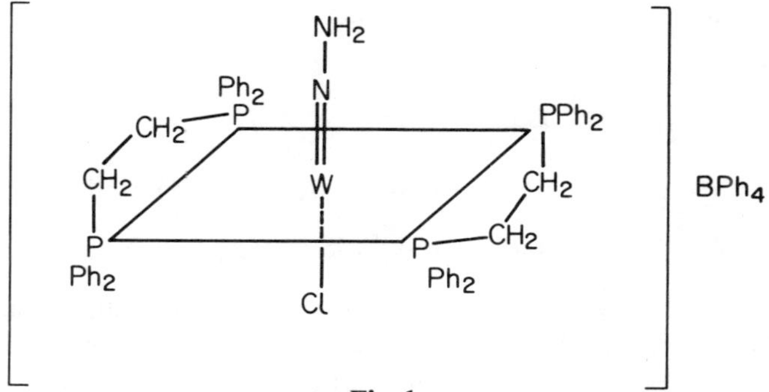

Fig. 1.

When sulphuric acid replaces halogen acids in equation (**5**), cationic, hydrazido(2—) complexes $[M(NNH_2)(SO_4H)(dppe)_2]HSO_4$ (M = Mo or W), are the only products.[10] If HBF_4 is used, fluoro-hydrazido(2—) compounds, $[MF(NNH_2)(dppe)_2]BF_4$ (M = Mo or W) result,[11] of which the molybdenum analogue has also been prepared by use of $[Et_3O]BF_4$ instead of HBF_4.[12]

The hydrazido(2—) complexes may be dehydrohalogenated with weak base to give diazenido complexes, equation (**6**):[11]

$$trans\text{-}[MX(NNH_2)(dppe)_2]BF_4 + NEt_3 \rightleftharpoons trans\text{-}[MX(N_2H)(dppe)_2] + [NEt_3H]BF_4 \qquad \textbf{(6)}$$

$(M = Mo \text{ or } W, X = F, Cl \text{ or } Br)$

The diazenido ligand is isoelectronic with nitrogen oxide and is quantitatively displaced by it, equation (**7**).

$$\textit{trans}\text{-}[MX(N_2H)(dppe)_2] + 2NO \rightleftharpoons [MX(NO)(dppe)_2] + \text{“HNO”} \qquad \textbf{(7)}$$

Treatment of the diazenido complexes with acids regenerates the hydrazido(2–) compounds. Thus the diazenido complexes may be regarded as the precursors to the hydrazido(2–) complexes in a step-wise protonation of ligating dinitrogen.

Attempts to reduce the $-N_2H_2$ grouping in the above complexes by a variety of reducing agents failed, but moderate yields of ammonia (0·24 mol/atom) were obtained when $[MoBr_2(N_2H_2)(dppe)_2]$ together with *trans*-$[Mo(N_2)_2(dppe)_2]$ was treated with aqueous HBr in *N*-methylpyrrolidone or propylene carbonate followed by removal of solvent and Kjeldahl distillation.[13] Protonation and reduction beyond the N_2H_2 stage is probably the result of phosphine ligand displacement at the relatively high temperatures of solvent removal. On the other hand, the complexes *cis*-$[M(N_2)_2(PMe_2Ph)_4]$ or *trans*-$[M(N_2)_2(PPh_2Me)_4]$ (M = Mo or W), on treatment with sulphuric acid in methanol at 20° C, give high yields of ammonia (with a little hydrazine in some cases). The reaction is illustrated in equation (**8**) and typical yields are given in Table 3.[10]

$$\textit{cis}\text{-}[M(N_2)_2(PR_3)_4] \xrightarrow{H_2SO_4/MeOH} N_2 + 2NH_3 + M(VI)\ \text{products} + 4[PR_3H]HSO_4 \qquad \textbf{(8)}$$

($M = Mo$ or W; $PR_3 = PMe_2Ph$ or $PMePh_2$)

The yield of ammonia, according to equation (**8**) is essentially quantitative for M = W, but only ~0·7 mol per metal atom for M = Mo. The final tungsten product appears to be essentially a W(VI) oxide species, but the nature of the molybdenum-containing product has not yet been elucidated. Some indication of probable intermediates in these reactions has been gained by variation of the reagents in reaction (**8**). Thus by using HX (X = Cl, Br or I) instead of sulphuric acid, dihalido-hydrazido(2–) complexes have been isolated (equation (**9**).[14]

$$\textit{cis}\text{-}[M(N_2)_2(PMe_2Ph)_4 + HX(\text{excess}) \xrightarrow{MeOH} [MX_2(NNH_2)(PMe_2Ph)_3] + N_2 + [PHMe_2Ph]X \qquad \textbf{(9)}$$

(M = Mo or W; X = Cl, Br or I)

Substitution reactions may be carried out on these hydrazido(2–) complexes using a variety of neutral or anionic ligands as shown in equations (**10–12**).[14]

Table 3. Yields of ammonia and hydrazine from dinitrogen, and hydrazido(2−) complexes[a]

Compound	Solvent	Reagent[b]	NH_3	N_2H_4	N_2[c]
cis-$[W(N_2)_2(PMe_2Ph)_4]$	thf	H_2SO_4(15)	0·89	0·20	0.98
cis-$[W(N_2)_2(PMe_2Ph)_4]$	MeOH	H_2SO_4(15)	1·86	0·02	0·97
cis-$[W(N_2)_2(PMe_2Ph)_4]$	MeOH/H_2O	H_2SO_4(15)	1·26	0·03	0·98
trans-$[W(N_2)_2(PMe_2Ph)_4]$	thf	H_2SO_4(19·5)	0·72	0·15	1.0
trans-$[W(N_2)_2(PMePh_2)_4]$	MeOH	H_2SO_4(20)	1·80	0·07	0·99
cis-$[Mo(N_2)_2(PMe_2Ph)_4]$	thf	H_2SO_4(7)	0·55	trace	1·06
cis-$[Mo(N_2)_2(PMe_2Ph)_4]$	MeOH	H_2SO_4(12)	0·64	trace	1·23(0·21)[d]
trans-$[Mo(N_2)_2(PMePh_2)_4]$	thf	H_2SO_4(12)	0·08	0·02	1·87
trans-$[Mo(N_2)_2(PMePh_2)_4]$	MeOH	H_2SO_4(12)	0·66	0	1·64
trans-$[Mo(N_2)_2(dppe)_2]$[e]	n.m.p.	HBr/H_2O(~100)	0·37	—	n.d.
trans-$[Mo(N_2)_2(dppe)_2]$[e]	p.c.	HBr/H_2O(~100)	0·19	—	n.d.
cis-$[W(N_2)_2(PMe_2Ph)_4]$	MeOh[f]	—	1·56	0·03	1·03(0·02)[d]
cis-$[W(N_2)_2(PMe_2Ph)_4]$	MeOH[g]	—	1·64	0·02	1·0(0·04)[d]
cis-$[Mo(N_2)_2(PMe_2Ph)_4]$	MeOH	—	0	0	1·9(1·1)[d]
$[WX_2(NNH_2)(PMe_2Ph)_3]$	MeOH	H_2SO_4(20)	1·54[h] (X = Cl)	0·06	n.d.
$[W(NNH_2)(8\text{-hq})(PMe_2Ph)_3]Cl$	MeOH	H_2SO_4(20)	0·0	0·39	n.d.
$[W(NNH_2)(8\text{-hq})(PMe_2Ph)_3]Cl$	H_2O	KOH(~250)	0·03	0·43	0·45
$[WCl_2(NNH_2)(PMe_2Ph)_3]$	H_2O	KOH(~250)	1·4	0·14	n.d.
$[MoCl_2(NNH_2)(PMe_2Ph)_3]$	MeOH	H_2SO_4(20)	0·67	0·01	0·61
$[MX_2(NNH_2)(PMe_2Ph)_3]$	H_2O	KOH(~250)	trace	trace	n.d.
$[WCl_3H(NNH_2)(PMePh_2)_2]$	MeOH	H_2SO_4(20)	0·38	0·66	0·63
$[WCl_3H(NNH_2)(PMePh_2)_2]$	H_2O	KOH(~250)	0·76	0·08	n.d.

[a] Refs. *10, 14, 16* and *17*, [b] No. Mol reagent in parentheses. [c] Gas evolved. [d] H_2 gas. [e] Ref. *161*. [f] Irradiated with 2 × 150 Watt tungsten filament lamps for 42 h. [g] At 60°C for 4 h. [h] 1·72, X = Br; 1·88, X = I.

n.d. = not determined. n.m.p. = *N*-methylpyrrolidone, p.c. = propylene carbonate, X = Cl, Br or I.

$$[MX_2(NNH_2)(PMe_2Ph)_3] + L \xrightarrow{CH_2Cl_2} [MX(NNH_2)(L)(PMe_2Ph)_3]X \quad \textbf{(10)}$$

(L = substituted pyridine or tertiary phosphine)

$$[MX_2(NNH_2)(PMe_2Ph)_3] + (8\text{-hqH}) \xrightarrow{CH_2Cl_2} [M(8\text{-hq})(NNH_2)(PMe_2Ph)_3]X + [8\text{-hqH}_2]X \quad \textbf{(11)}$$

$$[WBr_2(NNH_2)(PMe_2Ph)_3] + Na[S_2CNMe_2] \xrightarrow{CH_2Cl_2} [WBr(NNH_2)(S_2CNMe_2)(PMe_2Ph)_3] + NaBr + PMe_2Ph \quad \textbf{(12)}$$

(M = Mo or W; X = Cl, Br or I; 8-hqH = 8-hydroxyquinoline)

The X-ray structures of several of these complexes have been determined and the results are collected in Table 4. Important points to note are: the parameters of the N—NH_2 ligand confirm that it is of the four-electron-donor, hydrazido(2—) type in all cases, with an essentially linear M≡N=N system; the bond distances within this system appear to vary little with the nature of the co-ligands; extensive hydrogen bonding, both *inter*- and *intra*-molecular, between the NH_2 group and halide anions occurs. (An example is shown in Fig. 2 for $[W(NNH_2)(8\text{-hq})(PMe_2Ph)_2]I$).[15] In such complexes, the N—H stretching IR band is broad and split (~170 cm^{-1}) and the N—H resonance is at low field (8–10 ppm relative to $SiMe_4$) owing to the asymmetric hydrogen bonding. Thus these spectroscopic parameters are not diagnostic of a monodentate diazene complex and all complexes of the N_2H_2 ligand are best regarded as being hydrazido(2—) complexes.

The further protonation reactions of these hydrazido(2—) compounds are dependent upon the nature of the ligand environment. Thus on treatment with H_2SO_4/MeOH followed by base distillation, the compounds $[MX_2(NNH_2)(PMe_2Ph)_3]$ provide ammonia in essentially similar yields to those obtained from the parent dinitrogen complexes under the same conditions. These reactions are quicker and give higher yields in the presence of Tl_2SO_4 (Table 3). Thus the hydrazido(2—) ligand is probably an intermediate on the route to ammonia, common to both molybdenum and tungsten. Similar treatment of $[W(8\text{-hq})(NNH_2)(PMe_2Ph)_3]X$, however, gives no ammonia but only hydrazine.[16] Hydrazine is also produced on treatment of these complexes with strong base alone. Thus the ligand environment profoundly effects the relative yield of ammonia and hydrazine from hydrazine(2—) complexes. The solvent used in protonation reactions of the parent dinitrogen complexes also effects the relative yields of ammonia and hydrazine (Table 3). There are as yet insufficient data to indicate whether electronic or other factors influence the relative proportion of ammonia and hydrazine, but as noted above, the X-ray

Table 4. X-ray parameters for hydrazido(2–) complexes

Complex	M–N (Å)	N–N (Å)	M–N–N (Å)	Reference
$[MoF(NNH_2)(dppe)_2]BF_4$ [a]	1·762(12)	1·333(24)	176·4(13)	*12*
$[Mo(NNH_2)(8\text{-hq})(PMe_2Ph)_3]I$ [b]	1·743(4)	1·347(7)	172·3(5)	*14*
$[WCl(NNH_2)(dppe)_2]BPh_4$ [c]	1·73(1)	1·37(2)	171(1)	*9*
$[W(NNH_2)(8\text{-hq})(PMe_2Ph)_3]I$ [b]	1·753(10)	1·360(17)	174·7(9)	*14*
$WBr(NNH_2)(PMe_2Ph)_3(MeC_5N_4N)]Br$ [d]	1·75	1·34	177	*e*

[a] F *trans* to NNH_2. [b] O *trans* to NNH_2. [c] Cl *trans* to NNH_2. [d] Br *trans* to NNH_2. [e] M. B. Hursthouse and M. Montevali, unpublished results.

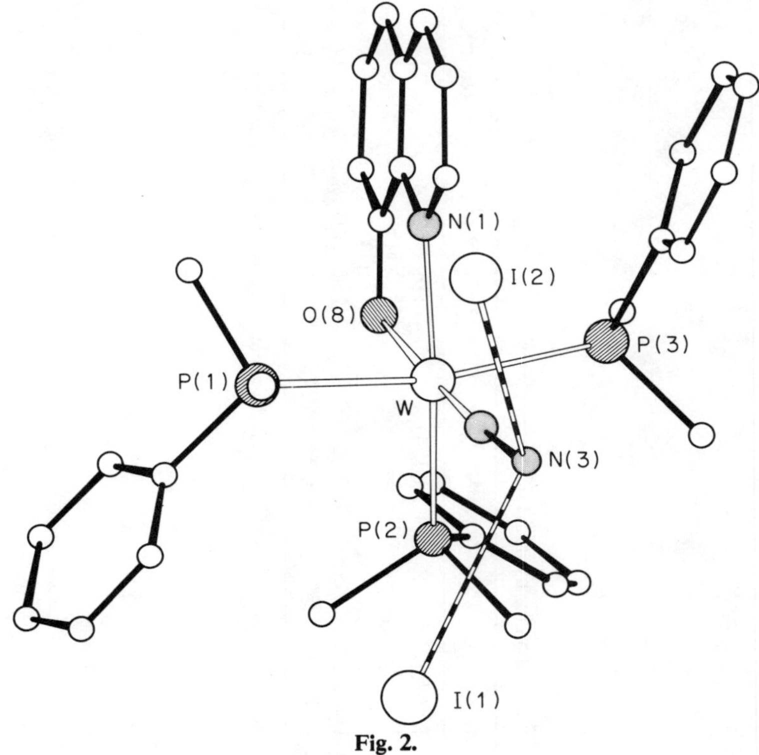

Fig. 2.

parameters of the M=M—NH_2 unit in a variety of complexes, albeit a rather insensitive measurement, show that it undergoes no gross change with variation in co-ligand. Protonation of dinitrogen complexes beyond the NNH_2 stage has been reported to give a complex of the hydrazido(1—) ($NHNH_2$) ligand (equation (**13**).[17]

$$trans)[W(N_2)_2(PMe_2Ph)_2] + 5HCl \xrightarrow{CH_2Cl_2} [WCl_3(NHNH_2)(PMe_2Ph_2)_2] + N_2 + 2[PMePh_2H]Cl \quad \mathbf{(13)}$$

A series of analogues of this complex has now been prepared by similar protonation of *cis*-$[W(N_2)_2(PMe_2Ph)_4]$ or of the corresponding hydrazido(2—) complexes, $[WX_2(NNH_2)(PMe_2Ph)_3]$ (X = Cl or Br).[16] All the products of further protonation show a relatively high-field $N_\beta H_2$ resonance (~5 ppm) relative to $SiMe_4$) which readily exchanges with D_2O and a low-field resonance (~12 ppm) which is inert to exchange. This latter resonance was originally assigned to $N_\alpha H$,[17] but it has recently been observed that the ^{15}N NMR spectrum of the most soluble member of the series (equation (**14**)) shows no

$^{15}N_{\alpha}H$ coupling.[16] This observation, together with the large coupling of the low-field resonance to phosphorus, is more compatible with the formation of these complexes as hydrides (equation (**14**)).

$$[WCl_2(NNH_2)(PMe_2Ph)_3] + \xrightarrow[HX]{thf} [WHX_3(NNH_2)(PMe_2Ph)_3]$$

$$\downarrow \text{thf} \mid \text{HX}$$

$$[PMe_2PhH]X + [WHX_3(NNH_2)(PMe_2Ph)_3] \qquad \mathbf{(14)}$$

(X = Cl or Br)

These complexes tend to give relatively high yields of hydrazine on further acid treatment (Table 3).

Treatment of dinitrogen and dinitrogen hydride complexes with strong acid is not necessary in every case to obtain ammonia. Thus ammonia is obtained from $[WX_2(NNH_2)(PMe_2Ph)_3]$ in high yield, on their distillation from 40% KOH alone, but under these alkaline conditions the molybdenum analogues give at best only traces of ammonia (see Table 3).[18] Moreover, *cis*-$[W(N_2)(PMe_2Ph)_4]$ gives almost as high a yield of ammonia (1·7 mol/W atom), but more slowly, on treatment with methanol alone, either at reflux (3–5 h) or under tungsten-filament irradiation (30 h at 20°C). A lower yield is obtained by reaction with ethanol (0·4 mol/W atom) but the molybdenum analogue gives virtually no ammonia with either alcohol (0·08 mol/Mo atom at best).[18]

III. Mechanism of Protonation of Ligating, Terminal Dinitrogen

The main features of the protonation reactions discussed above are: first, the production of ammonia and varying, but relatively minor, quantities of hydrazine from tungsten complexes, and production of ammonia but merely trace quantities of hydrazine from molybdenum complexes; second, the production of about two molecules of ammonia per tungsten atom but only at best one per molybdenum atom; third, the isolation of a number of stable complexes which contain dinitrogen in various intermediate stages of reduction. Although the latter complexes have not all been isolated with the same phosphine or anionic ligand they can be used as a reasonable basis for Scheme (**1**).

Some further comments on Scheme (**1**) are necessary in the light of the above observations. First, the hydride–hydrazido(2–) complexes have been isolated only when M = W, but this is not surprising since complexes of metals of the third long period are usually more robust than those of the second. The

$$[M(N_2)_2P_4] \xrightarrow[M = Mo \text{ or } W]{+HX, -N_2} [MX(N_2H)P_4] \xrightarrow[M = Mo \text{ or } W]{+2HX, -[PH]X} [MX_2(NNH_2)P_3]$$

$$[MX_2(NNH_2)P_3] \xrightarrow[M = W \text{ or } Mo]{H_2SO_4} 2NH_3 + M(VI)$$

$$[MX_2(NNH_2)P_3] \xrightarrow[M = W]{-[PH]X, +2HX} [MX_3H(NNH_2)P_2] \xrightarrow[M = W]{H_2SO_4} 2NH_3 + M(VI)$$

Scheme (1). Reduction of dinitrogen in molybdenum and tungsten complexes.

yield of ammonia for M = Mo is less than that obtained for M = W and close to 0·67 NH_3 per Mo atom. If the compounds $[MX_2(NNH_2)(PMe_2Ph)_3]$ (M = Mo or W) are treated with sulphuric acid in methanol, they give ammonia (~1·8 mol/W atom; ~0·67 mol/Mo atom together with ≃0·67 mol N_2/Mo atom) in similar yields to those obtained from the parent dinitrogen complexes.[14] Thus the mechanism appears to diverge for the two metals at the $N-NH_2$ stage. The yield of ammonia obtained for M = Mo is consistent with a disproportionation step as in equation (**15**), perhaps via release of isodiazene, whereas further protonation must occur when M = W to account for the very high yields (90% NH_3, 5% N_2H_4).

$$Mo{=}N{-}NH_2 \rightarrow \text{“Mo”} + \tfrac{2}{3}N_2 + \tfrac{2}{3}NH_3 \qquad \textbf{(15)}$$

The possibility that the N_2H_2 ligand is released from molybdenum but held by tungsten for further protonation and reduction may be a reflection of (a) the greater lability of molybdenum complexes and/or (b) the poorer reducing power of molybdenum in its lower oxidation states.

There does not now appear to be good evidence for the $M-NH-NH_2$ stage. If it does exist, however, it may be easily accommodated in the main scheme as in Scheme (**2**), which represents only the N—N bond-splitting part of the cyclic scheme.

$$M{-}N\begin{matrix}H\\NH_2\end{matrix} \xrightarrow{H^+} M{=}N\begin{matrix}H\\\overset{+}{N}H_3\end{matrix} \longrightarrow M{\equiv}NH + NH_3$$

Scheme (2).

Complex compounds containing the groupings M≡N and M≡NH have been prepared indirectly (e.g. $[MoCl_2(N)(PMePh_2)_2]$ prepared from an azide) and have been shown to hydrolyse readily to ammonia.[19]

The involvement of dinitrogen-bridged complexes has often been invoked to explain the reduction of dinitrogen to ammonia or hydrazine in metal complexes or by metal complexes as catalysts. The high yields of ammonia obtained from the bis-dinitrogen tungsten complexes and the isolation of monomeric reduced complexes up to the $N{-}NH_2$ stage, make the involvement of dimeric species highly improbable in their case. Similarly monomeric species are probably involved in all the stages of the reactions of the bis-dinitrogen molybdenum compounds, but as yet the involvement of dimeric species cannot be completely excluded.

Table 5. ^{15}N NMr data for N_2 and N_2H_2 complexes

Complex	Chemical shifts (ppm)[a]		Coupling constants
	N_α[b]	N_β[b]	(Hz)
trans-[$Mo(^{15}N_2)_2(dppe)_2$]	−46·6	−46·3	$^1J(NN) = 5{\cdot}0$
cis-[$Mo(^{15}N_2)_2(PMe_2Ph)_4$]	−42·6	−34·9	$^1J(NN) = 6{\cdot}3$
trans-[$W(^{15}N_2)_2(dppe)_2$]	−63·5	−52·0	$^1J(NN) = 5{\cdot}4$
cis-[$W(^{15}N_2)_2(PMe_2Ph)_4$]	−61·2	−35·9	$^1J(NN) = 6{\cdot}2$
trans-[$MoF(^{15}N_2H_2)(dppe)_2$]BF_4	−83·3	−243·9	$\lvert ^1J(NN) + {}^3J(FMoN)\rvert = 23{\cdot}9$ $^2J(FMoN) = 77{\cdot}4$
trans-[$WF(^{15}N_2H_2)(dppe)_2$]BF_4	−101·4	−255·1	$\lvert ^1J(NN) + {}^2J(FWN)\rvert = 58{\cdot}6$ $^2J(FWN) = 43{\cdot}7$
[$WCl(^{15}N_2H_2)(C_5H_5N)(PMe_2Ph)_3$]Cl	−90·6	−240·6	$^1J(NN) = 10{\cdot}5$ $^1J(NH)^c = 83{\cdot}0$ $^1J(WN) = 124{\cdot}5$
[$W(^{15}N_2H_2)$(8-hq)$(PMe_2Ph)_3$]Cl	−82·1	−241·6	$^1J(NN) = 8{\cdot}8$
$CH(CO_2Et)N_2$[d]	−112·6	−3·6	$^1J(WN) = 114{\cdot}0$
N_2	−75·3		
NH_4^+	−351		

[a] Relative to $C^2H_3NO_2$, negative values to high field. [b] Tentative assignment for N_2 complexes based on relative size of $J(NP)$ and analogy with $CH(CO_2Et)N_2$. [c] From ^1H undecoupled spectrum. [d] R. L. Lichter, P. R. Srinvasan, A. B. Smith, R. K. Dieter, C. T. Denny and J. M. Schulman, *J. C. S. Chem. Commun.* (1977) 366.
dppe = $Ph_2PCH_2CH_2PPh_2$, 8-hq = 8-oxoquinolate.

As yet, this mechanism stands upon the basis of isolation of stable complexes which represent intermediate stages of reduction, and analysis of the kinetics and mechanism of the reaction (**8**) in solution has not yet been attempted in detail. Recent ^{15}N NMR studies of the reaction of *cis*-[$W(N_2)_2(PMe_2Ph)_4$] with sulphuric acid, however, have clearly shown the formation of intermediate $W{=}N{-}NH_2$ species prior to the formation of NH_4^+.[16] This demonstration makes use of the data shown in Table 5 which clearly distinguish $M{-}N_2$, $M{=}N{-}NH_2$, and NH_4^+ on the basis of their different ^{15}N chemical shifts.[20]

IV. Relevance to Nitrogenase Action

The recent advances in the determination of the structure and function of nitrogenase have been adequately reviewed elsewhere[21] (see Chapter 2) and will not be repeated in this article, save to note the following important features. First, the larger protein of the two which comprise nitrogenase contains two molybdenum atoms and appears to be concerned in the uptake of dinitrogen. Second, it contains iron atoms (18–32 depending on the source), together with about an equal number of sulphide ions with which at least half the iron forms cluster units. The smaller protein contains only four iron atoms, which are also in a cluster with four sulphides. The cluster units, of the types $[Fe_4S_4(SR)_4]^{n-}$ (R = cysteinyl residue, n lies between 1 and 4) and possibly also $[Fe_2S_2(SR)_2]^{n-}$ ($n = 1$ or 2), are concerned with transport of electrons to the dinitrogen binding site which is probably molybdenum. At the site the electrons are passed to dinitrogen, together with protons from the environment, to give ammonia. Until very recently no intermediate had been observed between dinitrogen and ammonia, but it has now been shown that rapid quenching, with acid or base, of the functioning enzyme yields hydrazine.[22] It appears that the intermediate is not enzyme-bound hydrazine because of its failure to accumulate in solution at low enzyme concentrations, which would be expected since it is a poor substrate for nitrogenase.[22]

These observations and the stepwise protonation observed for dinitrogen in molybdenum and tungsten complexes can be accommodated by such a process as is shown in Scheme (**3**) for the reduction of dinitrogen on the nitrogenase site.[23]

$$\mathrm{Mo}\cdots\mathrm{N}\cdots\mathrm{N} \xrightarrow[e^-]{H^+} \mathrm{Mo}\cdots\mathrm{N}\equiv\mathrm{N}{-}\mathrm{H} \xrightarrow[e^-]{H^+} \mathrm{Mo}\equiv\mathrm{N}\cdots\mathrm{NH_2} \xrightarrow[\text{acid or base}]{\text{quenching}} \mathrm{N_2H_4}$$

$$\uparrow N_2 \qquad\qquad\qquad\qquad \downarrow H^+$$

$$\mathrm{Mo} + \mathrm{NH_3} \xleftarrow[3e^-]{3H^+} \mathrm{Mo}\equiv\mathrm{N} + \mathrm{NH_3} \xleftarrow[e^-]{} \mathrm{Mo}\cdots\mathrm{N}{-}\overset{+}{\mathrm{N}}\mathrm{H_3}$$

Scheme (3). Possible reduction cycle of dinitrogen on molybdenum in nitrogenase.

It is envisaged in this scheme that molybdenum is maintained in some intermediate oxidation state, capable of binding dinitrogen and of transferring electrons to it from the Fe–S centres of the enzyme. The dinitrogen is then degraded in a similar manner to that shown in Scheme (**3**).

In the artificial complex systems, the hydrazido(2–) state is then the only one which gives hydrazine under both acid and base conditions, in the complex cation $[W(8\text{-hq})(NNH_2)(PMe_2Ph)_3]^+$. The hydrazido(2–) ligand represents a particularly stable stage of reduction of dinitrogen and if it occurs in the natural system, it is the most likely species to be a sufficiently long-lived intermediate to be caught by the acid or alkali quench. Its state in the above tungsten complexes, where it gives about equal yields of N_2H_4 whether it is treated with acid or alkali, may well represent its state in nitrogenase.

Other schemes involving the reduction of dinitrogen binding two metal atoms, or sidewise-bound dinitrogen on one metal atom, leading to diazene intermediates, have been proposed (see Part 2). Obviously such intermediates could lead to hydrazine as the result of acid or base quenching and none of our experiments excludes them. On the other hand Scheme (**3**) is the first to offer a detailed sequence of reduction steps and is the one which is preferred by the author at the present time.

References

1. J. Chatt, C. M. Elson, N. E. Hooper and G. J. Leigh, *J. C. S.* (*Dalton*) (1975) 2392.
2. J. Chatt, J. R. Dilworth and G. J. Leigh, *J. C. S.* (*Dalton*) (1973) 612.
3. D. F. Harrison, E. Weissberger and H. Taube, *Science* (1968) **159**, 320.
4. J. Chatt and R. L. Richards, *J. Less Common Metals* (1977) **54**, 477.
5. R. H. Crabtree, M. Mercer and R. L. Richards, *J. C. S. Chem. Commun.* (1973) 808; M. Mercer, *J. C. S.* (*Dalton*) (1974) 1637.
6. J. Chatt, R. H. Crabtree, E. A. Jeffery and R. L. Richards, *J. C. S.* (*Dalton*) (1973) 1167.
7. J. Chatt, G. A. Heath and R. L. Richards, *J. C. S.* (*Dalton*) (1974) 2079 and references therein.
8. Y. Miura and A. Yamamoto, *Chem. Letters* (1978) 937.
9. G. A. Heath, R. Mason and K.M. Thomas, *J. Amer. Chem. Soc.* (1974) **96**, 259.
10. J. Chatt, A. J. Pearman and R. L. Richards, *J. C. S.* (*Dalton*) (1977) 1853.
11. J. Chatt, A. J. Pearman and R. L. Richards, *J. C. S.* (*Dalton*) (1976) 1520.
12. M. Hidai, T. Kodoma, M. Sato, M. Harakawa and Y. Uchida, *Inorg. Chem.* (1976) **15**, 2694.
13. C. R. Brûlet and E. E. Van Tamelen, *J. Amer. Chem. Soc.* (1975) **97**, 911.
14. J. Chatt, A. J. Pearman and R. L. Richards, *J. Organometallic Chem.* (1975) **101**, C45; *J. C. S.* (*Dalton*) (1978) 1766.
15. D. Hughes and I. R. Hanson, personal communication.
16. J. Chatt, M. E. Fakley and R. L. Richards, unpublished results.
17. J. Chatt, A. J. Pearman and R. L. Richards, *J. C. S.* (*Dalton*) (1977) 2139.
18. J. Chatt, A. J. Pearman and R. L. Richards, *Nature* (*London*) (1976) **259**, 204.
19. J. Chatt and J. R. Dilworth, *J. C. S. Chem. Commun.* (1975) 983; *J. Indian Chem. Soc.* (1977) **LIV**, 13.
20. J. Chatt, M. E. Fakley, J. Mason, R. L. Richards and I. A. Stenhouse, *J. Chem. Research* (1979) 3701, 0873 and unpublished results.

21. W. H. Orme-Johnson, L. C. Davis, M. T. Henzel, B. A. Averill, N. R. Orme-Johnson, E. Munck and R. Zimmerman, *In* "Recent Developments in Nitrogen Fixation" (Ed. W. Newton, J. R. Postgate and C. Rodriguez-Barrueco), p. 131. Academic Press, London and New York (1977); R. R. Eady and B. E. Smith, "Dinitrogen Fixation" (Ed. R. W. F. Hardy), Vol. II, p. 399. Wiley–Interscience, New York (1978).
22. R. R. Eady, D. J. Lowe and R. N. F. Thorneley, *Nature* (*London*) (1978) **272**, 557.
23. J. Chatt, R. R. Eady, D. J. Lowe, J. R. Postgate, R. L. Richards, B. E. Smith and R. N. F. Thorneley, "Proceedings of the Steenbock International Symposium on Nitrogen Fixation, June 12–16, 1978", University of Wisconsin, Madison, Wisconsin, U.S.A.

8

The Formation of Nitrogen–Carbon Bonds from Dinitrogen

G. J. LEIGH

A.R.C. Unit of Nitrogen Fixation The University of Sussex, Brighton, England

AS WELL as the efforts to find ways of activating dinitrogen to form ammonia under mild conditions, there have been parallel efforts to produce amines directly from N_2 without intermediate steps involving ammonia, nitric acid, or similar materials. Commercially the only process employed to form carbon–nitrogen bonds directly from N_2, the cyanamide process, suffered from the same drawback as the arc process for ammonia, namely high energy consumption, and is obsolescent for the same reason. Metal complexes of dinitrogen have been involved in the investigations aimed at producing amines, just as they have in the ammonia work, and so have metal nitrides, but there are also reports of organic species reacting directly with N_2, and these are discussed first.

I. Reactions of N_2 with Species which Do Not Contain Metal Atoms

There are no stable neutral organic compounds capable of combining with dinitrogen at 1 atm. pressure and at ca. 25°C. However several reactive species do react with N_2 and will be discussed here. Reactions of stable organic compounds with "active nitrogen" (which contains excited nitrogen molecules as well as nitrogen atoms) have been reviewed[1] and will not be further mentioned.

Aryldiazonium ions, ArN_2^+, decompose on warming to yield $Ar^+ + N_2$. Several attempts were made to reverse this reaction and bind N_2 into organic molecules by using, for example, highly strained ring systems to incorporate N_2, but apparently none was successful. However, thermal decomposition of

$Ph^{15}N_2^+$ at 30°C in CF_3CH_2OH under 300 atm. $^{14}N_2$ has shown that Ph^+ does react with N_2. When the thermal decomposition is 70% complete, 2·46 ± 0·40% of the nitrogen in the undecomposed PhN_2^+ is unlabelled, that is, derived from the $^{14}N_2$.[2] Now Ph^+ is a short-lived and reactive species, but the degree of reaction with N_2 is rather small.

Another organic cation, the phenylsulphenium ion, PhS^+, has also been shown to react with N_2.[3] This cation is also a highly reactive species. When generated by the reaction of phenylsulphenium bromide with silver benzene-2,4,6-trisulphonic acid in dichloromethane under N_2 it gives rise to a cream-coloured solid which has a weak band at 2260 cm^{-1} in the IR spectrum ($\nu(N_2)$?). If He is used in place of N_2, no such band is observed. The presence of combined N_2 in the reaction product has been confirmed by typical diazonium coupling reactions and by analysis,[3] and the product has been formulated as $PhSN_2^+$ or $(PhSN{=}NPh)H^+$.

Diazoalkanes are, like aryldiazonium compounds, materials which decompose readily to yield N_2, but the organic product is in this case not an ion but a carbene, in both singlet and triplet states.[4] Both singlet and triplet CH_2, whether produced by thermolysis or by photolysis of diazomethane, can react with N_2[4,5] in the gas phase, despite earlier reports[6] that such a reaction can occur only in an N_2 matrix at 20 K. The technique used was to decompose $CH_2{}^{14}N_2$ in the presence of $^{15}N_2$ and to stop the decomposition before it was complete. Thus, at 383°C, after 65% decomposition of $CH_2{}^{14}N_2$ in $^{15}N_2$, the remaining diazomethane contained 0·114% $CH_2{}^{15}N_2$. After 75% photolytic decomposition, the $CH_2{}^{15}N_2$ content was 1·00%. Apparently only CH_2:, CNCH: and CF_3CH: react with N_2; CH_3CH:, CH_3CH_2CH:, $CH_2{=}CHCH$:, C_2H_5OCOCH: and $(CH_2)_4C$: do not.[4]

There is no report of alkyl or aryl radicals reacting with N_2. CH radicals do react, but the products have not been identified.[7]

We have been unable to confirm[8] the only report[9] of a stable organic compound which can react with N_2. According to the literature,[9] the tetrakis(tertiary phosphine), $C(CH_2PPh_2)_4$, forms a sodium derivative $Na_4\{C(CH_2PPh)_4\}$ which with CS_2 under Ar at 60°C forms a yellow product,

$$C(CH_2\overset{\overset{\displaystyle S^-}{|}}{\underset{\underset{\displaystyle Ph}{|}}{P}}{=}C{=}S\cdot Na^+)_4,$$

and this in turn reacts with N_2 during 24 h to yield a solid which contains nitrogen by analysis and has a band in its IR spectrum at 2090 cm^{-1}. The tentative structure suggested is as below.

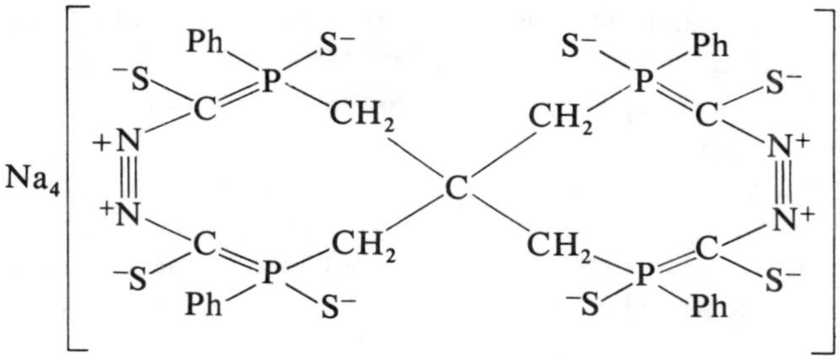

Ph S⁻ S⁻ Ph
⁻S P P S⁻
C CH₂ CH₂ C
+N N+
Na₄ C
+N N+
C CH₂ CH₂ C
⁻S P P S⁻
Ph S⁻ ⁻S Ph

Confirmation and a full publication are lacking.

All other species which have been reported to react with dinitrogen are metal-containing species. We consider first those in which dinitrogen complexes have not been directly implicated. These are normally "nitriding" systems.

II. Reactions with Metal Compounds Leading to N—C Bonds and Not Involving Stable N_2 Complexes

The principal work in this area has been Russian.[10] As early as 1966, Vol'pin and Shur reported that at 20°C and 100 atm., N_2 reacts with either $[(C_5H_5)_2TiPh_2]$ or $[(C_5H_5)_2TiCl_2]$ and PhLi in ether to yield, after hydrolysis of the reaction mixture, NH_3 (0·65 mol) and $PhNH_2$ (0·15 mol).[10,11] This kind of reaction had, however, been hinted at even earlier by the observation that the preparation of particular Grignard reagents in ether consistently gives higher yields under He or Ar than under N_2.[12] Thus the reaction of cyclohexyl bromide (CyBr) and Mg gives a clear solution of CyMgBr (68% yield) under He but a cloudy solution with reduced yield (31%) under N_2.[12]

The production of aniline is also observed in the titanium systems at lower pressures of N_2. For example at 1 atm., NH_3 (0·17 mol) and $PhNH_2$ (0·03 mol) are found. Other titanium compounds, such as $[C_5H_5TiCl_3]$, $TiCl_4$, and $Ti(OBu^n)_4$ are capable of producing aniline from PhLi under N_2.[13]

Originally two possible routes to aniline formation were suggested, one the insertion of N_2 into a titanium–phenyl bond to form a group T—N=N—Ph.[10,11,13] Since neither $[(C_5H_5)_2TiPh_2]$ nor PhLi reacts with N_2,[13] this seems unlikely. However, the evidence is not clear cut.

The reaction of $[(C_5H_5)_2TiPh_2]$ with PhLi is believed to proceed as shown:[14]

$$[(C_5H_5)_2TiPh_2] + 2PhLi \rightarrow TiPh_2 + 2C_5H_5Li + 3Ph\cdot + Li$$

In normal circumstances, the Ph· radicals attack the solvent and appear as benzene. However, small amounts of triphenylene have also been isolated from this reaction, leading to the suggestion[14] that benzyne is also involved. When the three isomeric bis(tolyl) compounds are used in place of $[(C_5H_5)_2TiPh_2]$ in the fixation experiment, $[(C_5H_5)_2Ti(CH_3\text{-}3\text{-}C_6H_4)_2]$ gives *m*-toluidine, $[(C_5H_5)_2Ti(CH_3\text{-}4\text{-}C_6H_4)_2]$ gives *p*-toluidine but $[(C_5H_5)_2Ti(CH_3\text{-}2\text{-}C_6H_4)_2]$ gives a mixture of *o*-toluidine (40–77%, *m*-toluidine (52–21%) and *p*-toluidine (8–2%).[10,13] The production of *m*- and *p*-toluidines, as well as the comparatively low yields of amines when alkyl lithium reagents are used (Ti–alkyl bonds are less stable than Ti–aryl bonds) is consistent with an insertion mechanism, sketched schematically below:

$$\text{Ti—Ph} + N_2 \rightarrow \text{Ti—N=N—Ph} \xrightarrow{\varepsilon} PhN_2^- \xrightarrow{H_2O} PhNH_2$$

The results with the *o*-tolyl derivative suggest that

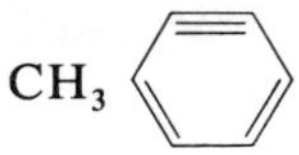

CH_3

must be generated, perhaps leaving a Ti—H system. The "toluyne" could then reinsert into the Ti—H bond generating an isomeric tolyl derivative.[10] The generation of *o*-aminodiphenyl from the $[(C_5H_5)_2TiPh_2]$ could arise from the reaction of

with PhLi. A suggested[10] route from benzyne to amine is shown below.

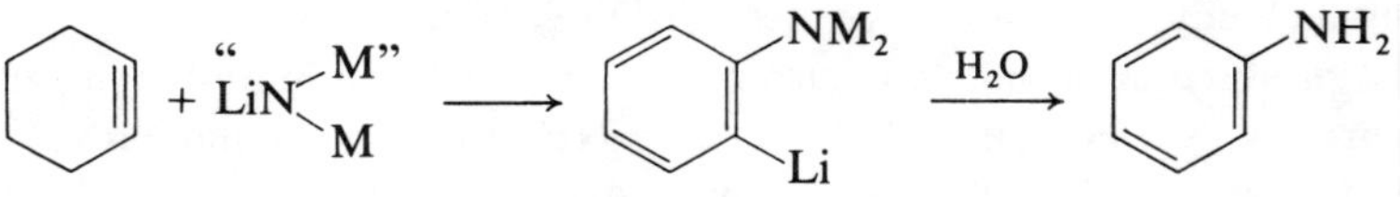

$$+ \text{"LiN}\begin{matrix}\text{M}\\\text{M}\end{matrix}\text{"} \longrightarrow \ \ (NM_2,\ Li) \ \xrightarrow{H_2O} \ (NH_2)$$

(M = Li or Ti)

This leaves obscure how the fragment "$LiNM_2$" might form in the first place.

There are various modifications of this system, generally due to the same Russian group. $[(C_5H_5)_2TiPh_2]$ decomposes thermally at 80–130°C in benzene or ether under high pressures of N_2 (80–100 atm.) to yield, after hydrolysis, aniline, ammonia, and, apparently, phenylhydrazine.[15] Again, insertion of N_2 into a Ti—aryl bond is plausible, but decomposition of the *o*-tolyl titanium compound yields *o*- and *m*-toluidine, of the *p*-tolyl compound gives *m*- and *p*-toluidine, and of the *m*-compound gives, unexpectedly, *m*- and *p*-toluidines only, and not a mixture of all three toluidines. This is felt to be most consistent with a benzyne scheme such as:

$$[(C_5H_5)_2TiPh_2] \longrightarrow \left[(C_5H_5)_2Ti\,\|\,C_6H_4\right] + C_6H_6$$

$$\downarrow N_2$$

$$\left[(C_5H_5)_2Ti{-}N{=}N{-}C_6H_4\right] \xrightarrow[H_2O]{\varepsilon}$$

$$PhNH_2 + PhNHNH_2 + NH_3$$

A route via a titanium nitride is not favoured, and, indeed, evidence for a benzyne complex has been obtained independently from a study[16] of the thermal decomposition of $[(C_5H_5)_2TiPh_2]$.

Yet another variant is the discovery that at 100 atm. N_2 in ether at 20°C, metals such as lithium, sodium, or magnesium can convert $[(C_5H_5)_2TiPh_2]$ to a material which hydrolyses to give aniline (0·15–0·18 mol) and ammonia (1·1 mol). Because the *p*- and *m*-tolyl derivatives yield *p*- and *m*-toluidine, respectively, an N_2 insertion into the metal–aryl bond seems likely.[17]

A system not involving a well defined organometallic compound is that comprising N_2, lithium, and naphthalene.[18] In tetrahydrofuran at 20°C and 100 atm. this can yield 0·2–0·3% *α*-naphthylamine after hydrolysis. The yield is a function of the hydrocarbon concentration and the ratio of Li to hydrocarbon. Apparently, the presence of $TiCl_4$ increases the yield. This fixation could proceed via lithium nitride since lithium is well known to react with N_2 under conditions such as those pertaining here,[19] but it is claimed[18] that in the absence of naphthalene almost no nitrogen is fixed by the lithium.

Finally, there are two systems which almost certainly do involve metal nitrides as intermediates. Magnesium and $[(C_5H_5)_2TiCl_2]$ in tetrahydrofuran at 23°C under 1 atm. N_2 absorb considerable amounts of N_2 during 1 h.[20] The mixture so produced reacts slowly with Et_2CO to produce, after hydrolysis, a 2:1 mixture of 3-pentylamine and di(3-pentyl)amine in 25–50% yield based upon nitrogen absorbed. In a similar fashion PhCOCl can be converted to PhCN, and $PhCO_2Et$ to $PhCH_2NH_2$. One of the two routes suggested involves the formation of a titanium N_2 complex which yields a nitride, which in turn reacts with the ketone. If this is so, it is strongly reminiscent of the longer known reactions of Li_3N with acid chlorides,[21] and of the similar reaction with acid anhydrides known for over a century.[22]

In the second system,[23] the material obtained from the mixture $MCl_4/Mg/N_2$/tetrahydrofuran (M = V or Ti) is allowed to react with CO_2. This material is not properly characterized, but is reproducible and has the stoichiometric formula Mg_2Cl_2N(tetrahydrofuran) and is probably a nitride. It takes up CO_2 rapidly to yield a product containing the NCO group, and further reaction with MeI yields MeNCO. A parallel series of experiments using $[(C_5H_5)_2TiCl_2]$ in place of MCl_4 gives $[(C_5H_5)_2TiNCO]$.

The systems described above are generally characterized by working best under high N_2 pressures, by the presence of strong reducing agents or conditions, and by the formation of ill defined intermediates. Since the discovery of N_2 complexes capable of undergoing reactions under relatively mild conditions, the general emphasis of research work into the formation of N—C bonds has changed. The next section discusses relevant reactions of N_2 complexes, and some of the properties of the products.

III. Formation of N—C Bonds from Dinitrogen Complexes

The seminal observation in this area was that acetyl chloride reacts with *trans*-$[W(N_2)_2(dppe)_2]$ (dppe = $Ph_2PCH_2CH_2PPh_2$) in benzene at room temperature to form an acetyldiazenido derivative, $[WCl(N_2COCH_3)(dppe)_2]$, which can be isolated in high yield.[24] The acetyldiazenido ligand can be protonated reversibly to give a hydrazido(2—) ligand, but no further:

$$-N{=}NCOCH_3] + H^+ \underset{\text{base}}{\overset{\text{acid}}{\rightleftharpoons}} \left[{=}N{-}N\begin{smallmatrix} H \\ COCH_3 \end{smallmatrix} \right]^+$$

The evidence for the formation of a nitrogen–carbon bond was originally circumstantial, but the correctness of the structure has been confirmed by X-ray analysis of the molybdenum analogue.[25] Some of the molecular dimensions are shown in Table 1. The numbers in the table are to be referred to the molecular skeleton at its head. For most cases, X = halogen, M = Mo or W, R_1 = alkyl and R_2 = alkyl or hydrogen, or is absent.

It was subsequently shown[26] that it is possible to acylate and aroylate N_2 in $[ReCl(N_2)(PMe_2Ph)_4]$ with the appropriate carboxylic acid chloride in an essentially similar reaction which yields $[ReCl_2(N_2COR)(PMe_2Ph)_3]$ (R = alkyl or aryl). However $[OsCl_2(N_2)(PMe_2Ph)_3]$ and acid halides do not react.[26] This was taken to show a drop in reactivity of coordinated N_2 towards attack in the series W > Re > Os. It is not now altogether clear what this series implies.

At that time (about 1972) it seemed probable to us that acetylation and aroylation reactions were of the same type as the protonations of N_2 which were also being investigated in our laboratory (see §3.1., Ch. 7). We believed

Table 1. Some molecular dimensions of some diazenido, hydrazido, and diazoalkane complexes

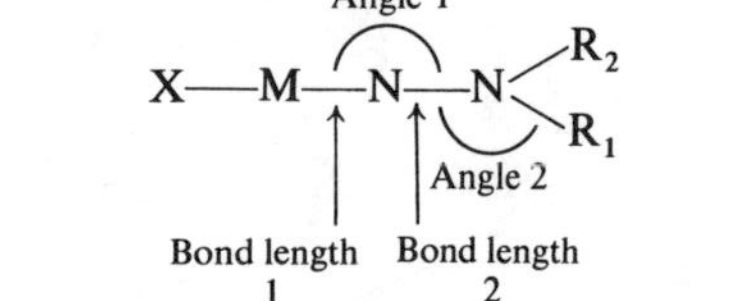

Complex	Bond lengths (Å)		Angles (deg)		Ref.
	1	2	1	2	
$[MoCl(N_2COPh)(dppe)_2]$	1·813(7)	1·255(10)	172·1(6)	116·7(7)	*25*
$[MoI(N_2C_6H_{11})(dppe)_2]$	1·95(1)	0·91(1)	176(1)	142(2)	*28*
$[MoI(N_2HC_8H_{17})(dppe)_2]I$	1·801(11)	1·259(14)	174(1)	120(1)	*29*
$[WBr(N_2HMe)(dppe)_2]Br$	1·768(14)	1·32(2)	174(1)	121	*55*
$[ReCl_2(N_2Ph)(PMe_2Ph)_3]$	1·80(1)	1·23(1)	172	118	*56*
$[ReCl_2(N_2COPh)(PMe_2Ph)_3]$	1·74(2)	1·22(3)	170(2)	124(2)	*57*
$[ReCl_2(NH_3)(N_2HPh)(PMe_2Ph)_2]Br$	1·75(1)	1·28(2)	172(1)	119(2)	*58*
$[WBr\{N_2CH(CH_2)_3OH\}(dppe)_2]PF_6$	1·78(1)	1·31(2)	172	116	*38*
$[WBr(N_2CMe_2)(dppe)_2]Br$	1·71(1)	1·34(2)	171(1)	124(1)	*42*
$[WF(N_2CMeCH_2COMe)(dppe)_2]BF_4$	1·770(17)	1·317(25)	173·8(15)	125·2(19)	*44*

that all the reactions represented an electrophilic attack of a cation (RCO^+ or H^+) on a nucleophilic nitrogen of dinitrogen polarized in the sense $M{-}\overset{\delta+}{N}{-}\overset{\delta-}{N}$. This led us to investigate the properties of other electrophilic reagents which might conceivably behave similarly. In the event, our mechanistic ideas concerning acetylation and aroylation were probably wrong (see below). They did, however, stimulate us to attempt alkylation of coordinated N_2 using alkyl halides, attempts which proved very fruitful.[27] Independently, other workers have also synthesized compounds related to those we discovered.[28–30]

Alkyl bromides and iodides RX react with complexes $[M(N_2)_2(dppe)_2]$ in an inert solvent, such as benzene, at room temperature during several hours to form alkyldiazenido complexes, $[MX(N_2R)(dppe)_2]$.[27–29] Despite some confusion,[31] the rule appears to be that reactions for M = Mo proceed in the dark whereas those for M = W require irradiation. In general, alkyl chlorides do not react under these conditions, though "activated" chlorides such as $ClCH_2COOEt$[30] and ClCOOEt[32] do, and reactivity with normal chlorides can be stimulated thermally. Thus Bu^nCl and $[Mo(N_2)_2(dppe)_2]$ do not react at 20°C, but they react normally at 40°C.[33] Aryl halides do not give nitrogen-containing products upon reaction with $[M(N_2)_2(dppe)_2]$ under these or similar conditions.[34]

The diazenido complexes thus produced react with protic acids to yield hydrazido(2–) compounds which revert to diazenido complexes upon reaction with base:

$$[MX(N_2R)(dppe)_2] \underset{\text{Base}}{\overset{\text{HX}}{\rightleftharpoons}} [MX(N_2HR)(dppe)_2]X$$

However, there is no firm indication of further protonation of these compounds to yield amines. In this, the properties parallel those of other diazenido and hydrazido(2–) derivatives. The structures of representative alkyldiazenido and hydrazido(2–) complexes have been determined by X-ray structure analysis and representative structural parameters are shown in Table 1. These will be discussed later.

One N_2 in $[M(N_2)_2(dppe)_2]$ can be replaced by other donors, such as PhCN,[35,36] MeCN,[35,36] NCS^-,[36] CN^-,[36] and N_3^-.[36] The products, whether of the form $[ML(N_2)(dppe)_2]$ (L = neutral ligand) or of the form $[ML(N_2)(dppe)_2]^-$ (L = anionic ligand) can also be alkylated on the remaining dinitrogen. Thus $[Mo(PhCN)(N_2)(dppe)_2]$ and PhCOCl produce $[MoCl(N_2COPh)(dppe)_2]$, mentioned above.[25] In this reaction as in all others discussed so far the halide from the attacking reagent is included in the complex product. However, the reaction of $[Mo(N_2)(SCN)(dppe)_2]^-$ with Bu^nI produces $[Mo(N_2Bu^n)(SCN)(dppe)_2]$, which suggests that the mechanism of reaction must differ in detail from the previously described reactions.

When the reaction of $[M(N_2)_2(dppe)_2]$ with MeBr is carried out in a solvent such as tetrahydrofuran (thf) rather than benzene, then another product is isolated, rather than a methyldiazenido complex. This product was initially and erroneously described as a tetrahydropyridazido complex,[37] but subsequent X-ray structure analysis[38] showed it to be a diazobutanol complex, $[MBr\{N_2CH(CH_2)_3OH\}(dppe)_2]^+$. Evidently the solvent becomes incorporated into the complex in a way subsequently elucidated,[39] and similar solvent incorporation was observed with tetrahydropyran, tetrahydrothiophen, 1,4-dioxan and *N*-methylpyrrolidine.[33,39]

Diazoalkane complexes are a considerable novelty. In general, they cannot be synthesized by the direct interaction of a diazoalkane with a metal complex, because the complex will normally catalyse the decomposition of the diazoalkane. The reaction products are usually varied and unexpected.[40] However we have discovered two further ways of synthesizing diazoalkane complexes from dinitrogen, in addition to that mentioned above, which involves thf. In both these cases, the diazoalkane is actually assembled *in situ*, from complexed dinitrogen and an organic moiety. The first synthesis was from $[M(N_2)_2(dppe)_2]$ and a *gem*-dibromide.[41]

$$[M(N_2)_2(dppe)_2] + R^1R^2CBr_2 \xrightarrow{C_6H_6} [MBr(N_2CR^1R^2)(dppe)_2]Br + N_2$$

The NMR characteristics of such complexes (e.g., the very low downfield chemical shift of the unique diazoalkane carbon) indicated that they are of the same general type as the diazobutanol complex discussed above. The ^1H NMR spectrum of the compounds with $R^1 = R^2 = Me$ suggested that these methyl groups are in different environments, and this was confirmed by an X-ray structural analysis.[42]

The compounds do not react with acids, but do react with nucleophilic reagents such as LiMe to yield diazenido complexes:

$$[MBr(N_2CR^1R^2)(dppe)_2]^+ + LiMe \rightarrow [MBr(N_2CR^1R^2Me)(dppe)_2] + Li^+.$$

A wider range of diazoalkane complexes, some of which cannot be obtained from reactions of *gem*-dibromides, can be obtained from the acid catalysed condensation of hydrazido(2–) complexes with aldehydes or ketones. The hydrazido(2–) complexes are themselves obtained by protonation of dinitrogen complexes.[42,44]

$$[M(N_2)_2(dppe)_2] + 2HX \rightarrow [MX(NNH_2)(dppe)_2]X + N_2$$

$$[MX(NNH_2)(dppe)_2]X + R^1R^2CO \rightarrow [MX(N_2CR^1R^2)(dppe)_2]X + H_2O$$

The structure of one diazoalkane complex so obtained has been confirmed by X-ray structural analysis.[44]

The reaction of complexes $[M(N_2)_2(PR_3)_4]$ (PR_3 = monotertiary phosphine) with alkyl halides leads to general decomposition and loss of all dinitrogen as N_2. In particular, diazoalkane and alkyldiazenido complexes are not obtainable directly from these bis(dinitrogen) complexes. However, prior protonation, to yield $[MX_2(NNH_2)(PR_3)_3]$ followed by condensation with ketones does yield diazoalkane complexes, generally of the form $[WX_2(N_2CR^1R^2)(PMe_2Ph)_3]$.[45] These complexes react with proton acids as below.

$$[WBr_2(N_2CMe_2)(PMe_2Ph)_3] \xrightarrow{HBr} [WBr_3(NHNCMe_2)(PMe_2Ph)_3] \xrightarrow{HBr}$$

$$[WBr_3(NHNHCMe_2Br)(PMe_2Ph)_3] \xrightarrow{-PMe_2Ph}$$

$$[WBr_3(NHNHCMe_2Br)(PMe_2Ph)_2] \xrightarrow{HBr}$$

$$[WBr_4(PMe_2Ph)_2] + NH_2NHCMe_2Br \xrightarrow{-HBr}$$

$$NH_2N{=}CMe_2 \rightarrow N_2H_4 + Me_2CNNCMe_2$$

Evidently, this does not provide a very versatile route to organonitrogen compounds.

One final class of alkyl halide reagents deserves mention here, α,ω-dibromides, $Br(CH_2)_nBr$. These react with $[M(N_2)_2(dppe)_2]$ to yield products which are a function of n in the dibromide formula. For $n = 1$, a diazoalkane complex is formed,[41] but for M = W, the complex $[\{WBr(dppe)_2\}_2(N_2CH_2N_2)]$ was also isolated. In general, more than one complex is always formed,[34,41,43] and the products are indicated in Table 2. Evidently the long-chain bromoalkyldiazenido complexes behave as alkyl bromides and attack further bis(dinitrogen) complexes. The mechanism of this and all the other similar reactions belonging to N–C bond formation will now be discussed.

Table 2. Products of the reactions of $[M(N_2)_2(dppe)_2]$ with $Br(CH_2)_nBr$ ($n = 1$–12)

n	Products
1	$[MBr(N_2CH_2)(dppe)_2]Br$, $[\{WBr(dppe)_2\}_2(N_2CH_2N_2)]$
2	$[MBr_2(dppe)_2] + C_2H_4$
3	$[MBr\{N_2(CH_2)_3Br\}(dppe)_2]$
4	$[MBr\{NN(CH_2)_4\}(dppe)_2]Br$
5	$[MBr\{NN(CH_2)_5\}(dppe)_2]Br$
6–12	$[MBr\{N_2(CH_2)_nBr\}(dppe)_2]$, $[\{MBr(dppe)_2\}_2\{N_2(CH_2)_nN_2\}]$

Conditions; benzene solution, room temperature, tungsten-filament irradiation, several hours reaction time.

Despite some suggestions to the contrary,[31,46] the alkylation reactions of coordinated N_2 are not necessarily photocatalysed, at least on Mo, although light accelerates them. However, the belief that the reactions are photocatalysed has led to some erroneous theorizing.[46,47] In the dark in benzene solution, it has been shown that a wide range of substitution, exchange, and alkylation reactions proceed at the same rate, and that, under pseudo first-order conditions, the reactions are first-order in molybdenum complex. The same also apparently applies to tungsten compounds, although for these irradiation is also necessary.[48] This implies a common rate-controlling step for all these reactions, which can only be the loss of dinitrogen:

$$[M(N_2)_2(dppe)_2] \rightleftharpoons [M(N_2)(dppe)_2] + N_2$$

In substitution and exchange reactions there is competition between the new ligands and N_2 for the vacant coordination site. In all the reactions discussed above, there is persuasive evidence[48] that the alkyl halide (mono- or di-halide) also competes with N_2, forming an unstable intermediate $[M(N_2)(XR)$-$(dppe)_2]$ which homolyses at room temperature ($X = Br$ or I) or at slightly elevated temperatures (in general for $X = Cl$). This generates an M(I) species, $[MX(N_2)(dppe)_2]$, and a radical $R\cdot$, which, if RX is a dibromide, may be a bromoalkyl radical $Br(CH_2)_{n-1}CH_2\cdot$.

If the solvent is generally inert to homolytic attack, and the radical is reasonably stable, then the radical appears to stay within the solvent cage containing it and the M(I) species until it comes into reaction with the remaining coordinated N_2:

$$[MX(N_2)(dppe)_2] + R\cdot \rightarrow [MX(N_2R)(dppe)_2]$$

A radical which is too unstable (e.g. $\cdot CH_2CH_2Br$) will decompose before this can occur (see Table 2), and a radical which is too stable will not attack the N_2. Benzyl bromide, for example, yields $[MBr_2(dppe)_2]$ and $PhCH_2CH_2Ph$. Methylene bromide generates the radical $BrCH_2\cdot$, which apparently attacks N_2 as described above, and the bromomethyldiazenido product then ionizes to yield the diazolkane complex:

$$[MBr(N_2CH_2Br)(dppe)_2] \rightarrow [MBr(N_2CH_2)(dppe)_2]Br.$$

Our evidence suggests that alkylation and acetylation of coordinated N_2 have similar mechanisms.[48]

Support for this general mechanism has come from a study of the flash-photolysis of $[W(N_2)_2(dppe)_2]$.[49] This shows that under photolysis conditions, both dinitrogens are lost, but that the $[W(dppe)_2]$ residue can then bind two more molecules of N_2 stepwise. If Bu^nBr is present this competes with N_2 for the remaining site on $[W(N_2)(dppe)_2]$, as discussed above.[48]

It would be unwise to attempt to generalize this reaction sequence for complexes other than $[M(N_2)_2(dppe)_2]$ for at least two reasons. Firstly, it may be that a given metal–nitrogen–carbon bonding system is not stable enough to survive even under mild conditions. Thus, $[Ru(NH_3)_5(N_2O)]^{2+}$ and $N_2CHCOOEt$ react in neutral water at 20°C to yield $[Ru(NH_3)_5(N_2)]^{2+}$ and a mixture of ethyl esters.[50] This reaction is said to proceed via reductive cleavage of the nitrogen–carbon bond. A more obvious explanation is that the diazoacetic ester displaces water from the penta(ammine) to yield $[Ru(NH_3)_5(N_2CHCOOEt)]^{2+}$ in which the N–C bond homolyses. Hence, one would not expect to form diazoalkane and diazenidocomplexes of ruthenium by a route analogous to that we have demonstrated for molybdenum and tungsten.[48]

Secondly, we have now shown that in at least one case an alternative initiation step occurs.[36] The reaction of $[M(NCS)(N_2)(dppe)_2]^-$ with Bu^nI is first order in each reactant, and the product is $[M(NCS)(N_2Bu^n)(dppe)_2]$, so that the halogen from the alkyl halide is not incorporated. There are two possible rate-controlling steps consistent with this.

$$[M(NCS)(N_2)(dppe)_2]^- + Bu^nI \longrightarrow M{-}N{\equiv}N \longrightarrow \begin{matrix} C_3H_7 \\ | \\ CH_2 \\ | \\ I \end{matrix}$$

$$\longrightarrow \text{products}$$

$$[M(NCS)(N_2)(dppe)_2]^- + Bu^nI \longrightarrow [M(NCS)(N_2)(dppe)_2] + C_4H_9\cdot + I^-$$

$$\longrightarrow \text{products}$$

We favour the latter step at present on two grounds. The first is that the redox potential of $[M(NCS)(N_2)(dppe)_2]^-$ is about a volt lower than that of $[M(N_2)_2(dppe)_2]$, so that electron transfer to the halide is much more feasible from the anion than from the bis(dinitrogen) complex.[51] The second is that the electron transfer step itself generates an M(I) species plus a radical, which is exactly what results from metal-assisted homolysis of the alkyl halide. We do not yet know how general this alternative initiation is, nor have we yet elucidated the details of the mechanism.

We have so far described how the detailed alkylation mechanism is a function of both metal complex and halide. It is evident from the data cited above that it is also a function of solvent. If the solvent contains hydrogen atoms which are easily removable by radicals (and tetrahydrofuran is a case in point) then attack on the solvent may also occur.[39,48]

$$C_4H_8O + R\cdot \longrightarrow C_4H_7O\cdot + RH$$

The radical thus generated can then attack the M(I) species, yielding a diazenido complex:

$$[MX(N_2)(dppe)_2] + \text{C}_4\text{H}_7\text{O}^{\bullet} \longrightarrow [MX(N_2\text{—C}_4\text{H}_7\text{O})(dppe)_2]$$

In certain cases (e.g. N-ethylpyrrolidine) this kind of diazenido complex is isolated.[3] With tetrahydrofuran there is a subsequent, and reversible, reaction with proton acids to yield a diazoalkane complex, as indicated:

$$M\text{—}N\text{=}N\text{—C}_4\text{H}_7\text{O} \; (+\,H^+) \longrightarrow [M\text{=}N\text{—}N\text{=}CH(CH_2)_3OH]^+$$

The synthetic utility of this pathway is considerable.

Extensive as this chemistry has now become, it has evidently not yet reached its limits. Very recently, an entirely different kind of reactivity has been demonstrated.[52] This stemmed from the suggestion that coordinated N_2 is polarized in the sense metal—$\overset{\delta+}{N}$—$\overset{\delta-}{N}$. Hence the nitrogen adjacent to the metal should be susceptible to nucleophilic attack. The complex $[(C_5H_5)Mn(CO)_2(N_2)]$ reacts with organolithium reagents LiR to form diazenido complexes of a new type, but which decompose above −10°C.

$$[(C_5H_5)Mn(CO)_2(N_2)] + PhLi \rightarrow [(C_5H_5)Mn(CO)_2\text{—}N(Ph)\text{=}N^-]Li^+$$

Methyl lithium produces a similar reaction. Both products (R = Ph and Me) should now be susceptible to electrophilic attack at the terminal nitrogen, and they react with protons at −30°C to give diazene derivatives.

$$[(C_5H_5)Mn(CO)_2NRN]Li + H^+ \rightarrow [(C_5H_5)Mn(CO)_2(NRNH)] + Li^+$$

That with R = Me also reacts with $(Me_3O)(BF_4)$ at −30°C to give $[(C_5H_5)Mn(CO)_2(NMeNMe)]$. Finally, the dimethyldiazene complex thus formed reacts with N_2 (100 atm., 20°C) to regenerate $[(C_5H_5)Mn(CO)_2(N_2)]$ and release MeN=NMe. This is a potentially cyclic system, but it is not very useful since the many side-reactions quickly lead to loss of activity. Nevertheless, the work is significant on two counts. Firstly it demonstrates a new kind of reactivity of N_2, and secondly it is the first example of a system which allows the formation of an organonitrogen compound and retention of the integrity of the metal complex system.

IV. Formation of Amines from Complexes Formed via Coordinated Dinitrogen

In contrast to the manganese systems[52] discussed above, production of organonitrogen compounds from the diazoalkane, diazenido and hydrazido(2–) complexes derived from coordinated N_2 only occurs with complete breakdown of the metal complex. Thus, $[MoBr(N_2Bu^n)(dppe)_2]$ reacts with a ten-fold excess of sodium borohydride in methanol–benzene in a sealed tube at 100°C during 10 h to yield volatile base (58% of that expected). The yields of individual bases were 51% (NH_3) and 54% (Bu^nNH_2).[53] Butylmethylamine was also detected.

The mechanism of this reaction is not clear. No molybdenum species was isolated, except when ethanol was used as solvent, when $[MoH_4(dppe)_2]$ was recovered in ~25% quantitites.[53]

Hydrazido(2–) complexes can be converted to amines by a variety of destructive methods,[54] including treatment with sulphuric acid in propylene carbonate, reaction with $Li[AlH_4]$, and distillation from strong aqueous or alcoholic base. The yields of amine varied from very low to almost quantitative, depending upon the conditions used. No ammonia was detected (i.e. at least half of the combined nitrogen was lost) and the metal complex was generally completely destroyed, although $[WH_4(dppe)_2]$ and $[WBr_2H_2(dppe)_2]$ were isolated on occasion. As an example, treatment of $[WBr(N_2Me_2)(dppe)_2]Br$ with $Li[AlH_4]$ in methanol yielded 0·95 mole Me_2NH after appropriate work-up.[54]

Finally, diazoalkane complexes also give amines upon treatment with $Li[AlH_4]$,[45] $[WBr_2(N_2CMe_2)(PMe_2Ph)_3]$ yielding approximately quantitative amounts of Pr^iNH_2, but only about 60% of the expected NH_3. Again, the metal complex is degraded.

V. Conclusions

Despite all the work described above, our understanding of the chemistry of N_2 is still in its infancy. We have discovered how to trap N_2 on a metal site, but we do not fully appreciate why one metal site will bind N_2 and another will not. We can convert some kinds of bound N_2 to a variety of complexed organonitrogen moieties, but we do not know why other kinds of N_2 are not reactive. We can displace the organonitrogen moiety to yield organonitrogen compounds in some cases, but it is not possible to do this such that the metal complex can be satisfactorily re-used. We are some distance from a catalytic cycle.

It is not immediately obvious where the source of the differences in behaviour of these complexes lies. Quite clearly, Table 1 implies that the metal–nitrogen and nitrogen–nitrogen bonds in the complexes are rather similar, both quantitatively and qualitatively wherever the N—N—C system is found and whatever the stoichiometry of the complexes. The one apparent exception, $[MoI(N_2C_6H_{11})(dppe)_2]$ should be treated with caution.[28] It is probably consistent with all the other compounds.

Nevertheless, the partial answers we do have must reassure us that the questions we are asking are sensible ones. As experience accumulates we shall be able to phrase them more precisely and ultimately obtain an answer to the problem of the construction of a catalytic system for converting N_2 to organonitrogen compounds.

References

1. O. K. Fomin, *Uspekhi Khim.* (1967) **36**, 1701 (p. 725 in translation); R. E. March and H. I. Schiff, *Can. J. Chem.* (1967) **16**, 1891.
2. R. G. Bergstrom, R. G. M. Landells, G. H. Wahl and H. Zollinger, *J. Amer. Chem. Soc.* (1976) **98**, 3301.
3. D. C. Owsley and G. K. Helmkamp, *J. Amer. Chem. Soc.* (1967) **89**, 4558.
4. A. E. Shilov, A. A. Shteinmann and M. B. Tjabin, *Tetrahedron Letters* (1968) 4177.
5. Yu. G. Borodko, A. E. Shilov and A. A. Shteinmann, *Doklady Akad. Nauk S.S.S.R.* (1966) **168**, 581 (p. 510 in translation).
6. T. B. Wilson and G. B. Kistiakowsky, *J. Amer. Chem. Soc.* (1958) **80**, 2934; C. B. Moore and G. C. Pimentel, *J. Chem. Phys.* (1964) **41**, 3504.
7. W. Brown, J. McNesby and A. M. Bass, *J. Chem. Phys.* (1967) **46**, 2071.
8. J. Chatt and G. J. Leigh, unpublished observations.
9. J. Ellermann, F. Poersch, R. Kunstmann and R. Kramalowsky, *Angew. Chem.* (*Internat. Edn.*) (1969) **8**, 203.
10. M. E. Volpin and V. B. Shur, *Organometallic Reactions* (1970) **1**, 55.
11. M. E. Volpin and V. B. Shur, *Izvest. Akad. Nauk S.S.S.R.* (1966) 1873 (p. 1819 in translation).
12. F. H. Owens, H. P. Fellman and F. E. Zimmerman, *J. Org. Chem.* (1960) **25**, 1808.
13. M. E. Volpin, V. B. Shur, R. V. Kudryavtsev and L. A. Prodayko, *J.C.S. Chem. Commun.* (1968) 1038.
14. V. N. Latyaeva, L. I. Vyshinskaya, V. B. Shur, L. A. Fedorov and M. E. Volpin, *Doklady Akad. Nauk S.S.S.R.* (1968) **179**, 875 (p. 303 in translation); *J. Organometallic Chem.* (1969) **16**, 103.
15. V. B. Shur, E. G. Berkovich and M. E. Volpin, *Izvest. Akad. Nauk S.S.S.R.* (1971) 2358 (p. 2248 in translation); V. B. Shur, E. G. Berkovich, L. B. Vasilyeva, R. V. Kudryavtsev and M. E. Volpin, *J. Organometallic Chem.* (1974) **78**, 127.
16. J. Dvorak, R. J. O'Brien and W. Santo, *J. C. S. Chem. Commun.* (1970) 411.
17. V. B. Shur, E. G. Berkovich, Ṡ. M. Yunusev, Sh. M. Mamedov, I. L. Nizker and M. E. Volpin, *Izvest. Akad. Nauk S.S.S.R.*, (1977) 2841 (p. 2630 in translation).
18. M. E. Volpin, A. A. Belyi, N. A. Katkov, P. V. Kudryavtsev and V. B. Shur, *Izvest. Akad. Nauk S.S.S.R.* (1969) 2858 (p. 2697 in translation).

19. For a review of metal/N_2 reactions, see G. J. Leigh in "*The Chemistry and Biochemistry of Nitrogen Fixation*" (Ed. J. R. Postgate). Plenum Press, London and New York (1971).
20. E. E. Van Tamelen and H. Rudler, *J. Amer. Chem. Soc.* (1970) **92**, 5253.
21. P. E. Koenig, J. M. Morris, E. J. Blanchard and P. S. Mason, *J. Org. Chem.* (1961) **26**, 4777.
22. F. Briegleb and A. Geuther, *Liebigs Ann.* (1862) **123**, 237.
23. P. Sobota, B. Jezowska-Trzebiatowska and Z. Janas, *J. Organometallic Chem.* (1976) **118**, 253.
24. J. Chatt, G. A. Heath and G. J. Leigh, *J. C. S. Chem. Commun.* (1972) 444.
25. M. Sato, T. Kodama, M. Hidai and Y. Uchida, *J. Organometallic Chem.* (1978) **152**, 239.
26. J. Chatt, G. A. Heath, N. E. Hooper and G. J. Leigh, *J. Organometallic Chem.* (1973) **57**, C67.
27. J. Chatt, A. A. Diamantis, G. A. Heath, N. E. Hooper and G. J. Leigh, *J. C. S.* (*Dalton*) (1977) 688.
28. V. W. Day, T. A. George and S. D. A. Iske, *J. Amer. Chem. Soc.* (1975) **97**, 4127.
29. V. W. Day, T. A. George, S. D. A. Iske and S. D. Wagner, *J. Organometallic Chem.* (1976) **112**, C55.
30. D. C. Busby and T. A. George, *J. Organometallic Chem.* (1976) **118**, C16.
31. T. A. George and S. D. A. Iske, "Proceedings of First International Symposium on Nitrogen Fixation" (Ed. W. E. Newton and C. J. Nyman), p. 27. Washington State University Press, Washington (1975).
32. G. Butler, J. Chatt, G. J. Leigh and D. L. Hughes, *Inorg. Chim. Acta* (1978) **30**, L287.
33. J. Chatt, R. A. Head and G. J. Leigh, unpublished observations.
34. J. Chatt, G. J. Leigh and F. P. Terreros, unpublished observations.
35. T. Tatsumi, M. Hidai and Y. Uchida, *Inorg. Chem.* (1975) **14**, 2530.
36. J. Chatt, G. J. Leigh and C. J. Pickett, unpublished observations.
37. A. A. Diamantis, J. Chatt, G. A. Heath and G. J. Leigh, *J. C. S. Chem. Commun.* (1975) 27.
38. P. C. Bevan, J. Chatt, R. A. Head, P. B. Hitchcock and G. J. Leigh, *J. C. S. Chem. Commun.* (1976) 509.
39. P. C. Bevan, J. Chatt, A. A. Diamantis, R. A. Head, G. A. Heath and G. J. Leigh, *J. C. S.* (*Dalton*) (1977) 1711.
40. See, for example, M. L. Ziegler, K. Weidenhammer, H., Zeuner, P. S. Skell and W. A. Herrmann, *Angew. Chem.* (*Internat. Edn.*) (1976) **15**, 695.
41. R. Ben-Shoshan, J. Chatt, W. Hussain and G. J. Leigh, *J. Organometallic Chem.* (1976) **112**, C9.
42. J. Chatt, R. A. Head, P. B. Hitchcock, W. Hussain and G. J. Leigh, *J. Organometallic Chem.* (1977) **133**, C1.
43. J. Chatt, W. Hussain and G. J. Leigh, unpublished observations.
44. M. Hidai, Y. Mizobe and Y. Uchida, *J. Amer. Chem. Soc.* (1976) **98**, 7824; M. Hidai, Y. Mizobe, M. Sato, T. Kodama and Y. Uchida, *J. Amer. Chem. Soc.* (1978) **100**, 5740.
45. P. C. Bevan, J. Chatt, M. Hidai and G. J. Leigh, *J. Organometallic Chem.* (1978) **160**, 165.
46. D. L. Dubois and R. Hoffmann, *Nouveau J. Chim.* (1977) **1**, 79.
47. J. Chatt, G. J. Leigh, C. J. Pickett, R. L. Richards and A. J. L. Pombeiro, *Nouveau J. Chim.* (1978) **2**, 541.

48. J. Chatt, R. A. Head, G. J. Leigh and C. J. Pickett, *J. C. S. Chem. Commun.* (1977) 299; *J. C. S.* (*Dalton*) (1978) 1638.
49. R. J. W. Thomas, G. S. Laurence and A. A. Diamantis, *Inorg. Chim. Acta* (1978) **30**, L353.
50. I. A. Tikhonova, V. B. Shur and M. E. Volpin, *Izvest. Akad. Nauk S.S.S.R.* (1976) p. 221 (p. 225 in translation).
51. J. Chatt, G. J. Leigh, H. Neukomm, C. J. Pickett and G. J. Leigh, unpublished observations.
52. D. Sellmann and W. Weiss, *Angew. Chem.* (*Internat. Edn.*) (1977) **16**, 880; (1978) **17**, 269; *J. Organometallic Chem.* (1978) **160**, 183.
53. D. C. Busby and T. A. George, *J. Organometallic Chem.* (1978) **29**, L273.
54. P. C. Bevan, J. Chatt, G. J. Leigh and E. G. Leelamani, *J. Organometallic Chem.* (1977) **139**, C59.
55. F. C. March, R. Mason and K. M. Thomas, *J. Organometallic Chem.* (1975) **96**, C43.
56. V. F. Duckworth, P. G. Douglas, R. Mason and B. L. Shaw, *J. C. S. Chem. Commun.* (1970) 1083.
57. V. F. Duckworth, G. J. Gainsford and R. Mason, unpublished work, quoted in ref. 58.
58. R. Mason, K. M. Thomas, J. A. Zubieta, P. G. Douglas, A. R. Galbraith and B. L. Shaw, *J. Amer. Chem. Soc.* (1974) **96**, 260.

9

Reactions of Ligating Dinitrogen in Dinuclear Complexes

JAN H. TEUBEN

Laboratorium voor Anorganische Chemie, Rijksuniversiteit Groningen, The Netherlands

I. Introduction

DINUCLEAR dinitrogen complexes have been reported for all transition metals with the exception of Hf, V, Ta, Tc and Ir. The field has been extensively reviewed recently,[1,2,3] and this chapter will be limited to the reactivity of the dinitrogen ligand in complexes in which the units M—N_2—M, M—N_2—M′ (M and M′ = transition metal) are present and also to complexes M—N_2—A where A = Lewis acid. The reactivity of the dinitrogen ligand in dinuclear complexes has been subject of intense research for only a small number of complexes and is limited mainly to Ti, Zr, and Fe.

In Section II aspects of bonding and reactivity of the dinitrogen ligand are discussed, while in Section III an account is given of successful and unsuccessful attempts to cause this ligand to react.

II. Bonding and Reactivity

The theoretical considerations for the binding of dinitrogen in dinuclear complexes are in principle the same as those which apply for mononuclear complexes. Dinitrogen, being a weak σ-donor and moderate π-acceptor, will build up electronic charge as a consequence of bonding and this will lead to a rather unstable dinuclear complex if there is not a secondary effect that helps to relieve this charge build-up. An indication for the flow of charge to dinitrogen can be inferred from the XPE spectrum of the asymmetric dinuclear complex $[MoCl_4(OMe)\{(N_2)ReCl(PMe_2Ph)_4\}]$.[4] The position of the N(1s) bond energy (398·6 eV) indicates that the electron density in both N atoms in this complex is about the same as found for the non-coordinated and more negatively charged N atom in the mononuclear complex $[Mo(N_2)_2(dppe)_2]$ (dppe = $Ph_2PCH_2CH_2PPh_2$), which can be protonated or alkylated quite easily.[1] Unfortunately other XPE data on the bridging dinitrogen molecule are not known at the moment.

More evidence for the activation of the dinitrogen ligand is obtained from X-ray structural and vibrational spectroscopic data. In Table 1 the data for a number of relevant, dinuclear complexes are given. With the exception of the

Table 1. X-ray structural and vibrational spectroscopic data for some dinuclear dinitrogen complexes

Complex	N–N (Å)	M–N–N (deg)		$\nu(N_2)$ $(cm^{-1})^a$	Ref.
$[\{Ti(\eta^5\text{-}C_5Me_5)_2\}_2(N_2)]$	1·160(14)[b]	178·8(4)[b]		—	*5*
$[\{Ti(\eta^5\text{-}C_5H_5)_2R\}_2(N_2)]$	1·162(12)	176·5(5)		—	*6*
$[\{Zr(\eta^5\text{-}C_5Me_5)_2(N_2)\}_2(N_2)]$	1·182(5)	176·7(3)		—	*7, 8*
$[\{Mo(\eta^6\text{-}C_6H_3Me_3)(dmpe)\}_2(N_2)]$	1·145(7)	175·6(4)		1989	*9, 10*
$[MoCl_4(OMe)\{(N_2)ReCl(PMe_2Ph)_4\}]$	1·18(3)	179·6(14)	(M = Re)	1660	*11*
		178·7(9)	(M′ = Mo)		
$[MoCl_4\{(N_2)ReCl(PMe_2Ph)_4\}_2]$	1·154(29)	177·1(22)	(M = Re)	1800	*12, 13*
		178·6(21)	(M′ = Mo)		
$[\{Mn(\eta^5\text{-}C_5H_4Me)(CO)_2\}_2(N_2)]$	1·118(7)	176·5(4)		1971	*14*
$[\{Ru(NH_3)_5\}_2(N_2)](BF_4)_4$	1·124(15)	178·3		2100	*15*
$[\{Co(PMe_3)_3(N_2)\}_2Mg(thf)_4]$	1·18	linear	(M = Co)	2068	*16*
		158	(M′ = Mg)		
$K[Co(N_2)(PMe_3)_3]$	1·16–1·18	linear	(M = Co)	1795	*17*
				1758	
$[\{Ni(PCy_3)_2\}_2(N_2)]$	1·12	178·2		—	*18*
$[\{(PhLi)_6Ni_2(N_2)(Et_2O)_2\}_2]$	1·35(2)	—[c]		—[d]	*19*
N_2	1·0976[e]	—		2331	
MeN=NMe	1·23[f]	—		—	
$H_2N–NH_2$	1·46[f]	—		—	

[a] $\nu(N_2)$ is normally not IR-active in the centrosymmetric complexes; Raman spectroscopy is frequently impossible due to decomposition of the samples during irradiation. [b] Average value from two independent molecules in the unit cell. [c] The dinitrogen ligand is "edge-on" bonded to a Ni–Ni moiety. [d] $\nu(N_2)$ is possibly below 1550 cm^{-1}. [e] P. G. Wilkinson and N. B. Houk, *J. Chem. Phys.* (1956) **24**, 528. [f] "Interatomic Distances", The Chemical Society, London (1958).

Mn, Ru and Ni–PCy_3 complexes, which show N–N distances very similar to those of mononuclear dinitrogen complexes,[1] all the complexes show a substantial lengthening of the N–N bond compared with free dinitrogen to about 1·14–1·18 Å. The M–N–N–M moiety is essentially linear, with exception of Co–N–N–Mg–N–N–Co, where the transition metal–dinitrogen grouping is linear, but the magnesium–dinitrogen group is bent. In the very complicated molecule $[\{(PhLi)_6Ni_2(N_2)(Et_2O)_2\}_2]$, the dinitrogen is "edge-on" bonded to a Ni–Ni moiety and interacts simultaneously with the Li atoms. The result of this multiple interaction is a very long N–N distance (1·35 Å).

The $\nu(N_2)$ data clearly are in agreement with the substantial decrease in bond order of the dinitrogen molecule as follows from the long N–N distances. The shifts of $\nu(N_2)$ relative to free dinitrogen are large (of the order of 350–650 cm^{-1} to lower energy).

To explain these observations, a number of qualitative MO schemes have been suggested [for detailed discussions of these theories see the references given in the review by Chatt *et al.*[1] and Chapter 6.] One essential aspect however should be emphasized at this point. The weak σ-donation from the dinitrogen to the metal and the relatively stronger back-donation into the π^* orbitals of dinitrogen could easily lead to a too high charge build-up on dinitrogen. This unfavourable situation can be relieved by delocalization into empty π-type orbitals of the transition metal. It is evident that this mechanism will work effectively for complexes with few d electrons, e.g. early transition metals like Ti and Zr[8] or acceptor-type molecules like $[MoCl_3(thf)_3]$ which can function both as π-donors and as π-acceptors.

A consequence of this type of bonding may be an additional weakening of the N≡N bond and an increase in multiple bond character of the M–N bond, with ultimately cleavage of the N–N bond and formation of a transition metal nitride. Other ligands on the metal may participate in the bonding scheme and can even play an essential role in the bonding of the bridging dinitrogen ligand. This is the case in the complexes $[\{(\eta^5\text{-}C_5Me_5)_2M(N_2)\}_2(N_2)]$, (M = Ti, Zr).[5,8]

Based on the foregoing arguments one may expect quite a high reactivity of the bridging dinitrogen ligand. It is however impossible to predict whether this reactivity will be towards electrophilic (π-donor properties of the metal dominating) or nucleophilic reagents (π-acceptor properties dominating). The nature of the additional ligands may be decisive. Moreover, in most dinuclear complexes the dinitrogen ligand is in the inside of the complex, and reactions that are feasible electronically may be sterically blocked. It is also possible that in reactions initially aimed at the dinitrogen ligand, attack on peripheral ligands takes place and this may change the electronic conditions for the bonding of the dinitrogen so that the N–N bond order increases and dinitrogen is liberated.

III. Reactions

A. Unsuccessful experiments

Very few publications on dinuclear dinitrogen complexes of the transition metals mention explicitly experiments exploring the reactivity of the ligated dinitrogen molecule. The very long N—N distance (1·35 Å) in $[\{(PhLi)_6Ni_2(N_2)(Et_2O)_2\}]$ suggests a strong increase in the electron density of the dinitrogen molecule and consequently a high reactivity of this ligand to protons or other electrophilic agents, but up to now no such reactions have been reported for this and a related complex.[20,21] From these publications a rather disappointing picture emerges. In practically all of the experiments reported, the dinitrogen is simply substituted by other ligands and is evolved as dinitrogen gas or simply remains bound to the metal.

The molybdenum complex $[\{Mo(\eta^6\text{-}C_6H_6)(PPh_3)_2\}_2(N_2)]$ does not produce any N-containing products on reaction with H_2O, $LiAlH_4$ or BuLi.[22] The analogous complex $[\{Mo(\eta^6\text{-}C_6H_3Me_3)(dmpe)\}_2(N_2)]$ (dmpe = $Me_2PCH_2CH_2PMe_2$) is protonated at the metal with HBF_4 giving $[\{Mo(\eta^6\text{-}C_6H_3Me_3)(H)(dmpe)\}_2(N_2)]\ (BF_4)_2$, which on further protonation (with HCl) or reaction with $FeSO_4/H_2SO_4$ also forms no nitrogen derivatives.[9] This protonation reaction suggests that in the Mo—dinitrogen complex, the electron density at the metal is higher than at the dinitrogen molecule.

The iron complexes $[\{Fe(\eta^5\text{-}C_5H_5)(L_2)\}_2(N_2)]$ (L_2 = dmpe or dppe) also do not give N—H bond formation on reaction with reducing agents (e.g. $LiAlH_4$ or $NaBH_4$), or on protonation.[23,24] The dinitrogen ligand in $[\{RuL_5\}_2(N_2)]^{2+}$ (L = NH_3, H_2O) does not react with oxidizing and reducing agents and attempts to protonate it also fail.[25] An unsuccessful attempt to reduce the dinitrogen ligand in $K_2[\{Rh(NO_2)_3(NH_3)(OH)\}_2(N_2)]$ with sodium dithionate has been reported.[26]

B. Reduction and protonation of the ligated dinitrogen molecule

The discovery by Vol'pin and Shur[27] in 1964, that combinations of transition metal compounds and an excess of reducing agents take up and reduce dinitrogen under mild conditions and give substantial amounts of ammonia on hydrolysis, started off intense research on these and related systems. Titanium compounds were especially effective and systems based on $[Ti(\eta^5\text{-}C_5H_5)_2Cl_2]$ have been investigated by a number of research groups, varying such reaction conditions as temperature, dinitrogen pressure, and nature of the reducing agents. There is substantial evidence that in these systems, which are very complicated and difficult to study systematically dinitrogen is reduced to titanium nitrides via dinuclear dinitrogen complexes (see Chapter 3). A number of mechanisms have been suggested to explain the complicated reactions that take place in the $[Ti(\eta^5\text{-}C_5H_5)_2]$-based systems and much confusion has resulted.

A survey of this early work will not be given here; this has already been covered in the review literature.[2,28]

C. Systems based on dicyclopentadienyl compounds of Ti and Zr

Since about 1970, research on the reduction of dinitrogen by Ti- and Zr-based systems has been limited to $M(\eta^5\text{-}C_5H_5)_2$-derivatives. Although considerable progress concerning the mechanism of complexation and reduction of the dinitrogen molecule has been made, the results obtained so far do not permit discussion in terms of an all-inclusive mechanism.

The reason for this may in part be that although various related dinitrogen complexes may have almost identical physicochemical properties, they quite often show very different chemical behaviour. This is well known for mononuclear dinitrogen complexes and it may also be true for dinuclear dinitrogen complexes.[29] Secondly, although the intermediates in the system under discussion here have not been fully characterized, it is quite possible that due to the high reactivity of cyclopentadienyl groups in low-valent Ti and Zr compounds, various different structures may exist which are not easily distinguishable but may have different chemical properties. In this respect the reported structures of species formerly referred to as "titanocene" have to be mentioned here[30,31,32] (Fig. 1).

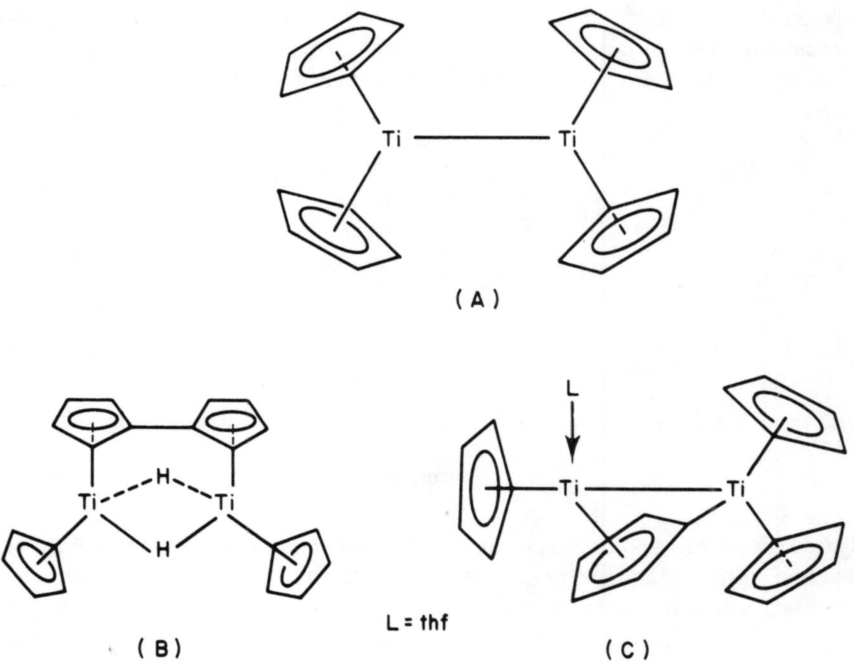

Fig. 1. Some forms of "titanocene", (A) metastable dimer[30] (structure unknown), (B) fulvalene bridged dimer,[31] (C) cyclopentadienyl bridged dimer.[32]

Authentic titanocene $[Ti(\eta^5\text{-}C_5H_5)_2]$ is most probably generated as an intermediate in the alkali metal (amalgam, naphthalene anion or metal) reduction of $[Ti(\eta^5\text{-}C_5H_5)_2Cl_2]$. It is not stable as such, but Brintzinger *et al.*[5,30] showed that it can dimerize to give the metastable $[\{Ti(\eta^5\text{-}C_5H_5)_2\}_2]$, which forms a dinuclear dinitrogen complex $[\{Ti(\eta^5\text{-}C_5H_5)_2\}_2(N_2)]$. The dinitrogen ligand is not sufficiently activated to produce ammonia or hydrazine on protonation, but the addition of a strong reductor like lithium–naphthalene (1 Li/Ti) is sufficient to give a stoichiometric yield of NH_3 (0·95 mol NH_3 per Ti atom) after protonation.

A complex with the same stoichiometry but quite different properties was obtained by Shilov *et al.*[33] from $[\{Ti(\eta^5\text{-}C_5H_5)_2Cl\}_2]$ and MeMgI in ether under dinitrogen. In contrast to the former complex, this complex shows a band at 1280 cm^{-1}, which is assigned to $\nu(N_2)$. This low value of $\nu(N_2)$ suggests that the complex is a diazene derivative and consequently a non-linear Ti–N=N–Ti unit is proposed.[34] Protonation studies support this view. Reaction with HCl (in methanol) at −60°C give dinitrogen and hydrazine (1:1 molar ratio) and reaction with HCl (in ether) at the same temperature produces dinitrogen and ammonia (30% based on the complexed dinitrogen). The difference in product formation is ascribed to disproportionation of the initially formed diazene under the different reaction conditions. The presence of the dicyclopentadienyl-titanium moiety follows from the formation of $[Ti(\eta^5\text{-}C_5H_5)_2Cl_2]$ on reaction with HCl (Scheme (**1**)). Complete reduction of the

$$Cp_2TiCl \xrightarrow[\text{Et}_2\text{O, N}_2,\ -70°\text{C}]{\text{MeMgI}} Cp_2Ti\text{–}N\text{=}N\text{–}TiCp_2$$

$(Cp = \eta^5\text{-}C_5H_5)$

−70°C

HCl/MeOH, −60°C: $Cp_2TiCl + N_2 + N_2H_4$

HCl/Et_2O, −60°C: $Cp_2TiCl + N_2 + NH_3$

Scheme (1)

ligated dinitrogen and formation of a nitrido–titanium complex takes place on reaction with PhLi (in ether). Conversion of this complex to a hydrazine precursor under the action of Pr^iMgCl was claimed in a reduction scheme.[35] It is not clear from the paper whether this observation was made on the complex itself or whether it has been assumed as an intermediate in the system $[Ti(\eta^5\text{-}C_5H_5)_2Cl_2]$/Pr MgCl/$N_2$ in ether.

Pez[32] reports that the dimeric titanocene-type compound $[(\eta^5\text{-}C_5H_5)_2\text{-}Ti\text{-}\mu\text{-}(\eta^1,\eta^5\text{-}C_5H_4)Ti(\eta^5\text{-}C_5H_5)]$, obtained by low-temperature potassium–naphthalene reduction of $[Ti(\eta^5\text{-}C_5H_5)_2Cl_2]$ in thf under argon, reacts with dinitrogen to give a deep-blue complex $[\{Ti_2(C_5H_5)_3(C_5H_4)\}_2(N_2)]$ (Fig. 2).[36] This complex shows no $\nu(N_2)$ in the infrared spectrum and therefore a centrosymmetric dinuclear structure is proposed with the dinitrogen bridging the titanocene parts of two dimers. In this respect the complex is very similar to $[\{Ti(\eta^5\text{-}C_5H_5)_2\}_2(N_2)]$ reported by Brintzinger *et al.*[30]

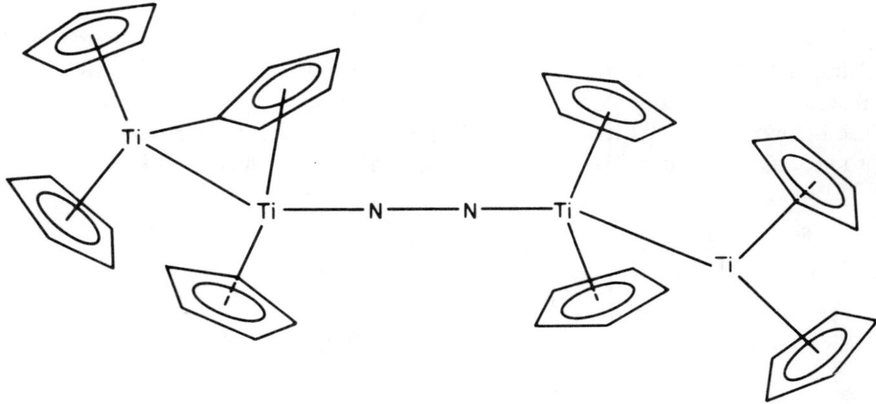

Fig. 2. Possible structure of dinitrogen complex of cyclopentadienyl bridged titanocene dimer.[36]

This similarity is also found in the reactivity of the dinitrogen ligand. It does not react with H_2, but a strong reducing agent (e.g. potassium–naphthalene) reduces it to an ammonia precursor. About 70% of the originally complexed dinitrogen is formed as ammonia after hydrolysis.[36]

The high reactivity of the cyclopentadienyl rings in low-valent dicyclopentadienyltitanium compounds prompted Bercaw *et al.*, to study the more stable permethylcyclopentadienyl analogues. $[Ti(\eta^5\text{-}C_5Me_5)_2]$ at 0°C forms $[\{Ti(\eta^5\text{-}C_5Me_5)_2\}_2(N_2)]$ with a linear bridging dinitrogen as was shown by X-ray structure determination (Fig. 3).[5] The long N—N distance (1·160 Å) indicates a reduction of the dinitrogen bond order, but this is not substantiated by an easy protonation of the dinitrogen; additionally, further reduction of this ligand (e.g. with Mg/Hg or a N_2/H_2 mixture) cannot be achieved.[37]

At low temperature (−80°C) the titanium complex takes up extra dinitrogen and an equilibrium is established:

$$[\{Ti(\eta^5\text{-}C_5Me_5)_2\}_2(N_2)] \xrightleftharpoons{N_2} [\{Ti(\eta^5\text{-}C_5Me_5)_2(N_2)\}_2(N_2)] \quad (1)$$

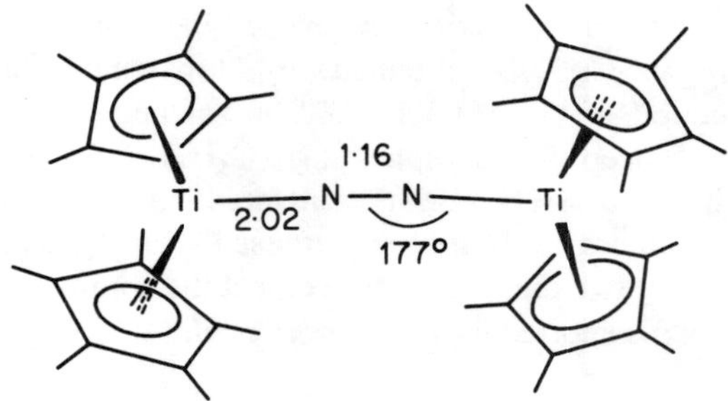

Fig. 3. Structure of $[\{Ti(\eta^5\text{-}C_5Me_5)_2\}_2(N_2)]$.

giving a dinuclear dinitrogen complex with terminal dinitrogen ligands. The structure was based on analogy with the corresponding zirconium complex (see below) and on ^1H and ^{15}N NMR spectra.[38] Surprisingly these complexes produce a nearly quantitative yield of hydrazine when treated with HCl.[5]

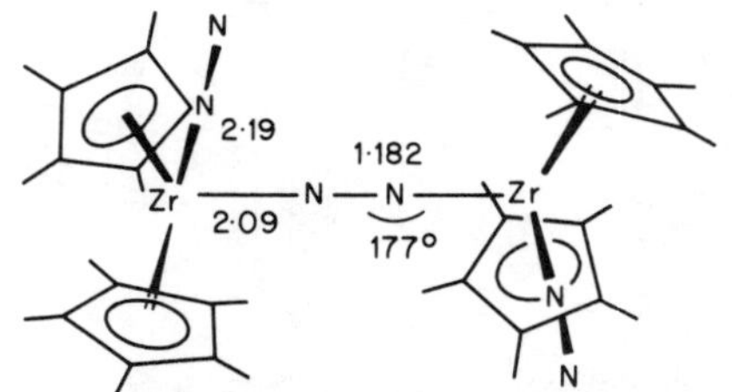

Fig. 4. Structure of $[\{Zr(\eta^5\text{-}C_5Me_5)_2(N_2)\}_2(N_2)]$.

Permethylzirconocene forms a dinitrogen-containing complex both terminal and bridging, $[\{Zr(\eta^5\text{-}C_5Me_5)_2(N_2)\}_2(N_2)]$, the structure of which has been determined[8] as that shown in Fig. 4. With HCl in toluene at −80°C it reacts to give hydrazine (0·86 mol)[8] (equation (**2**).

$$[\{Zr(\eta^5\text{-}C_5Me_5)_2(N_2)\}_2(N_2)] + 4HCl \rightarrow 2[Zr(\eta^5\text{-}C_5Me_5)_2Cl_2] + 2N_2 + N_2H_4 \quad (2)$$

A study of the reaction of hydrogen chloride with the complex having terminal $^{15}N_2$ shows that half of the hydrazine arises from terminal dinitrogen and the other half from the originally bridging dinitrogen.[8] It is not likely that the bridging dinitrogen is attacked first, since the related complex $[\{Zr(\eta^5\text{-}C_5Me_5)_2(CO)\}_2(N_2)]$ does not give hydrazine on reaction with hydrogen chloride.[39] The mechanism proposed is one in which first a terminal dinitrogen is protonated, then the other terminal ligand is lost and a symmetric bis-diazenido complex $[Zr(\eta\text{-}C_5Me_5)_2(N_2H)_2]$ is formed. Further protonation of the latter gives diazene, which disproportionates to give dinitrogen and hydrazine.[8,38]

Shilov *et al.*[33,35] have studied systems based on ethereal mixtures of $[Ti(\eta^5\text{-}C_5H_5)_2Cl_2]$ and alkyl-Grignard reagents at low temperatures (−80°C) under N_2. With MeMgI (Mg:Ti = 2) the previously discussed dinitrogen complex $[\{Ti(\eta^5\text{-}C_5H_5)_2\}_2(N_2)]$ was obtained. With Pr^iMgCl (Mg:Ti = 4) a Mg- and Cl-containing complex of stoichiometry $[\{Ti(\eta^5\text{-}C_5H_5)_2\}(N_2MgCl)]$ having $\nu(N_2)$ at 1255 cm^{-1}, was isolated. Treatment of this complex with HCl (in MeOH) at −60°C gave hydrazine in about 80% yield. Structures with a MgCl fragment bound to two titanium atoms bridging the dinitrogen molecule were proposed. The reduction of dinitrogen in these systems is suggested to proceed according to Scheme (**2**). In this scheme the conservation of the dicyclopentadienyltitanium unit throughout the reduction process is essential.

$$Cp_2TiCl \xrightarrow{RMgX} Cp_2TiR \xrightarrow{N_2} (Cp_2TiR)_2N_2 \xrightarrow{\Delta} (Cp_2Ti)_2N_2$$

$$(Cp_2TiR)_2N_2 \xrightarrow{H^+} N_2 \qquad (Cp_2Ti)_2N_2 \xrightarrow{H^+} N_2H_2$$

$$(Cp_2Ti)_2N_2 \xrightarrow{RMgX} (Cp_2Ti)_2(N_2MgX) \xrightarrow{RMgX} Cp_2TiN(MgX)_2$$

$$Cp_2TiN(MgX)_2 \xrightarrow{H^+} NH_3 \qquad (Cp_2Ti)_2(N_2MgX) \xrightarrow{H^+} N_2H_4$$

$Cp = \eta^5\text{-}C_5H_5$; R = alkyl; X = halide.

Scheme (2)

This Shilov mechanism, however, is probably not the only one operative in these systems. Van der Weij *et al.*[40,41] studied related systems $[\{Ti(\eta^5\text{-}C_5H_5)_2R\}(N_2)]$ with R = aryl, CH_2CMe_3 and gave evidence for a reduction mechanism in which loss of a $(\eta^5\text{-}C_5H_5)$ group is essential (see below). They observed that for the system $[Ti(\eta^5\text{-}C_5H_5)_2Cl_2]$/4 Pr^iMgCl in ether under dinitrogen at −80°C this mechanism also works and can at least account for part of the ammonia and hydrazine obtained. The loss of cyclopentadienyl groups was confirmed by the isolation of $[Mg(\eta^5\text{-}C_5H_5)_2]$ from the reaction mixture. The Pr^iMgCl system is very labile and the results are strongly dependent on the reaction conditions.[41] Based on experimental evidence and by analogy with the sodium–naphthalene reduction of $[\{Ti(\eta^5\text{-}C_5H_5)_2R\}_2(N_2)]$ they suggest the mechanism shown in Scheme (**3**).

The complexity of the systems under discussion here, and the variety of products that can be expected under various reaction conditions is also

$$Cp_2TiCl \xrightarrow{RMgX} Cp_2TiR \xrightarrow{N_2} (Cp_2TiR)_2N_2$$

$$(Cp_2TiR)_2N_2 \xrightarrow{RMgX} (CpTiR)_2N_2 \xrightarrow{RMgX} (CpTiR)_2N_2(MgX)_2 \xrightarrow{\Delta} (CpTiR)NMgX$$

$$(CpTiR)_2N_2(MgX)_2 \xrightarrow{H^+} N_2H_4 \qquad (CpTiR)NMgX \xrightarrow{H^+} NH_3$$

Cp = η^5-C_5H_5; R = alkyl; X = halide.

Scheme (3)

illustrated by the work of Bercaw *et al.*[30] They isolated [{Ti(η^5-C_5H_5)$_2$-(Et)}$_2$·6$MgCl_2$·7Et_2O] from a mixture of [Ti(η^5-C_5H_5)$_2$$Cl_2$] and EtMgCl in ether at 20°C under Ar. The structure of this complex is unclear. It reacts directly with dinitrogen (230 atm. in 1,2-dimethoxyethane) and gives a nitride formulated as [{Ti(η^5-C_5H_5)$_2$}$_3$(N)$_2$], which gives ammonia and [Ti(η^5-C_5H_5)$_2$$Cl_2$] (2:3) on reaction with HCl gas.

It is clear that further work under carefully controlled reaction conditions is necessary to solve the problem of the mechanism of dinitrogen reduction in these systems.

Unlike the alkyldicyclopentadienyltitanium dinitrogen complexes, the analogous complexes [{Ti(η^5-C_5H_5)$_2$R}$_2$(N_2)] (R = Ph, *o*-, *m*-, *p*-$CH_3C_6H_4$, C_5F_5 and CH_2Ph) are completely characterized crystalline compounds.[42] Recently an X-ray structure determination on [{Ti(η^5-C_5H_5)$_2$(*p*-CH_3C_6-H_4)}$_2$(N_2)] (Fig. 5) showed the presence of a linear Ti—N—N—Ti unit as has been found for the other well characterized dinuclear dinitrogen complexes. The N—N distance (1·162 Å) is the same as reported for [{Ti(η^5-C_5-Me_5)$_2$}$_2$(N_2)],[5] and slightly shorter than that (1·182 Å) in the complex [{Zr(η^5-C_5Me_5)$_2$(N_2)}$_2$(N_2)].[8] The dinitrogen ligand in [{Ti(η^5-C_5H_5)$_2$R}$_2$-(N_2)] cannot be protonated directly. It would be interesting to know the behaviour towards protonation of the bridging dinitrogen in the recently published[43] complex [{Zr(η^5-C_5H_5)$_2$R}$_2$(N_2)] (with R = CH($SiMe_3$)$_2$), but no such study has yet been reported.

Further activation of the dinitrogen ligand in [{Ti(η^5-C_5H_5)$_2$R}$_2$(N_2)] can be achieved by the use of reducing agents (e.g. sodium–naphthalene, BuLi or Pr^{i}MgCl). Sodium–naphthalene is the most reactive, Pr^{i}MgCl the least,[44] but its effectiveness increases for the alkyl complexes (e.g. R = CH_2CMe_3).[41] Van der Weij *et al.*[40] have thoroughly investigated the system [{Ti(η^5-C_5-H_5)$_2$R}$_2$(N_2)]/sodium–naphthalene (R = *m*-$CH_3C_6H_4$) at low temperature (−78°C) in thf. They propose a mechanism (Scheme (**4**)) in which the bridging

$$(Cp_2TiR)_2N_2 \xrightarrow{NaC_{10}H_8} (CpTiR)_2N_2 \xrightarrow{NaC_{10}H_8} [(CpTiR)_2N_2]^{2-}$$

$$(CpTiR)_2N_2 \xrightarrow{H^+} N_2H_2$$

$$[(CpTiR)_2N_2]^{2-} \xrightarrow{H^+} N_2H_4$$

$$[(CpTiR)_2N_2]^{2-} \xrightarrow{\Delta} [(CpTiR)N]^- \xrightarrow{H^+} NH_3$$

$Cp = \eta^5\text{-}C_5H_5$; R = aryl; $NaC_{10}H_8$ = sodium naphthalene.

Scheme (4)

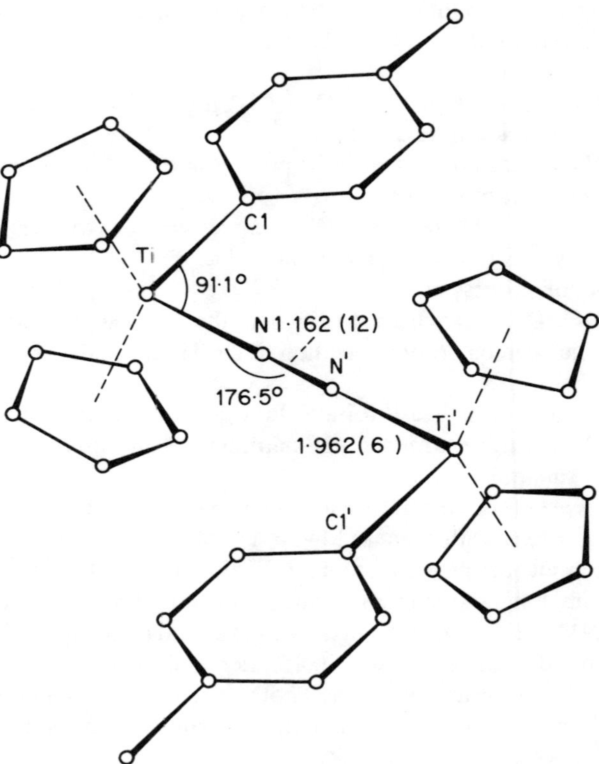

Fig. 5. Structure of $[\{Ti(\eta^5\text{-}C_5H_5)_2(p\text{-}MeC_6H_4)\}_2(N_2)]$.

dinitrogen is reduced stepwise via a diazene precursor to a dianion which gives hydrazine when treated with HCl. In the first step the essential reaction is the abstraction of a cyclopentadienyl group from titanium and the formation of $(C_5H_5)Na$. In the second step electrons of the naphthalene anion are transferred into the $[\{Ti(\eta^5\text{-}C_5H_5)R\}_2(N_2)]$ system and a dianion is generated. Both reactions proceed quite easily even at −78°C. In the hydrazine precursor the dinitrogen molecule is so far modified (reduced) that it is thought to possess an essentially single N—N bond. Both intermediates are very reactive and react easily with a variety of substrates (e.g. solvent or slightly acidic hydrogen on organic molecules) or extra sodium–naphthalene to give a very complicated mixture which is very difficult to unravel. Warming of the hydrazine precursor to room temperature results in the splitting of the N—N bond and formation of a nitrido compound. Ammonia is produced when this nitrido species reacts with protons. Reaction with extra sodium–naphthalene proceeds with degradation of the cyclopentadienyltitanium structure, without increasing the total yield of reduced dinitrogen. The various intermediates have not been isolated or detected in solution. The evidence for their existence and identity is indirect and based on the following observations:

(i) Substantial amounts of $[Ti(\eta^5\text{-}C_5H_5Cl_3]$ are obtained after reacting the reduced mixtures with HCl/ether.
(ii) $(C_5H_5)Na$ was demonstrated to be present in the solutions (by IR and ^1H NMR spectroscopy).
(iii) Ferrocene was obtained in yields corresponding to the amounts of $(C_5H_5)Na$ present; it was demonstrated that $[FeCl_2]$ did not interfere with the reduction process itself.
(iv) When DCl/D_2O was used to quench the reduced mixtures RD was formed, thus demonstrating retention of the Ti—R bond.

Other reducing agents like BuLi and Pr^iMgCl give reaction mixtures which are very similar to those with sodium–naphthalene and it is assumed that the same mechanism applies.

From the work discussed in this section it may be inferred that in dinuclear dinitrogen complexes with a linear M—N—N—M unit (M = Ti, Zr), the ligated dinitrogen cannot be protonated directly despite the fact that the N—N distance is quite long and is intermediate between that found in free dinitrogen and MeN=NMe. Further reduction is an essential condition for obtaining N—H compounds. On the other hand, terminal dinitrogen in Ti and Zr complexes may be just as reactive as in other transition metal compounds[1] as is suggested by the smooth formation of hydrazine from the complexes $[\{M(\eta^5\text{-}C_5Me_5)_2(N_2)\}_2N_2]$ M = Ti, Zr.

C. Carbon–nitrogen bond formation

A number of publications have appeared in which the (mostly low yield) formation of carbon–nitrogen bonds is reported. The systems under investigation normally consist of Ti compounds mixed with an excess of strong reducing agent and are normally considered as containing nitrido–metal species. It is very unlikely that in these systems the N—N group is still present and therefore they fall outside the scope of this section, and are discussed in Chapter 3.

D. Dinuclear complexes of niobium, chromium and iron

The formation of a dinuclear dinitrogen complex of niobium, $[\{NbCl(dmpe)_2\}_2(N_2)]$, has recently been reported. It gives a small amount of hydrazine (10%) on reaction with HCl in thf. Details of structure and the reactivity of the dinitrogen ligand are not yet known.[1]

Sobota *et al.*[45] report the formation of the complexes $[\{Cr(dppe)_2\}_2(N_2)]$ and $[Cr_2N_2Mg_4Cl_4(thf)_5]$ in the system $CrCl_2/Mg/N_2$ in thf (in the presence or absence of dppe respectively). Hydrolysis of the first complex gives small amounts of ammonia and hydrazine, whereas the latter gives these compounds in higher yields (60% and 25% respectively). These complexes are poorly characterized and the presence of bridging dinitrogen was not clearly established. However, the formation of hydrazine from the thf complex indicates that although the major part of the complex mixture probably is a nitride, at least a minor part contains N—N bonds. Whether this is a dinitrogen ligand in a reduced form or a hydrazido-derivative is not clear, but the system is very interesting.

Other systems in which the reduction of dinitrogen is observed (probably via dinuclear dinitrogen complexes) are formed by $FeCl_3$ and reducing agents in ether under dinitrogen. A dinuclear complex $[(PPh_3)_2(Pr^i)HFe(N_2)Fe(Pr^i)(PPh_3)_2]$ is formed from $[(PPh_3)_2FeCl_3]$ with Pr^iMgCl in ether.[46] A band at $1761\ cm^{-1}$ is attributed to $\nu(N_2)$ and indicates a low N—N bond order. This is illustrated in its reactivity towards HCl (in ether), where a low yield of hydrazine is obtained (10% based on the complexed dinitrogen). The mechanism shown in Scheme (5) is postulated. Kinetic studies on the system

$$L_n\overset{H}{Fe}\cdots N\equiv N\cdots FeL_n \xrightarrow{H^+} L_nFe{-}\overset{H}{N}{=}\underset{+}{\overset{H}{N}}{-}FeL_n \longrightarrow$$

$$L_n\overset{+}{Fe}{-}NH{-}NH{-}FeL_n \xrightarrow{H^+} N_2H_4$$

Scheme (5)

$FeCl_3/PhLi/N_2$ in ether suggest the formation of a dinuclear iron–dinitrogen complex.[47] The nature of this complex is still unknown, EPR spectra indicate that the iron part of the complex is probably an Fe(I) derivative. On treatment with HCl, part of the dinitrogen is liberated as hydrazine (about 20% of the complexed dinitrogen) and the remainder as dinitrogen gas. In the presence of LiBr this system also forms ammonia on hydrolysis while the yield of hydrazine drops; in the absence of LiBr only hydrazine (and dinitrogen) is obtained.[48]

References

1. J. Chatt, J. R. Dilworth and R. L. Richards, *Chem. Rev.* (1978) **78**, 589.
2. A. E. Shilov, *Russ. Chem. Rev.* (1974) **43**, 378.
3. D. Sellmann, *Angew. Chem.* (1974) **86**, 692.
4. J. Chatt, C. M. Elson, N. E. Hooper and G. J. Leigh, *J. C. S.* (*Dalton*) (1973) 612.
5. R. D. Sanner, D. M. Duggan, T. C. McKenzie, R. E. Marsh and J. E. Bercaw, *J. Amer. Chem. Soc.* (1976) **98**, 8358.
6. J. D. Zeinstra, J. H. Teuben and F. Jellinek, *J. Organometallic Chem.* (1979) **170**, 39.
7. J. M. Manriquez and J. E. Bercaw, *J. Amer. Chem. Soc.* (1974) **96**, 6229.
8. J. M. Manriquez, R. D. Sanner, R. E. Marsh and J. E. Bercaw, *J. Amer. Chem. Soc.* (1976) **98**, 8351.
9. M. L. H. Green and W. E. Silverthorn, *J. C. S.* (*Dalton*) (1974) 2164.
10. R. A. Forder and K. Prout, *Acta Crystallogr* (1974) **30B**, 3778.
11. M. Mercer, *J. C. S.* (*Dalton*) (1974) 1637.
12. P. D. Cradwick, J. Chatt, R. H. Crabtree and R. L. Richards, *J. C. S. Chem. Commun.* (1975) 351.
13. P. D. Cradwick, *J. C. S.* (*Dalton*) (1976) 1934.
14. M. L. Ziegler, K. Weidenhammer, H. Zeiner, R. S. Skell and W. A. Herrmann, *Angew. Chem.* (*Internat. Edn.*) (1976) **15**, 695.
15. I. M. Treitel, M. T. Ford, R. E. Marsh and H. B. Gray, *J. Amer. Chem. Soc.* (1969) **91**, 6512.
16. R. Hammer, H.-F. Klein, U. Schubert, A. Frank and G. Huttner, *Angew. Chem.* (1976) **88**, 648.
17. R. Hammer, H.-F. Klein, P. Friedrich and G. Huttner, *Angew. Chem.* (1977) **89**, 499.
18. P. W. Jolly, K. Jonas, C. Kruger and Y.-H. Tsay, *J. Organometallic. Chem.* (1971) **33**, 109.
19. C. Kruger and Y.-H. Tsay, *Angew. Chem.* (1973) **95**, 1051.
20. K. Jonas, *Angew. Chem.* (1973) **85**, 1050.
21. K. Jonas, D. J. Brauer, C. Krüger, P. J. Roberts and Y.-H. Tsay, *J. Amer. Chem. Soc.* (1976) **98**, 74.
22. M. L. H. Green and W. E. Silverthorn, *J. C. S.* (*Dalton*) (1973) 301.
23. W. E. Silverthorn, *J. C. S. Chem. Commun.* (1971) 1310.
24. D. Sellmann and E. Kleinschmidt, *J. Organometallic Chem.* (1977) **140**, 211.
25. J. Chatt, A. B. Nikolsky, R. L. Richards, J. R. Sanders, J. E. Fergusson and J. L. Love, *J. Chem. Soc.* (*A*) (1970) 1479.

26. L. S. Volkova, V. M. Volkov and S. S. Chernikov, *Zh. Neorg. Khim.* (1971) **9**, 2594.
27. M. E. Volpin and V. B. Shur, *Doklady Akad. Nauk SSSR* (1964) **156**, 1102.
28. E. E. van Tamelen, *Accounts Chem. Res.* (1970) **3**, 361.
29. D. Sellmann, A. Brandl and R. Endell, *J. Organometallic Chem.* (1975) **97**, 229.
30. J. E. Bercaw, R. H. Marvich, L. G. Bell and H. H. Brintzinger, *J. Amer. Chem. Soc.* (1972) **94**, 1219.
31. A. Davison and S. S. Wreford, *J. Amer. Chem. Soc.* (1974) **96**, 3017.
32. G. P. Pez, *J. Amer. Chem. Soc.* (1976) **98**, 8072.
33. Yu. G. Borodko, I. N. Ivleva, L. M. Kachipina, S. I. Salienko, A. K. Shilova and A. E. Shilov, *J. C. S. Chem. Commun.* (1972) 1178.
34. I. N. Ivleva, A. K. Shilova, S. I. Salienko and Yu. G. Borodko, *Dokl. Akad. Nauk. SSSR.* (1973) **213**, 116.
35. Yu. G. Borodko, I. M. Ivleva, L. M. Kachipina, E. F. Kvashina, A. K. Shilova and A. E. Shilov, *J. C. S. Chem. Commun.* (1973) 169.
36. G. P. Pez and S. C. Kwan, *J. Amer. Chem. Soc.* (1976) **98**, 8079.
37. J. E. Bercaw, *J. Amer. Chem. Soc.* (1974) **96**, 3017.
38. J. M. Manriquez, D. M. McAllister, E. Rosenberg, A. M. Shiller, K. L. Williamson, S. I. Chan and J. E. Bercaw, *J. Amer. Chem. Soc.* (1978) **100**, 3078.
39. J. M. Manriquez, R. D. Sanner, R. E. Marsh and J. E. Bercaw, *J. Amer. Chem. Soc.* (1976) **98**, 3042.
40. F. W. van der Weij and J. H. Teuben, *J. Organometallic Chem.* (1976) **120**, 223.
41. F. W. van der Weij, J. H. Teuben and H. Scholtens, *J. Organometallic Chem.* (1977) **127**, 299.
42. J. H. Teuben, *J. Organometallic Chem.* (1973) **57**, 159.
43. M. J. S. Gynane, J. Jeffery and M. F. Lappert, *J. C. S. Chem. Commun.* (1978) 34.
44. F. W. van der Weij and J. H. Teuben, *J. Organometallic Chem.* (1976) **105**, 203.
45. P. Sobota and B. Jezowska-Trzebiatowska, *J. Organometallic Chem.* (1977) **131**, 341.
46. Yu. G. Borodko, M. O. Broitman, L. M. Kachipina, A. E. Shilov and L. Yu. Ukhin, *J. C. S. Chem. Commun.* (1971) 1185.
47. G. Le Ny, A. E. Shilov, A. K. Shilova and B. Tchoubar, *Nouveau J. De Chimie* (1977) **1**, 397.
48. B. Tchoubar, A. E. Shilov and A. K. Shilova, *Kinet. Katal.* (1975) **16**, 1079.

10

Reactions of Ligands Analogous to Dinitrogen on Dinitrogen Binding Sites

ARMANDO J. L. POMBEIRO

Complexo Interdisciplinar, Instituto Superior Técnico, Lisboa, Portugal

Abbreviations

Bu^i = i-butyl; Bu^n = n-butyl; Bu^t = t-butyl; COD = cyclo-octa-1,5-diene; Cp = η^5-C_5H_5; Cy = cyclohexyl; depe = $Et_2PCH_2CH_2PEt_2$; diars = 1,2-$(Me_2As)_2C_6H_4$; dmpe = $Me_2PCH_2CH_2PMe_2$; dppe = $Ph_2PCH_2CH_2PPh_2$; Et = ethyl; HOMO = highest occupied MO; IR = infrared; M = metal atom; Me = methyl; MO = molecular orbital; NMR = nuclear magnetic resonance; Ph = phenyl; PR_3 = monotertiary phosphine; Pr^i = i-propyl; Pr^n = *n*-propyl; thf = tetrahydrofuran; Tol = tolyl; UV = ultraviolet; X = halide.

I. Introduction

LIGANDS which are isoelectronic with N_2 (such as isonitriles, nitriles, aryldiazonium, carbonyl, cyanide, acetylenes and NO^+) are known to ligate dinitrogen binding sites, as expected. Aside from the inherent interest in the chemistry of these ligands, the study of the derived species helps to gain an insight into the nature of the dinitrogen binding site and the properties of this ligand, and also provides suggestive pathways to induce its reactivity. Moreover, in building up a model of nitrogenase activity, the other substrates of this enzyme (isonitriles, nitriles, acetylene, cyanide, azide, nitrous oxide) must also be taken into account. Hence it is not surprising that in various groups most interested in dinitrogen activation, a parallel study of other isoelectronic or related species has been undertaken.

The expected strong electron-donating ability of the dinitrogen binding site (see Chapter 6) has been substantiated by a variety of experimental data such as the usually observed electrophilic character of the site (for example, easy protonation at the metal often occurs) which ligates dinitrogen and the enhancement upon coordination of the basic character of related ligands. Spectroscopic data have also often been rationalized in terms of a strong electron-donating capacity of the metal centre (for example carbon monoxide, isonitriles and nitriles exhibit usually very low $\nu(CO)$ and $\nu(CN)$ values when bonded to a dinitrogen binding centre; in a similar way, $\nu(N_2)$ of coordinated N_2 generally falls well below the value of the free ligand).

Chemical behaviour of coordinated aryldiazonium species has already been referred to and this section deals with the reactivity of other ligands analogous to N_2 when ligating dinitrogen binding sites. A parallelism of chemical behaviour may often be observed as expected; moreover, a few examples of reactions which the former undergo at dinitrogen binding sites, although not yet paralleled by N_2, are also given as suggestions of possible N_2 behaviour which may yet be discovered.

II. Isonitriles

Displacement of coordinated dinitrogen by isonitriles yields *trans*-$[M(CNR)_2(dppe)_2]$ (**1**) (M = Mo, W; R = Me, Bu^t, Ph, p-$CH_3C_6H_4$, p-$MeOC_6H_4$, p-ClC_6H_4, 2,6-$Cl_2C_6H_3$)[1] and series of monophosphine species $[M(CNR)_xL_{6-x}]$ (**2**) (M = Mo, W; R = Me, Bu^t, p-$CH_3C_6H_4$; L = PMe_2Ph and/or $PMePh_2$; x = 2–4),[2a] which have been prepared from *trans*-$[M(N_2)_2(dppe)_2]$ (**3**) and from the monophosphine complexes *cis*-$[M(N_2)_2(PMe_2Ph)_4]$ (**4**) and *trans*-$[Mo(N_2)_2(PMePh_2)_4]$ (**5**), respectively. In these isonitrile complexes a great decrease in the $\nu(CN)$ value of the ligand is observed upon coordination (e.g., $\Delta\nu/\nu = -0{\cdot}146$ and $-0{\cdot}178$ respectively for *trans*-$[W(CNMe)_2(dppe)_2]$ in thf solution and *trans*-$[Mo(CNMe)_2(PMe_2Ph)_4]$ in benzene solution) corresponding to the lowering which $\nu(N_2)$ undergoes in the parent complexes. Moreover, a single-crystal X-ray crystallographic structure determination has been carried out[3] on *trans*-$[Mo(CNMe)_2(dppe)_2]$ (R = 0·049) and the most striking feature is the C—N—Me angle, which is 156(1)°, much smaller than the usual value (≃180°) found in terminal isonitriles. The Mo—C (isonitrile) bond length is 2·091(5) Å thus substantiating a considerable double bond character. The expected large π-backbonding from the low oxidation state metal in the electron-rich metal centre $M(dppe)_2$ accounts for these facts and the corresponding electronic effect may be represented, in a VB picture, by the resonance form

$$M{=}C{=}\ddot{N}\diagdown_{R}$$

or by a simplified π-MO scheme of the type in Fig. 1(a).[2b] The coordinated isonitriles then exhibit a considerable carbene character.

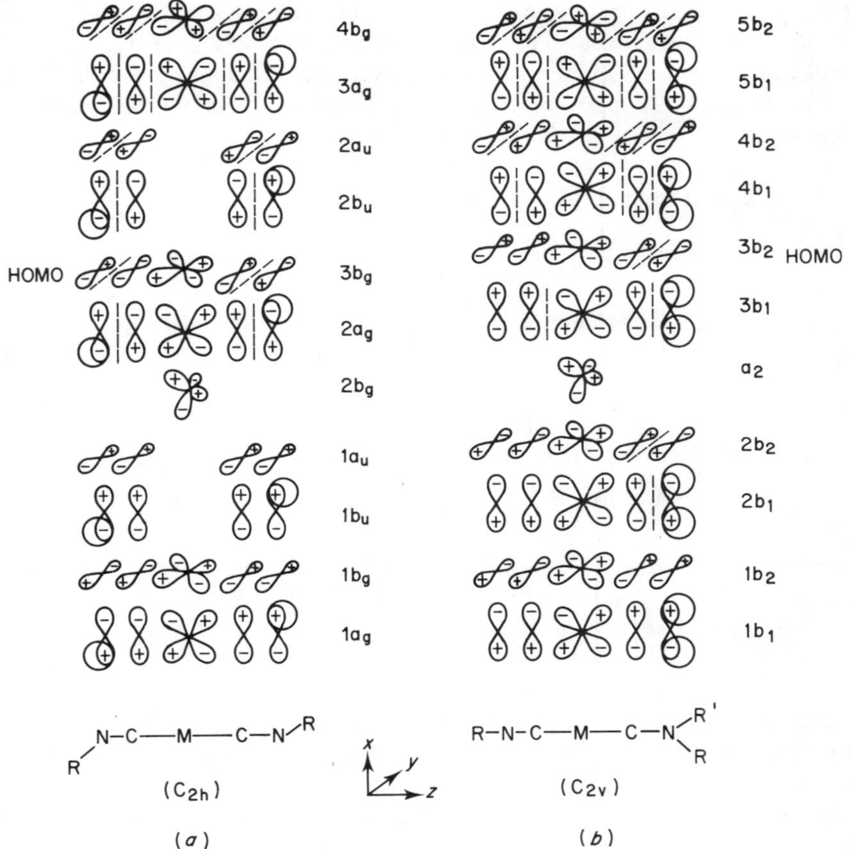

Fig. 1. Simplified π-MO schemes (*a, b*) for bisisonitrile *trans*-[M(CNR)$_2$(dppe)$_2$] (**1**) and monocarbyne complexes *trans*-[M(CNR′R)(CNR)(dppe)$_2$] (**6**), respectively.

Oxidations have been performed[2b] on the isonitrile species and ν(CN) increases upon metal oxidation (e.g., 2127 cm^{-1} for *trans*-[Mo(CNMe)$_2$-(dppe)$_2$](BF$_4$)$_2$, whereas for *trans*-[Mo(CNMe)$_2$(dppe)$_2$] the value of 1857 cm^{-1} is observed, in Nujol) which is accounted for by electron removal from a ligand-based antibonding orbital. However, the most striking feature of the electron-rich isonitrile species is their susceptibility to attack by protonating and alkylating agents yielding carbyne-type ligands $\overset{+}{M}\equiv C-NR'R$, with R′ = hydrogen or alkyl group: see also the π-MO scheme of Fig. 1(*b*).[4] Hence the *dinitrogen* binding sites M(dppe)$_2$ appear to behave also as *carbene*- and *carbyne*-binding metal centres. The carbyne complexes *trans*-[M(CNR′R)-

$(CNR)(dppe)_2]$ (**6**) are thermodynamically unstable in solution and proton transfer to the metal occurs (R′ = H) yielding hydride complexes (**7**); when R′ = alkyl, a *trans* to *cis* isomerization appears to occur. Ligating isonitriles in (**6**) (R′ = H) and (**7**) may also undergo protonation reactions affording the dicarbene† (**8**) and the hydride-carbyne (**9**) species.[4]

trans-$[M(CNR)_2(dppe)_2]$ (**1**)

$$\downarrow R'A$$

trans-$[M(CNR'R)(CNR)(dppe)_2]A$ (**6**) $\xrightarrow{HA}$ *trans*-$[M(CNHR)_2(dppe)_2]A_2$ (**8**)

$$\downarrow$$

$[MH(CNR)_2(dppe)_2]A$ (**7**) $\xrightarrow{HA}$ $[MH(CNHR)(CNR)(dppe)_2]A_2$ (**9**)

$A = BF_4, SO_3F$, etc.

Ligating CNMe in complexes (**1**) also undergoes attack by the Lewis acid $AlEt_3$ yielding a very unstable bis-adduct which has not been isolated.

These protonation and alkylation reactions have a close resemblance to the N—H and N—C bond formation by electrophilic attack which N_2 undergoes when ligating the same metal site (see Part 3).

The final species derived by reactions of coordinated N_2 with organic halides and acids have a halide (σ-donor) *trans* to the attacked dinitrogen. However in the species (**6**) the isonitrile which is in the *trans* position relative to the carbyne has not been displaced by a halide but has adapted itself to the strong electron withdrawing ability of the carbyne by showing a decrease in its π-acceptance as evidenced by an increase of $\nu(CN)$ and rationalized by the π-MO scheme of Fig. 1(*b*).[4e]

The aforementioned reactions constitute the first reported example of electrophilic attack at terminal coordinated isonitriles which in most of the known reactions undergo nucleophilic attack at ligating C by amines, alcohols, thiols, hydrazine, borohydride, alkyl lithium, etc.[6] When the latter reactions occur they exhibit $\nu(CN)$ values which are usually greater than those of the free ligand and their reactivity is enhanced by the factors which cause a decrease of the electronic density at isonitrile C (reduction of π-backbonding from the metal and presence of electron-withdrawing groups at the isonitrile);

† These complexes may be better called dicarbene or carbene–carbyne complexes (according to $RH\overset{+}{N}{=}C{=}M{=}C{=}\overset{+}{N}HR$ and $RH\overset{+}{N}{=}C{=}\overset{+}{M}{\equiv}C{-}\ddot{N}HR$, respectively, where the dppe ligands have been omitted for the sake of clarity) rather than dicarbyne species.

the isonitriles are then behaving mainly as σ-donors. One important example of such a nucleophilic attack is at coordinated CNMe in

$$[(\eta^5\text{-}C_5H_4Me)Mn(NO)(CO)_{2-x}(CNMe)_x]^+PF_6^- \qquad (x = 1, 2) \qquad \textbf{(10)}$$

by nucleophiles such as C_6F_5Li leading to $[(\eta^5\text{-}C_5H_4Me)Mn(NO)(CO)_{1-y}\text{-}(CNMe)_y\{C(C_6F_5){=}NME\}]$($y$ = 0, 1) whose N-alkylimino group is susceptible to protonation at N yielding the carbene ligand $C(C_6F_5)NHMe$.[6b,7] These reactions parallel the only hitherto reported nucleophilic attack at coordinated N_2, followed by protonation at terminal nitrogen. Dinitrogen is then ligating a related metal site in $[(\eta^5\text{-}C_5H_5)Mn(CO)_2N_2]$ (**11**). $Li[(\eta^5\text{-}C_5H_5)Mn(CO)_2\text{-}\{N(C_6H_5){=}N\}]$ is the species derived upon reaction with PhLi and it undergoes protonation yielding the phenyldiazene complex $[(\eta^5\text{-}C_5H_5)Mn(CO)_2\text{-}\{N(C_6H_5){=}NH\}]$.[8]

Reduction of coordinated CNMe mainly to methylamine (which involves C—N bond cleavage) has been achieved by simple irradiation of methanolic solutions of monophosphine isonitrile complexes (**2**) (for $x = 2, 3$); ammonia and hydrocarbons (mainly methane with smaller amounts of ethylene and ethane) are also formed in much lower yields and a total of about three electrons per metal are used in the reduction.[5] Greater yields (about double) of amine and ammonia can be obtained without irradiation by using an excess of strong acid allowing use of nearly all the six available metal electrons.[5] These reactions are analogous to the formation of ammonia from N_2 ligating similar monophosphine metal sites.[9]

The enzyme nitrogenase which is capable of reducing N_2 to NH_3 can also reduce CNMe to $MeNH_2$ and CH_4 (both processes involving the transfer of six electrons) with small amounts of saturated and olefinic C-2 (ethane and ethylene) and C-3 hydrocarbons.

Reduction of isonitriles to these hydrocarbons has also been reported in aqueous medium for the oxomolybdate(IV) species $[Mo(O)Cl(CNMe)_4]PF_6$ (up to 30% of the total isonitrile is reduced by BH_4^-)[10a] and for some complex dinitrogen fixation systems (**12**) composed of molybdate and a thiol [usually L(+)-cysteine] with $NaBH_4$ or $Na_2S_2O_4$ as reducing agents.[10b] The isolation and characterization of any isonitrile intermediate species has not yet been achieved in the latter system; however, the presence of active mononuclear Mo(IV) species has been postulated. Reduction of the oxomolybdate(IV) species in the presence of acetylene yields ethylene catalytically. Formation of propylene also occurs and has been interpreted as the result of an insertion involving coordinated CNMe and C_2H_2.[10a] The formation of the C—C bond in the production of C-2 and C-3 hydrocarbons is also believed to occur by isonitrile insertion into a Mo—C bond. The isonitrile reduction in these systems is competitively inhibited by N_2 (which is reduced to NH_3, diazene being suggested as an intermediate) and appears to be stimulated by ATP and other phosphorylating agents, as in the enzymatic reduction, by a mechanism which has not yet been ascertained (see also Section V). Iron(II)–cysteine systems do not show similar reductive activity, but iron salts act as cocatalysts in the molybdothiol systems[10b] (see Chapter 4).

III. Carbonyl

No example of protonation at oxygen of CO binding a metal centre has yet been reported, even for carbonyl species which are analogous to dinitrogen complexes, although oxygen protonation of carbonyl groups in organic species is known to occur (e.g., aminocarboxylate ligands are protonated by super acid, $HFSO_3$–SbF_5–SO_2, at low temperature).[11] Hence, from attempted protonation at carbonyl oxygen of $[Mo(CO)_2(dppe)_2]$ (obtained upon N_2 replacement by CO) only oxidation[4d,12a] or protonation[12b,12c] at the metal have been detected. The same difficulty for CO protonation appears to be encountered by nitrogenase in that CO, which is isoelectronic with N_2, is not a substrate of the enzyme, although it causes a strong inhibition of nitrogen binding probably due to its coordination and blocking of the binding site. Ligating CO may be more difficult to protonate terminally than are the ligating CNR or N_2 because of the greater electronegativity of O as opposed to N. The greater basicity of N compared to O has also been proved in other well documented examples, such as preferential protonic attack at N of carbamoyl ligands C(NHR)=O;[13a] ready protonation of imino groups C(R′)=NR,[7] contrary to the acyl ligands C(R)=O and the usual greater stability of aminocarbenes compared to hydroxycarbenes.[13b] However, Lewis acid $[AlR_3(R = Me, Et, Ph \text{ or } Cl)]$ adduct formation has been reported for carbonyl complexes which are analogous to dinitrogen species, such as *cis*-$[Mo(CO)_2(PMe_2Ph)_4]$,[14a] *trans*- and *cis*-$[Mo(CO)_2(dppe)_2]$,[14] *cis*-$[W(CO)_2(dppe)_2]$[14b] and *trans*-$[ReCl(CO)(LL)_2][LL = dppe \text{ or } (PMe_2Ph)_2]$.[14b]

Due to limitations on petroleum resources, current interest has arisen in the Fischer–Tropsch synthesis[15] of hydrocarbons and alcohols from CO and H_2 which had its greatest development during the Second World War. Once analogies between that synthesis and the Haber process have been realized[16] it is not unexpected that dinitrogen coordination sites, or related ones, are among the effective metal centres which activate CO to hydrogenation. Hence, catalytic reduction of CO at room temperature and moderate pressures (1–4 atm.) to give, on hydrolysis, linear aliphatic alcohols has been reported[17] for the system $[Cp_2ZrCl_2]$ + Bu^i_2AlH, and species such as M–CH(R)O–(Al)[17] or M←C(H)O–(Al)[18a] (where (Al) denotes the Lewis acid or a derivative) are suggested as possible intermediates.

Carbonyl reduction to alcohol, probably via an aldehyde intermediate moiety, ReCHO, has been reported[18b] by reaction of $[CpRe(CO)_2(NO)]^+$ **(13)**

(which is related to complex (**11**) in which N_2 undergoes a nucleophilic attack) with $NaBH_4$ yielding [CpRe(CH_2OH)(CO)(NO)]. However, surprisingly, under the same conditions the Mn analogue (**14**) of (**13**) undergoes polymerization yielding [$Cp_2Mn_2(CO)_2(NO)_2$]. In both (**13**) and (**14**) ligating carbonyl also undergoes a nucleophilic attack by MeOH in the presence of NaCN, and the carbomethoxy species [CpM(COOMe)(CO)(NO)] (M = Mn, Re) are formed.[18b]

Reduction by H_2 of CO ligating a zirconocene type centre has been reported for [{(η^5-C_5Me_5)$_2$Zr(CO)}$_2$$N_2$] and [($\eta^5$-$C_5Me_5$)$_2$Zr(CO)$_2$] yielding the methoxide species [(η^5-C_5Me_5)$_2$ZrH(OMe)], via a mechanism which is suggested to involve CO insertion into the Zr–H bond in [(η^5-C_5Me_5)$_2$Zr(H)$_2$(CO)].[18c]

Carbon monoxide has been shown[18a] to react with hydride and alkyl derivatives of the types [(η^5-C_5Me_5)$_2$ZrH_2] (**15**) (which is derived from the reaction of [{(η^5-C_5Me_5)$_2$ZrN_2}$_2$$N_2$] with H_2), [(η^5-C_5Me_5)$_2$ZrR_2] [R_2 = Me_2 (**16**), –$CH_2(CH_2)_2CH_2$– (**17**)] or [(η^5-C_5Me_5)$_2$ZrH(CH_2CHMe_2)] (**18**); initial formation of a carbonyl adduct is followed by a facile insertion of CO into the Zr–C or Zr–H bond. A side-on bonded acyl or formyl group is postulated and the resulting carbenoid character of the carbonyl carbon

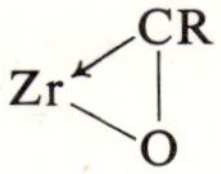

$$\mathrm{Zr}\overset{\swarrow\,\mathrm{CR}}{\underset{\searrow\,\mathrm{O}}{}}\quad(\text{CR–O bonded, both to Zr})$$

accounts for the observed behaviour, the final products having carbonyl oxygen bonded derivatives:

[{(η^5-C_5Me_5)$_2$ZrH}$_2$(OOCH=CHO)] and/or [(η^5-C_5Me_5)$_2$ZrH(OMe)] from (**15**),

[(η^5-C_5Me_5)$_2$ZrMe(MeCO)] and [(η^5-C_5Me_5)$_2$Zr(OMeC=CMeO)] from (**16**),

[(η^5-C_5Me_5)$_2$ZrH(OC=CH(CH_2)$_2$$CH_2$)] from (**17**) and

[(η^5-C_5Me_5)$_2$ZrH(OCH=CHCHMe_2)] from (**18**).[18a]

The above carbonyl insertion reactions (other examples involving olefins will be given later) may be suggestive of possible pathways to N–C bond formation, via N_2 insertion into M–C bonds, which are different from those known[19] at present for other dinitrogen sites. Although N_2 insertion into any isolated species with a transition metal–carbon bond has not yet been reported, this mechanism has been suggested for amine formation from N_2 in systems such as [Cp_2TiCl_2]–RLi (R = Ph; *o*-, *m*-, *p*-$CH_3C_6H_4$) which upon hydrolysis yield NH_3 and RNH_2, the latter formed via N_2 insertion into a Ti–R bond or through formation of a benzyne intermediate[20] (see also Chapter 3).

Hydrogenation of CO to methane (methanation) has also been reported in nitrogen fixation sites. Hence, from $[Cp_2Ti(CO)_2]$ + CO/H_2 at 150°C in homogeneous solution, methane is formed. The system is not catalytic but forms an octahedral cluster $Cp_6Ti_6O_8$ (probably derived from a reaction with the H_2O formed in the methanation reaction) which does not show catalytic activity.[21] Reduction of CO to methane has also been reported for the system $TiCl_3 \cdot 3thf$–Mg–thf; acetylene is formed upon hydrolysis of this reaction mixture, thus substantiating the formation of a metal carbide MC_2 (M = Mg, or Mg and Ti) or carbenoid species.[22] Also, the molybdate–thiol system (**12**) with $NaBH_4$ as the reducing agent, already referred to, reduces CO to methane.[23]

Carbon–nitrogen bond formation has been reported from the reaction of $[Cp_2TiNH]_2H$ (which is known to fix N_2 and to catalyse olefin hydrogenations) with CO and CO_2 leading to an isocyanate and carbamate species, respectively, which however cannot be liberated intact from the metal site, thus preventing catalytic behaviour.[24]

IV. Carbon Dioxide

A. Coordination at N_2 Binding Sites

The importance of the activation of CO_2 by transition metal elements, for example in catalytic CO_2 fixation into organic species and in the understanding of biological photosynthesis, has been recognized and the subject reviewed.[25] Carbon dioxide is neither isoelectronic with N_2 nor a substrate of nitrogenase. However, activation by nitrogen fixation sites, or related ones, has been reported and its importance justifies a reference in this chapter. A few examples will be given.

Although a few claims for *coordinated* CO_2 have been made, only a few examples have been unambiguously proved, namely involving electron-rich nitrogen fixation sites.

Hence, the complexes $[Ni(CO_2)(PR_3)_2]$ (R = Et, Bu^n, Cy)[26a] (which have been authenticated (R = Cy) by an X-ray structure determination, the CO_2 ligand co-ordinating through carbon and one of the oxygen atoms[26b]) have been prepared from reaction of $[Ni(PR_3)_4]$ with CO_2 via $[Ni(CO_2)(PR_3)_3]$ which is also four-coordinated (CO_2 bonded through the C atom) and analogous to $[Ni(N_2)(PR'_3)_3]$ (R'_3 = Et_3, Bu^n_3, Et_2Ph).[26c] Reactions of CO_2 with $[\{(PCy_3)_2Ni\}_2N_2]$, $[Ni(PCy_3)_3]$ or reduction of $[NiBr_2(PCy_3)_2]$ by sodium sand under CO_2 (preparative route analogous to the formation of $[Ni(N_2)(PR'_3)_3]$)[26c] yield the same C,O bonded carbon dioxide complex.[26b] A CO_2 dinickel species has also been reported from CO_2 addition to $[(PCy_3)_2Ni]$ (which is formed by displacement of N_2 from $[\{(PCy_3)_2Ni\}_2N_2]$).[26d]

$[IrCl(C_2O_4)(PMe_3)_3]$ containing the cyclic moiety $\overline{M{-}C(O)OC(O)O}$ has also been authenticated by X-rays and its formation from reaction of CO_2 with $[IrCl(C_8H_{14})(PMe_3)_3]$ is postulated to occur via a CO_2 electrophilic attack (through the C atom) to an oxygen atom of the ligating CO_2 in the proposed intermediate $[IrCl(CO_2)(PMe_3)_3]$.[26e] The cyclic moiety MC_2O_4 (which may undergo C—O or O—C bond cleavage within the grouping C—O—C) is suggested[26e] as a possible intermediate in the known conversion of ligating CO_2 into carbonyl and carbonate, e.g. in the reaction of CO_2 with *cis*-$[Mo(N_2)_2(PMe_2Ph)_4]$ yielding a μ-carbonate–carbonyl species[26f] (formation of the unstable intermediate $[Mo(CO_2)_2(PMe_2Ph)_4]$ followed by spontaneous conversion into the final species has been proposed by the original authors).[26f]

Other Ir(I)—CO_2 complexes have been reported, namely $[Ir(CO_2)(LL)_2]Cl$ (where the CO_2 ligand is suggested to bind through the C atom) which are prepared from CO_2 uptake by $[Ir(LL)_2]Cl$ (LL = dmpe, diars, depe). Heating the dmpe complex leads to carboxylation and metalation of the phosphine ligand yielding $[\overline{Ir(H)\{Me(CH_2CO_2})PC_2H_4PMe_2\}(dmpe)]Cl$.[26g]

Rhodium (I)—CO_2 complexes similar to the Rh(I)—N_2 species $[RhCl(N_2)(PR_3)_2]$[27a–27c] have also been reported[27d] with formulations $[RhCl(CO_2)L_2]$ (with C,O bonded CO_2) and $[RhCl(CO_2)L_3]$ (with C bonded CO_2) where L denotes a phosphorus ligand, and they have been prepared from reaction of CO_2 with $[RhClL_3]$ or $[RhCl(C_2H_4)_2]_2$ + L. Phosphine oxidation by CO_2 may occur and the phosphine oxide complex $[RhCl(CO)(OL)L]$ is formed from $[RhCl(CO_2)L_2]$, upon standing.[27d]

Carbon dioxide complexes may also be prepared by indirect routes such as oxidation of ligating CO by O_2 as in the preparation of $[Rh_2(CO)_2(CO_2)(PPh_3)_3]$ (plus $OPPh_3$) from reaction of $[Rh_2(CO)_4(PPh_3)_4]$ with molecular oxygen.[27e]

B. Insertion

Insertion reactions of CO_2 into a transition metal —H, —C, —N or —O bond are known to occur, they are believed to be key steps in the incorporation of CO_2 into organic species, and examples involving dinitrogen binding sites, or related ones, may be cited.

Carbon dioxide insertion into a metal–hydrogen (M—H) bond generally yields formate species M—OCHO, although a metal carboxylic acid moiety M—COOH (from which a carboxylic acid is derived upon alkylation) is formed in a few cases.

Carbon dioxide undergoes insertion into M—H bonds of $[CoHL(PPh_3)_3]$[28a–28c] and $[FeH_2L(PEtPh_2)_3]$[28d] (L = N_2, H_2) with N_2 (or H_2) liberation; formates HCOOR (R = Me, Et) are yielded upon alkylation of the derived species by RI.

A bidentate formate complex $[RuH(O_2CH)(PPh_3)_3]$ (whose structure has been authenticated by X-rays) has been reported from reversible insertion reactions of CO_2 into the Ru–H bond of $[RuH_2L(PPh_3)_3]$ (L = N_2, H_2) or $[RuH_2(PPh_3)_4]$.[28e,28f]

Formate is formed by alternate introduction of CO_2 and H_2 into the nitrogen fixation system $TiCl_4 \cdot 2thf$–Mg–thf (15 mol CO_2 fixed per mol Ti). The postulated mechanism involves formation of the intermediate $[\{(thf)Mg_2Cl_2Ti\}_2]$ (**19**) which adds H_2 leading to $[\{(thf)Mg_2H_2Cl_2Ti\}_2]$ which reacts with CO_2/Mg regenerating (**19**) and forming MgH_2 and $Mg(O_2CH)_2$.[29a,29b]

Formate may be further reduced as in the reaction of $[Cp_2ZrHCl]$ with CO_2 yielding CH_2O + $[(Cp_2ZrCl)_2O]$ or the methoxide $[Cp_2Zr(OMe)Cl]$ + $[(Cp_2ZrCl)_2O]$ which is believed to proceed via an intermediate formate species which is further reduced by an excess of $[Cp_2ZrHCl]$ (carbon monoxide undergoes a related reaction with this species yielding $[(Cp_2ZrCl)_2CH_2O]$.[29c]

Insertion of CO_2 into M—H is involved in the catalytic reduction of CO_2 by H_2 to formates or formamides according to the overall reactions: $CO_2 + H_2 + ROH \rightarrow HCOOR + H_2O$ and $CO_2 + H_2 + R_2NH \rightarrow HCONR_2 + H_2O$. Low-valent or hydride complexes of group VIII transition metals have been used as catalysts, namely, for the former reactions, neutral $[RuCl_2(PPh_3)_4]$ (in the presence of a Lewis acid, BF_3, as a cocatalyst),[29d] $[RuH_2(PPh_3)_4]$ + R'_3N (base cocatalyst),[29e] $[RhCl(PPh_3)_3]$ + BF_3,[29d] $[RhCl(PPh_3)_3]$ + R'_3N,[29e] $[IrH_3(PPh_3)_3]$ + R'_3N,[29e] or the anionic iron carbonyl hydrides $[HFe_3(CO)_{11}]^-$ and $[HFe(CO)_4]^-$,[29f] and, for formamide production, species such as $[RhCl(PPh_3)_3]$ and $[IrCl(CO)(PPh_3)_2]$.[29g]

Insertion of CO_2 into a *metal–carbon* bond of a phenylene titanium species is believed to occur in the reaction of CO_2 with $[Cp_2TiPh_2]$ where carboxylation of the phenyl ring and metallocycle formation occur yielding species (**20a**) whose structure has been authenticated by X-rays.[30a] A related

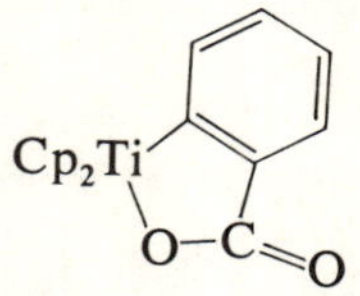

(**20a**) 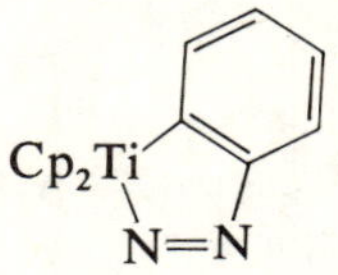 (**20b**)

species (**20b**) (derived from N_2 insertion into a Ti—C bond) is suggested as an intermediate in aniline formation from reaction of Cp_2TiPh_2 with N_2 upon hydrolysis.[25]

A carboxylate complex $[Rh(O_2CPh)(PPh_3)_3]$ whose structure has also been determined by X-rays has been prepared from CO_2 insertion into the Rh(I)—C bond of $[RhPh(PPh_3)_3]$.[30b] CO_2 insertion into the metal–carbon bond forming an ester species M—C(O)OR is also known to occur, although less commonly.[25]

As in the CO_2 insertion reactions into a M–H bond, the ligand moiety with incorporated CO_2 (via insertion into a M–C bond) may be released from the metal and benzoic acid or methylbenzoate for example are formed upon attack on (**20a**) by acid or BF_3 in methanol, respectively.[30a] Moreover, catalytic formation of lactones

e.g. [lactone structure]

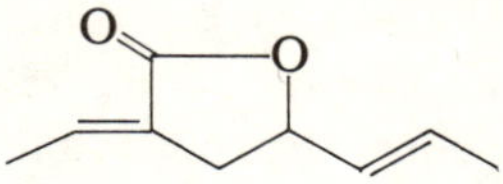

has been reported by CO_2 incorporation into polymerized butadiene, using Pd(0) tertiary phosphine complexes as catalysts; e.g. $[Pd(dppe)_2]$ in the formation of the above lactone. The reaction is believed to proceed via CO_2 insertion into a Pd–C bond (C of a ligating butadiene derived organic species).[30c,30d]

Coordinatively unsaturated Ni(0) complexes of the type $[NiL_2]$ (L = PPh_3, PCy_3, COD) catalyse the formation of alkylene carbonates from alkylene oxides and CO_2. Insertion of CO_2 into the metal–oxygen bond of $L_2NiCH_2CH_2O$ (postulated) is believed to be involved in the process. ($L_2NiCH_2CH_2OC(O)O$ is another postulated intermediate although

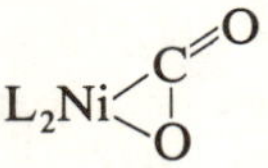

$L_2Ni(\eta^2\text{-}CO_2)$ [structure]

may also be considered).[30e]

The bidentate monoalkylcarbonate complexes $[Ru(O_2COR)L_4]PF_6$ (L = PMe_2Ph) have been reported from reaction of $[RuHL_5]PF_6$ with CO_2 and ROH (R = Me, Et) via CO_2 insertion into the Ru–O bond of a postulated alkoxide intermediate.[30f]

Ligating bicarbonate may be prepared by CO_2 insertion into metal hydroxo species: $[M(O_2COH)(CO)(PPh_3)_2]$ (M = Rh, Ir) from $[M(OH)(CO)(PPh_3)_2]$ for example.[30g] Other Co(III)[30h] and Pd(II)[30i] species also undergo similar reactions, which is of biological significance since the carbon dioxide hydration catalysed by the enzyme carbonic anhydrase may involve the formation of bicarbonate by a similar route.

CO_2 insertion into a metal–nitrogen bond has also been reported. Examples are the magnesium–nitrogen bond of the species $[MMg_2Cl_2N(thf)]$ (**21**) (isolated)[31a] and $[Cp_2TiNMgCl]$ (**22**) (postulated) formed in the systems MCl_4–Mg–thf–N_2 (M = Ti, V) and $[Cp_2TiCl_2]$–Mg–thf–N_2, respectively.[31b] The structural moiety M=$\ddot{N}$–MgCl is postulated in (**21**) and (**22**) and the fixation of CO_2 is believed to occur by donation of the free electron pair on the $2\pi_u$ antibonding CO_2 orbital. Insertion of CO_2 into the N–Mg bond followed by reduction leads to isocyanate derivatives ($[MMg_2Cl_2(NCO)(O)(thf)_3]$

isolated in the former systems) which yield $NH_3 + CO_2$ upon hydrolysis and methylisocyanate by reaction with methyl iodide.[31b]

Carbamate species may be derived from CO_2 reactions, with for example $[Cp_2TiNH]_2H$, as already mentioned[24] and with $[RuHL_5]PF_6$ ($L = PMe_2Ph$) + $HNMe_2$.[31c] $[Ru(O_2CNMe_2)L_4]PF_6$ (authenticated by X-rays) is formed in the latter case through reaction pathways which are suggested to involve CO_2 insertion into Ru–H or Ru–N of a ruthenium amide $Ru\text{–}NMe_2$ intermediate.[31c] CO_2 insertions into other amide complexes are known to occur, e.g. $[M(NMe_2)_n]$ (M = group IV, V or VI transition metal).[31d–31g]

Coordinated species derived from CO_2 insertions into M–N bonds may undergo further reactions yielding other organic species such as urethanes ($RR'NCO_2R''$) (via coordinated isocyanate + aklyl halide + alcohol[31h] or ligating carbamate + alkyl halide[31i]) as occurs in some Cu(I) systems which, however, do not bind dinitrogen.

V. Olefins and Acetylenes

Olefins and acetylenes ligating N_2 binding sites may be readily reduced. Hence, C_2H_4 ligating the dinitrogen binding centre $Mo(dppe)_2$ in *trans*-$[Mo(C_2H_4)_2(dppe)_2]$, obtained by displacement of N_2 by C_2H_4 or by a route analogous to the preparation of *trans*-$[Mo(N_2)_2(dppe)_2]$, may be protonated by CF_3COOH. Reversible proton exchange between one ethylene ligand and the metal occurs at 7°C: $MH(C_2H_4) \rightleftharpoons M(C_2H_5)$ where $M = Mo(C_2H_4)(dppe)_2$. It is slowed down by cooling and at −85°C the hydridic ^1H NMR resonance is observed. Similarly in $[HMo(\eta^3\text{-}C_3H_5)(dppe)_2]$ (prepared from $[MoCl_3thf_3]$ + Na/Hg + dppe + propylene) the hydridic ^1H NMR resonance is observed at +5°C, but reversible proton exchange between the metal and a terminal allyl carbon is detected on raising the temperature up to 101°C: $MH(\eta^3\text{-}C_3H_5) \rightleftharpoons M(\eta^2\text{-}CH_2{=}CHCH_3)$, where $M = Mo(dppe)_2$. Both the processes of olefin insertion–deinsertion into a metal–hydrogen bond and the exchange of the metal hydride with the terminal hydrogens of a π-allyl ligand are very important in olefin reaction catalysis.[32]

Also Rh–ethyl complexes are known[33] to be formed by HCl addition to Rh(I) ethylene or fluoroethylene species. Thus, the π-olefin complexes $[RhCl(C_2F_3X)(PPh_3)_2]$ (X = F, Cl, H), which are analogous to $[RhCl(N_2)(PPr^i_3)_2]$, readily react with HCl yielding the complex $[RhCl_2(C_2F_3HX)(PPh_3)_2]$. These reactions probably occur by initial metal protonation followed by addition of M–H across the coordinated olefin bond.[33]

Ethylene is formed by protonation of coordinated acetylene in $[Mo(CO)(C_2H_2)(S_2PPr^i_2)_2]$ which shows a very low field acetylene proton resonance of 12·33 ppm downfield from TMS, in $CDCl_3$,[34a] whereas $[\{Cr(dppe)_2\}_2C_2H_2]$ yields ethane.[34b] The latter is a paramagnetic dimeric species with a postulated edge-on bridging acetylene. It is analogous to $[\{Cr(dppe)_2\}_2N_2]$, although the bridging N_2 is suggested to be in a *cis* doubly bent structure

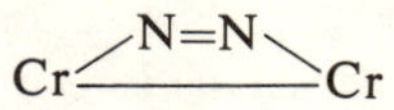

$$Cr\overset{N=N}{\longleftrightarrow}Cr$$

both being prepared by reaction of $CrCl_2$ wtih Mg and dppe in thf, the former under C_2H_2 and the latter under N_2; in this species the dinitrogen ligand is also reduced upon protonation to NH_3 (7%) and N_2H_4 (1%) (yields based on the fixed dinitrogen).[34b]

It has also been reported that the N_2 fixation system $[Cp_2MCl_2]$ + Na/Hg (M = Mo, W) (which yields dinitrogen species whose full characterization has been prevented by their thermal instability) can also fix olefins and acetylenes via a transient, coordinatively unsaturated and electron-deficient intermediate, Cp_2M, leading to

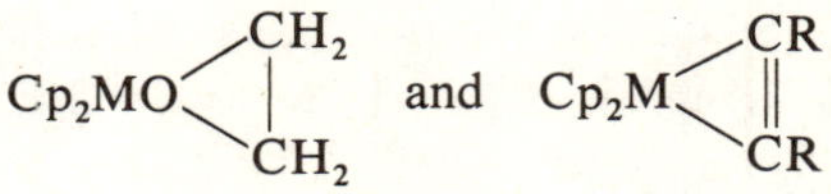

$$Cp_2Mo\Big<\begin{matrix}CH_2\\ |\\ CH_2\end{matrix} \quad \text{and} \quad Cp_2M\Big<\begin{matrix}CR\\ \|\\ CR\end{matrix}$$

($R = Ph$ or CF_3, M = Mo; R = Me, M = Mo or W).

The coordinated ethylene in $[Cp_2Mo(C_2H_4)]$ is reduced to ethane by reaction with HCl, $[Cp_2MoCl_2]$ also being formed. The ligating acetylenes in $[Cp_2M(RC{\equiv}CR)]$ display very low $\nu(CC)$ in their IR spectra (e.g., $\nu(C{\equiv}C) = 1750\ cm^{-1}$, $\Delta\nu = -563\ cm^{-1}$ for M = W, R = Me) and are reduced to *cis*-alkenes by reaction with acids (e.g., for M = Mo and R = Me, *cis*-2-butene and $[Cp_2MoCl_2]$ are formed from reaction with HCl).[35]

Photoreduction has been reported[36] of ethylene to a mixture of CH_4, C_2H_6, C_3H_8 and n-C_4H_{10} and of acetylene in low yield (<1%) to CH_4, C_2H_4, C_2H_6 and C-3 hydrocarbons upon UV irradiation on incompletely outgassed TiO_2 powder. Dinitrogen is also reduced in low yield to NH_3 and N_2H_4 by the same system.[36a] The hydrogenating agent is chemisorbed water and its presence is essential, hence the incomplete outgassing required.[36b] Water photolysis on TiO_2 yielding H_2 and O_2 occurs under argon; both C_2H_2 and N_2 inhibit H_2 evolution and they are believed to be reduced at the same site.[36a]

Catalytic activity of dinitrogen complexes in the hydrogenation, isomerization, polymerization and cyclization of olefins has already been referred to (Chapter 6) and other examples involving related complexes are also known. An important one which may be cited is the catalytic homogeneous hydroformylation of alkenes by CO/H_2 which has been reported by Wilkinson and co-workers[37a] using *trans*-$[RhX(CO)(PPh_3)_2]$ (X = halogen) (**23**), which is analogous to $[RhCl(N_2)(PPr^i_3)_2]$. The reaction has a lower rate than when

$[RhH(CO)(PPh_3)_3]$ (**24**)[37a,37b] is used and the principal active species is believed to be $[RhH(CO)_2(PPh_3)_2]$ (**25**). Insertion of a terminal olefin $RCH{=}CH_2$ into the Rh–H bond of (**25**) yields an alkyl derivative $[Rh(CH_2CH_2R)(CO)_2(PPh_3)_2]$ (**26**) which, upon CO insertion into the Rh–C bond, forms $[Rh(OCCH_2CH_2R)(CO)(PPh_3)_2]$. Oxidative addition of H_2 to this species yields a dihydride complex which rapidly degrades to aldehyde RCH_2CH_2CHO and $[RhH(CO)(PPh_3)_2]$ from which the catalyst (**25**) is regenerated by CO addition. Other pathways may be envisaged such as a dissociative one which involves a further phosphine loss from (**25**) followed by alkene insertion yielding $[Rh(CH_2CH_2R)(CO)_2(PPh_3)]$ which, upon phosphine addition, yields (**26**).[37a,37b] Catalytic homogeneous hydrogenation[37c] and isomerization[37d] of alkenes by (**24**) has also been reported and the principal active species is believed to be (**25**) or $[RhH(CO)(PPh_3)_2]$.

Vol'pin type systems for the reduction of dinitrogen via nitriding[20c] also provide catalysts for olefin reactions. Thus $[Cp_2TiCl_2]$ (or $\{Cp_2TiCl\}_2$) with sodium (or lithium naphthalenide) in thf catalyses the hydrogenation of olefins. Titanocene is believed to be the catalyst and the hydride species $[Cp_2TiH_2(olefin)]$ is postulated as an intermediate.[38] $TiCl_3$–Mg and VCl_3–Mg in thf catalyse the hydrogenation and dimerization of olefins, and hydridic species are suggested as the catalysts.[30b]

Catalytic reduction of acetylenes (and, in some cases, olefins) by nitrogen fixation systems has also been studied in aqueous media, mainly by two groups headed by A. E. Shilov (USSR) and G. N. Schrauzer (USA) and various nitrogenase features, such as reduction of various substrates, inhibition effects, stimulation by ATP, hydrogen evolution, have been reproduced by some of the reported systems, although the turnover numbers for the reductions are much lower than for the enzyme. For example a rate of $1{\cdot}7$ mol min^{-1} per mol Mo for C_2H_2 reduction by the molybdothiol system (**12**), whereas a value of 200 is observed for the enzyme.[39a]

Acetylene is catalytically reduced to ethylene by the dinitogen-fixing molybdothiol system (**12**), and related ones, with $NaBH_4$ or $Na_2S_2O_4$ as reducing agents.[39b,39c] The stereochemical course of the reduction is predominantly *cis*, leading in D_2O to *cis*-$C_2H_2D_2$ which is also formed stereospecifically in D_2O in the presence of nitrogenase. *Cis*-2-butene is formed exclusively from 2-butyne. A mononuclear side-on bonded acetylene complex is believed to be an intermediate which upon hydrolysis yields ethylene with concomitant oxidation of the metal. Under certain conditions reduction of acetylene to ethane (which is not formed by the enzyme) is also observed but this can be suppressed by modifying the experimental conditions. The catalytically active species is then suggested to involve a side-on bonded C_2H_2 bridging two Mo atoms. The C_2H_2 reduction appears to be stimulated by biological phosphorylating agents such as ATP as in the enzymatic reduction (to ethylene only); however the mechanism has not yet been elucidated. Assisted removal of OH in Mo—OH species (with enhancement of the rate of conversion to the active reduced form) has been suggested,[39d] but the liberation of H^+ from the hydrolysis of ATP is believed to play a key role.[40] Higher alkynes (propyne, 2-butyne, propargyl alcohol) are also reduced by the system. Simple iron salts enhance the catalytic activity of the system with $NaBH_4$.[39]

Ferredoxin models of the type $[Fe_4S_4(SR)_4]^{z-}$ ($z = 2$–4) have been shown to be efficient electron transfer catalysts from reductants such as $S_2O_4^{2-}$ and BH_4^- to molybdothiol catalysts.[39a] The reduced ferredoxin-like cluster species can also be employed as stoichiometric reducing agents for the molybdothiol-catalysed reduction of C_2H_2 to C_2H_4. In the absence of reducible substrates, H_2 is evolved by related systems containing Fe^{2+}, S^{2-} and RS^- in different molar ratios (1 : 2 : 6 for the most active).[41]

The active Mo reduced species is postulated[39] to involve Mo(IV) although electrochemical data[42] suggest the involvement of an Mo(III) species which is also favoured by other authors (see below).

Other aqueous dinitrogen-reducing molybdenum systems have been studied which are formed by a Mo(V or VI) compound (e.g., $MoOCl_3$, MoO_4^{2-}) plus reductant (Ti^{3+}, V^{2+} or Cr^{2+} salt) plus Mg^{2+} in aqueous/alcoholic alkaline medium.[43a] A yellow carbonyl cluster complex of Mo(III) is formed in the presence of CO (which is an inhibitor of the N_2 reduction) and it generates the catalyst upon CO displacement. Hence polynuclear Mo(III) (d^3) species are suggested to play a key role in the reduction.[43b] Acetylene is reduced to ethylene by these systems and also in the absence of any Mo compound, although only V^{2+} species have been able to reduce N_2 without molybdenum. Other aqueous reduction systems not involving molybdenum species may be cited, generally based on V(II) (d^3).

Alkaline suspensions of $V(OH)_2$, coprecipitated with $Mg(OH)_2$, reduce C_2H_2 to C_2H_4 (and also to C_2H_6, for low partial pressures of C_2H_2, due probably to a secondary reaction of C_2H_4 with V(II)), C_2H_4 to C_2H_6 and also N_2 to N_2H_4. A mononuclear diazene intermediate which further disproportionates to N_2H_4 and N_2 has been suggested by Schrauzer.[44] However, a different mechanism involving a tetranuclear V(II) species which reduces N_2 to N_2H_4 by a donation of one electron from each V atom has been postulated by Shilov.[43a,45a,45b] The stereochemistry of the reduction of acetylenes (C_2H_2 and 2-butyne) has been shown to be *cis*. A striking feature of this system is that CO is not a dinitrogen-fixation inhibitor. Two different mechanisms have been proposed for the reduction of C_2H_2 to C_2H_4 by the V(II) systems in analogy with N_2 reduction by the same metal site. One involves a tetranuclear V(II) species with C_2H_2 side-on bridging two V(II) atoms,[45c] and the other a mononuclear V(II) species with side-on bonding acetylene.[44] Ethylene is formed upon protonation by H_2O and oxidation of V(II).

Reduction of acetylene, ethylene and, to a lesser extent, N_2 has also been shown to occur in acidic solutions by V^{2+} (aq) on irradiation with UV light.[44a]

Acetylene may be reduced to ethylene only (*cis*-$C_2H_2D_2$ is formed from C_2D_2 in H_2O) by a dinitrogen reducing V(II)—catechol species in homogeneous and aqueous medium at $5 \leqslant pH < 14$. For $pH > 14$ some ethane is also formed, probably by subsequent reduction of the ethylene produced without leaving the coordination site. Deuterated ethane is not formed when C_2D_4 is added to acetylene. In this system (as in the presence of nitrogenase) acetylene is a strong inhibitor of N_2 reduction and H_2 evolution, and ethylene is not reduced to ethane.[45c] The mechanism of acetylene reduction is suggested to involve intermediate polynuclear V(II) species as in the above $V(OH)_2/Mg(OH)_2$ heterogeneous system.[45c]

Although the system *meso*-tetra (*p*-sulphonatophenyl)porphinato-cobalt(III)—$NaBH_4$ (**27**) catalytically reduces acetylene to *cis*-ethylene and also reduces other nitrogenase substrates,[46a] its claimed ability to reduce N_2 to NH_3 has been disputed.[46b]

Reductions of acetylenes and olefins when binding dinitrogen sites have been referred to, but those species may also undergo different reaction types in related systems. A striking example is the *oxidation* of terminal olefins to methyl ketones by molecular oxygen which has been reported to be promoted by the Rh(I) species $[RhCl(PPh_3)_3]$ and $[RhH(CO)(PPh_3)_3]$ at ambient temperature and pressure. Coordination of olefin and oxygen to the metal is believed to occur and one of the phosphines is oxidized to phosphine oxide, whereas the olefin is oxygenated by the other oxygen atom.[47] This reaction may suggest the possibility of activation of coordinated N_2 towards oxidation yielding species such as NO, NO_2^-, NO_2, NO_3^- which could be incorporated into organonitrogen derivatives. Within this context the reported[48] facile, aerial oxidation, in aqueous, basic medium, of coordinated NH_3 in $[Ru(NH_3)_6]^{3+}$ yielding $[Ru(NH_3)_5(NO)]^{3+}$ should also be mentioned.

VI. Cyanide

The cyanide ligand is very prone to electrophilic attack and alkylation reactions constitute a well established method for the preparation of isonitrile complexes. Even in $K[(\eta^5\text{-}C_5H_4Me)Mn(CO)_2(CN)]$ (**28**),[49a] $Na[(\eta^5\text{-}C_5H_5)Mn(CO)_2(CN)]$ (**29**)[49b] (with $\nu(CN)$ values lower than in free cyanide) which are analogous to $[(\eta^5\text{-}C_5H_5)Mn(CO)_2(N_2)]$ in which N_2 suffers nucleophilic attack, and in related species such as $K_3[(\eta^5\text{-}C_5H_4Me)Mn(CN)_3]$ (**30**)[49a] (with $\nu(CN)$ values which are higher than in free cyanide) the cyanide ligand may be alkylated or protonated. Such reactions yield $[(\eta^5\text{-}C_5H_4Me)Mn(CO)_2(CNR)]$ from (**28**), R = H or Me by reactions with phosphoric acid or MeI, respectively, $[(\eta^5\text{-}C_5H_5)Mn(CO)_2(CNH)]$ from (**29**) by reaction with phosphoric acid and

$K_2[(\eta^5\text{-}C_5H_4Me)Mn(CN)_2(CNEt)]$ and $K[(\eta^5\text{-}C_5H_4Me)Mn(CNEt)_2(CN)]$ from (**30**) by reaction with Et_3OBF_4.

Cyanide is enzymatically reduced by nitrogenase to NH_3, CH_4 and traces of C_2H_6, C_2H_4 and CH_3NH_2. The molybdothiol system (**12**) leads to the formation of the same products although in different proportions.[23] The presence of Fe^{2+} and substrate amounts of ATP appears to enhance the reduction activity. N_2 and CO are inhibitors of CN^- reduction. Also, coordinated CN^- in the molybdo(IV) cyanide complex $K_2[Mo(O)(H_2O)(CN)_4]$ (**31**) is reduced to the same products by reducing agents such as $NaBH_4$, $S_2O_4^{2-}$ or reduced ferredoxin model species,[50a] thus providing support for the involvement of Mo(IV) intermediate species in the molybdothiol system. The mechanism for cyanide reduction to methane has been postulated to be as follows:[39]

$$\text{Mo–CN} \xrightarrow{2e,\ 2H^+} \text{Mo–CH=NH} \xrightarrow{+H_2O,\ -NH_3}$$

$$\text{Mo–CH=O} \xrightarrow{2e,\ 2H^+} \text{Mo–CH}_2\text{OH} \xrightarrow{2e,\ 2H^+,\ -H_2O}$$

$$\text{Mo–CH}_3 \xrightarrow{H_2O} \text{Mo–OH} + \text{CH}_4.$$

Ethylene and ethane are said[39a] to be formed by cyanide insertion into the Mo–C bond of intermediates or by intramolecular alkylation of coordinated cyanide. During the cyanide reduction of (**31**) free coordination sites are formed which allow the reduction of other substrates of nitrogenase namely N_2[50b] and acetylene (yielding *cis*-$C_2D_2H_2$ in D_2O). Reduction of cyanide has been reported for other dinitrogen fixation systems such as (**27**) as already mentioned.[46]

The ambident nature of the cyanide may be shown by the formation of bridge structures through coordination to a second metal (e.g. a divalent metal ion such as Mn^{II}, Fe^{II}, Co^{II}, Ni^{II}, Zn^{II})[51] of the nitrogen lone pair of a CN group which is already ligating a first metal through carbon. These reactions resemble the formation of bridging dinitrogen binuclear complexes upon adduct formation from mononuclear dinitrogen species.

VII. Nitriles

In general, metal-bonded nitriles behave mainly as σ-donors. They exhibit an increase in ν(NC) from the free value and a linear geometry $M \leftarrow N{\equiv}C{-}R$ has been authenticated by X-ray studies. Hence these ligands can undergo nucleophilic attack by amines and alcohols for example.[52]

However, when ligating dinitrogen binding sites, a decrease of ν(NC) is observed upon coordination indicating extensive π-backbonding acceptance, as has been observed in *trans*-$[Mo(N_2)(NCR)(dppe)_2]$ (R = Me, Et, Ph, *p*-Tol, *p*-ClC_6H_4, *p*-MeC_6H_4 or *p*-$MeOC_6H_4$),[53a] $[CoH(NCMe)(PPh_3)_3]$[53b] and $[Ru(NH_3)_5(NCR)]^{2+}$ (R = Me, Et, Propyl, $CH{=}CH_2$, $CMe{=}CH_2$, *p*-ClC_6H_4, *p*-$MeOC_6H_4$ or *o*-, *m*- or *p*-NC_5H_4).[54]

Nitriles are also known to coordinate in an edge-on fashion as in the dihapto nitrile complex $[(\eta^5\text{-}C_5H_5)_2Mo(\eta^2\text{-}NCCF_3)]$ which is related to the dinitrogen species $[(\eta^5\text{-}C_5Me_5)_2Mo(N_2)]$. An electron pair is then localized at N,

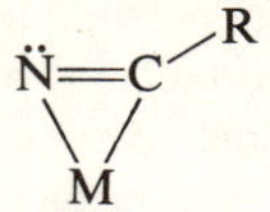

```
        R
N̈==C/
 \  /
  M
```

and double protonation by HCl at this atom has been carried out yielding $[(\eta^5\text{-}C_5H_5)_2ClMo\{C(CF_3){=}NH_2\}]Cl$.[55] The electron-deficient character of the iminium nitrogen has been used to obtain further reduction to ammonia, although in low yield, by reaction with sodium borohydride.[55]

Some dinitrogen binding sites can also reduce nitriles such as the porphyrin catalytic system (**27**).[46] The molybdothiol system (**12**) also reduces aliphatic nitriles to alkanes and ammonia in the presence of $NaBH_4$.[56] Olefins are formed in the reductions of unsaturated nitriles which occur at higher rates than those of saturated ones. The same products are obtained upon reduction by nitrogenase. The reaction is inhibited by CO and N_2. Mononuclear molybdenum–thiol species with a side-on bonded nitrile are postulated as intermediates.[56]

The complex $[CoH(NCR)(PPh_3)_3]$, which is analogous to $[CoH(N_2)(PPh_3)_3]$, being derived from the latter by N_2 displacement by nitrile,[53b] has been shown[57] to catalyse the isomerization of *cis*-2-pentene to the *trans* isomer although at a slower rate than the parent N_2 species. The catalytic rate decreases in the order $N_2 > PhCN > MeCN$, which is the reverse of the basicity of these ligands.[57]

VIII. Nitrosyl

Nitrogen oxide is known to be an inhibitor of nitrogenase activity, and is not reduced by the enzyme. However, nitrosyl ligating an inorganic dinitrogen fixation site may be reduced to NH_3 as in the reaction of the Ru(II) species $[Ru(NH_3)_5(NO)]^{3+}$ (**32**) with Cr^{2+} in aqueous acidic medium which yields the six-electron reduction product $[Ru(NH_3)_6]^{2+}$ and Cr(III). The mechanism of this reaction is believed to proceed by successive one-electron steps because only monomeric Cr(III) species have been produced (two-electron steps would lead to dimeric Cr(III) species).[58]

Complexes with ligands formed by successive degrees of reduction of ligating NO have been isolated. Hence, in alkaline medium, reduction of NO in species (**32**) to N_2 occurs yielding $[Ru(NH_3)_5(N_2)]^{2+}$. The mechanism involves reaction between NO and a coordinated amide, NH_2^-, formed by proton abstraction from NH_3 by OH^- (see Chapter 6).[59] Such a nucleophilic attack is not unexpected (see below).

Reduction of M–NO to M–NHO has been achieved by protonation at N of bent NO, e.g. in $[OsCl(CO)(NO)(PPh_3)_2]$,[60a] and to M–NHOH by protonation of $[Os(NO)_2(PPh_3)_2]$.[60a] Reduction to M–NH_2OH (coordinated hydroxylamine) occurs by protonation with excess of HX of linear NO in $[Ir(NO)(PPh_3)_3]$ with displacement of phosphine by halide affording $[IrX_3(NH_2OH)(PPh_3)_2]$ (X = Cl, Br)[60a,60b] or by reaction of probably bent NO in $[Cr(NH_3)_5(NO)]^{2+}$ with Cr^{2+} (reducer) in aqueous acidic medium.[61] Isomeric $[RhCl_3(NOH)(PPh_3)_2]$ and $[RhCl_3(HNO)(PPh_3)_2]$ have been reported[60c] from monoprotonation of $[Rh(NO)(PPh_3)_3]$ by HCl, but their reformulation as hydroxylamine complexes has been suggested.[60b]

The bonding of the nitrosyl group to a transition metal has been the subject of various approaches[62] and some controversy still exists. However, it is common practice to consider a linear coordinated nitrosyl as formally NO^+ (three-electron donor and isoelectronic with N_2), whereas a NO^- group (one-electron donor) is representative of a bent geometry. In a valence bond representation, linear NO may be considered a hybrid of the resonance forms

$$M \leftarrow N{\equiv}O{:}^+ \leftrightarrow M^+ \rightleftharpoons N{=}\ddot{O}{:} \leftrightarrow M^+{\equiv}N^+{-}\ddot{\underset{..}{O}}{:}^-,$$

the last one with a lower weighting. For bent NO the following may be considered:

$$M{\leftarrow}\ddot{N}{=}\underset{..}{O}{:} \longleftrightarrow M{=}\ddot{N}{-}\ddot{\underset{..}{O}}{:}^- .$$

As the nitrosyl σ-orbital involved in σ-bonding to the metal has bonding NO character, a lowering in ν(NO) upon coordination is observed for both linear and bent NO. The linear structure has higher M–N and N–O bond orders than the bent one and "corrected" ν(No) values (ν'(NO) is derived from ν(NO) by application of a number of empirically derived rules)[63] have been used as a diagnostic tool for assigning the structure of ligating NO. Values of ν'(NO) higher than 1620 cm^{-1} correspond generally to a linear geometry, and those lower than 1610 cm^{-1} to a bent one.

Bent nitrosyl is an electron-rich group and can undergo electrophilic attack at the N atom, as mentioned above, although some claims of attack at O have been reported in other systems.[64] It has been suggested from X-ray studies[63] that linear nitrosyls may be good π-acceptors, which shows how artificial it is to consider them as formally NO^+. A decrease in ν(NO) has been shown to follow an increase in the negative charge at both N and O of linear NO, the nitrogen atom carrying the greatest negative charge.[65] Hence it is not surprising that linear nitrosyls without too high a $\nu'(NO)$ value may also undergo protonation at N (see above). However, for sufficiently high $\nu'(NO)$ value, the NO ligand is electron-deficient and acquires a positive charge with the N atom having the highest value;[65] hence such nitrosyl can undergo nucleophilic attack at N[66] and examples are known involving dinitrogen binding sites. So for example the linear NO in *trans*-$[RuCl(NO)(diars)_2]^{2+}$ (**33**) with high $\nu'(NO)$ (1713 cm^{-1} corresponding to $\nu(NO)$ = 1883 cm^{-1}) reacts with bases according to the following reactions:[67a,67b]

(**33**) + $3H_2NNHR \rightarrow [RuCl(NONNHR)(diars)_2]$ as initial product;

(**33**) + $2OH^- \rightarrow$ *trans*-$[RuCl(NO_2)(diars)_2] + H_2O$;

(**33** + $2N_3^- \rightarrow [Ru(N_3)Cl(diars)_2]$ (**34**) + $N_2 + N_2O$.

The related species *cis*-$[RuCl(NO)(bipy)_2]^{3+}$ (bipy = 2,2′-bipyridine) undergoes a similar reaction with OH^- yielding $[RuCl(NO_2)(bipy)_2]^+$, whereas $[RuCl(S)(bipy)_2]^+$ species are formed from reactions of $[RuCl(NO)(bipy)_2]^{2+}$ with N_3^- in solvent S (S = water, acetonitrile and acetone).[67c] As a result of a study of the azide reaction with the labelled complex $[RuCl(^{15}NO)(diars)_2]^{2+}$, a cyclic intermediate, $Ru{-}^{15}N{-}O{-}^{14}N{-}^{14}N{-}^{14}N$ (ring closed between ^{15}N and terminal ^{14}N), is believed to be formed by azide attack at the nitrosyl.[67b] The coordinated nitrosyl in *trans*-$[OsCl(NO)(diars)_2]^{2+}$ ($\nu'(NO) \approx 1725\ cm^{-1}$) also undergoes nucleophilic attack by phenylhydrazine leading to *trans*-$[OsCl(NONNHPh)(diars)_2]$ and *trans*-$[OsCl(N_2)(diars)_2]^+$.[67d]

Disproportionation of NO ligating dinitrogen-binding or related metal sites is known to occur. $[Co(NO)_2(NO_2)(PPh_3)]$ and $[RhCl(NO)(NO_2)L_2]$ (L = PPh_3, $AsPh_3$) are formed from reaction of NO with $[Co(NO)(PPh_3)_3]$[68a] and $[RhClL'L_2]$ (L′ = PPh_3, $AsPh_3$, CO),[68b] respectively. Disproportionation of NO to $NO^+NO_2^- + N_2$ followed by oxidative addition of $NO^+NO_2^-$ yields the final nitrosylnitrito complexes.

Whereas the Lewis basicity of the terminal atom in coordinated N_2, CO and CN has been frequently reported and bridging structures have been authenticated by X-rays in a number of cases, only scant information[69] exists on the analogous NO ligand probably due to the weak basicity of its oxygen atom.

IX. Azide

Azide and N_2 are the only substrates which are reduced specifically only to NH_3 by nitrogenase (N_2 is also formed from azide).

Like nitrosyl, coordinated azide is known to be a ligating dinitrogen generator *in situ*, as it has been mentioned in Chapter 6, and the mechanism of some of these reactions (preparation of Ru(II)–N_2 complexes from Ru(III)–N_3 in an acidic medium) is believed to proceed via protonation at the coordinated nitrogen of the azide followed by N_2 loss and formation of a protonated nitrene species.[68–70] This mechanism is borne out by the isolation[71] of a species with ligating protonated azide, $[Co(NHN_2)(NH_3)_5](ClO_4)_3$ (from reaction of $[Co(N_3)(NH_3)_5](ClO_4)_2$ with $HClO_4$) whose IR data is best explained by the predominance of the canonical structure

$$Co \leftarrow \bar{N}(H){-}\overset{+}{N}{\equiv}N: \quad \text{over} \quad Co \leftarrow N(H){=}\overset{+}{N}{=}\bar{\ddot{N}}:.$$

The weak NH–NN bond accounts for the subsequent N_2 loss. The site of protonation is explained by the considerable weight of the canonical structure

$$M \leftarrow \bar{\ddot{N}}{-}\overset{+}{N}{\equiv}N: \quad \text{although} \quad M \leftarrow \bar{\ddot{N}}{=}\overset{+}{N}{=}\bar{\ddot{N}}: \text{ must also be considered).}$$

Related mechanisms involving the formation of electrophilic coordinated nitrene NH are supposed to occur in the reduction of azide to NH_3 by treating $[MoN_3(S_2CNEt_2)_2(Me_2SO)(NO)]$ with sulphuric acid followed by sodium hydroxide,[72] or by reaction of $[Ir(NH_3)_5N_3]^{2+}$ with acids[73] leading to $[Ir(NH_3)_6]^{3+}$ or $[Ir(NH_3)_5NH_2OH]^{3+}$. The mechanism of acid-catalysed hydrolysis reactions of a number of azide complexes (e.g. $[Cr(H_2O)_5N_3]^{2+}$,[74a] $[Co(NH_3)_5N_3]^{2+}$ [74b,74c]) has also been studied kinetically and a protonated azide intermediate has been postulated. Protonation at terminal nitrogen of coordinated azide has also been suggested, e.g. the formation of *trans*-$[RuCl(N_2)(diars)_2]^+$ (**35**) by reaction of (**34**) with HCl, although protonation at coordinated azide nitrogen cannot be ruled out. Reaction of (**34**) with NO^+ and NO_2^+ also leads to the dinitrogen complex (**35**) (see Chapter 6).[67a,67b]

Ammonia may also be formed from azide by a different route via a coordinated nitride. Reaction of $MoCl_5$ (or $MoCl_4$) with Et_4NN_3 yields a nitridomolybdate(VI) species $[Et_4N][MoNCl_4]$ which on hydrolysis gives NH_3 in ca 40% yield. The nitride ligand is believed to be formed from coordinated azide by N_2 evolution and reduction of the resulting N^- ligand by Mo^V (or Mo^{IV}).[75]

Hence at least two different types of mechanism for reduction of ligating azide to ammonia are known which may also resemble those used in N_2 reduction to ammonia.

Azide may also be catalytically reduced in aqueous medium by systems such as (**27**)[46] and (**12**).[23]

X. Final Remarks

The study of the reactivity of ligands analogous to N_2 at dinitrogen binding sites has shown the generally electron-rich character of these sites and their ability to stabilize by coordination such unstable species as the carbene and carbyne ligands derived from electrophilic attack at ligating isonitriles.

Such an electrophilic attack is paralleled by the reactions of ligating N_2 (Part 3) but other reactions, such as insertions and oxidations undergone by ligands at these sites have yet to be reproduced by N_2 in isolated complexes.

The precise conditions to force N_2 to undergo the latter types of reaction are yet to be attained, but detailed examination of the reaction of dinitrogen analogues may well point the way to the design of systems which will achieve this aim.

References

1. J. Chatt, C. E. Elson, A. J. L. Pombeiro, R. L. Richards and G. H. D. Royston, *J. C. S.* (*Dalton*) (1978) 165.
2. (a) J. Chatt, A. J. L. Pombeiro and R. L. Richards, *J. Organometallic Chem.*, in press; (b) A. J. L. Pombeiro and R. L. Richards, *J. Organometallic Chem.* (1979) **179**, 459.
3. J. Chatt, A. J. L. Pombeiro, R. L. Richards, G. H. D. Royston, K. W. Muir and R. Walker, *J. C. S. Chem. Commun.* (1975) 708.
4. (a) A. J. L. Pombeiro and R. L. Richards, *Transition Metal Chem.*, (1980) **5**, 55; (b) J. Chatt, A. J. L Pombeiro and R. L. Richards, *J. C. S.* (*Dalton*) (1979), 1585; (c) *J. Organometallic Chem.* (1980) **184**, 357; (d) *J. C. S.* (*Dalton*), in press; (e) A. J. L. Pombeiro, *Per. Port. Chem.*, in press.
5. (a) J. Chatt, A. J. L. Pombeiro and R. L. Richards, *Proc. Ninth Intern. Conf. Organometallic Chem.*, D47, Dijon, France, 1979; (b) A. J. L. Pombeiro and R. L. Richards, *Transition Metal Chem.*, in press.
6. (a) F. Bonati and G. Minghetti, *Inorg. Chim. Acta* (1974) **9**, 95; (b) P. M. Treichel, *Adv. Organometallic Chem.* (1973) **11**, 21; (c) L. Calligaro, P. Uguagliati, B. Crociani and U. Belluco, *J. Organometallic Chem.* (1975). **92**, 399; (d) B. Crociani, T. Boschi, M. Nicolini and U. Belluco, *Inorg. Chem.* (1972) **11**, 1292.
7. P. M. Treichel, J. J. Benedict, R. W. Hess and J. P. Stenson, *Chem. Commun.* (1970) 1627, and references therein.
8. D. Sellman and W. Weiss, *Angew. Chem.* (*Internat. Edn*) (1977) **16**, 880.
9. J. Chatt, A. J. Pearman and R. L. Richards, *J. Organometallic Chem.* (1975) **101**, C45.
10. (a) E. L. Moorehead, B. J. Weathers, E. A. Ufkes, P. R. Robinson and G. N. Schrauzer, *J. Amer. Chem. Soc.* (1977) **99**, 6089; (b) G. N. Schrauzer, P. A. Doemeny, G. W. Kiefer and R. H. Frazier, *J. Amer. Chem. Soc.* (1972) **94**, 3604.
11. J. L. Sudmeier, K. E. Schwarts and A. J. Senzel, *Inorg. Chem.* (1969) **8**, 2815.
12. (a) P. F. Crossing and M. R. Snow, *J. Chem. Soc.* (*A*) (1971) 610; (b) S. Datta, T. J. McNeese and S. S. Wreford, *Inorg. Chem.* (1977) **16**, 2661; (c) S. Datta, B. Dezube, J. K. Kouba and S. S. Wreford, *J. Amer. Chem. Soc.* (1978) **100**, 4404.

13. (a) T. Sawai and R. J. Angelici, *J. Organometallic Chem.* (1974) **80**, 91 and references therein; (b) M. L. H. Green and C. R. Hurley, *J. Organometallic Chem.* (1967) **10**, 188.
14. (a) M. Aresta, *Gazz. Chim. Ital.* (1972) **102**, 781; (b) J. Chatt, R. H. Crabtree, E. A. Jeffery and R. L. Richards, *J. C. S.* (*Dalton*) (1973) 1167.
15. G. Henrici-Olivé and S. Olivé, *Angew. Chem.* (*Internat. Edn*) (1976) **15**, 136.
16. A. Jones and B. D. McNicol, *J. Catalysis* (1977) **47**, 384.
17. L. I. Shoer and J. Schwartz, *J. Amer. Chem. Soc.* (1977) **99**, 5831.
18. (a) J. M. Manriquez, D. R. McAlister, R. D. Sanner and J. E. Bercaw, *J. Amer. Chem. Soc.* (1978) **100**, 2716; (b) A. N. Nesmeyanov, K. N. Anisimov, N. E. Kolobova and L. L. Krasnoslobodskaya, *Izvest. Akad. Nauk S.S.S.R., Ser. Khim.* (1970) 860 (English translation, p. 807); (c) J. M. Manriquez, D. R. McAlister, R. D. Sanner and J. E. Bercaw, *J. Amer. Chem. Soc.* (1976) **98**, 6733.
19. J. Chatt, R. A. Head, G. J. Leigh and C. J. Pickett, *Chem. Commun.* (1977) 299.
20. (a) M. E. Vol'pin and V. B. Shur, *Izvest. Akad. Nauk S.S.S.R., Ser. Khim.* (1966) 1873 (English translation, p. 1819); (b) M. E. Vol'pin, V. B. Shur, R. V. Kudryavtsev and L. A. Prodayko, *Chem. Commun.* (1968) 1038; (c) M. E. Vol'pin and V. B. Shur, *In* "Organometallic Reactions, I", p. 55. (Ed. E. I. Becker and M. Tsutsui). Interscience, New York, (1970).
21. J. C. Huffman, J. G. Stone, W. C. Krusell and K. G. Gaulton, *J. Amer. Chem. Soc.* (1977) **99**, 5829.
22. B. Jezowska-Trzebiatowska and P. Sobota, *J. Organometallic Chem.* (1973) **61**, C43.
23. G. N. Schrauzer, G. W. Kiefer, P. A. Doemeny and H. Kisch, *J. Amer. Chem. Soc.* (1973) **95**, 5582.
24. J. N. Armor, *Inorg. Chem.* (1978) **17**, 213.
25. (a) M. E. Vol'pin and I. S. Kolomnikov, *Pure Appl. Chem.* (1973) **33**, 567; "Organometallic Reactions" (Ed. E. I. Becker and M. Tsutsui) (1975), **5**, 313.
26. (a) M. Aresta, C. F. Nobile, *J. C. S.* (*Dalton*) (1977) 708; (b) M. Aresta, C. F. Nobile, V. G. Albano, E. Forni and M. Manassero, *J. C. S. Chem. Commun.* (1975) 636; (c) G. A. Tolman, D. H. Gerlach, J. P. Jesson and R. A. Schunn, *J. Organometallic Chem.* (1974) **65**, C23; (d) P. W. Jolly, K. Jonas, C. Kruger and Y. H. Tsay, *J. Organometallic Chem.* (1971) **33**, 109; (e) T. Herskovitz and L. J. Guggenberger, *J. Amer. Chem. Soc.* (1976) **98**, 1615; (f) J. Chatt, M. Kubota, G. J. Leigh, F. C. March, R. Mason and D. J. Yarrow, *J. C. S. Chem. Commun.* (1974) 1033; (g) T. Herskovitz, *J. Amer. Chem. Soc.* (1977) **99**, 2391.
27. (a) H. L. M. van Gaal, F. G. Moers and J. J. Steggerda, *J. Organometallic Chem.* (1974) **65**, C43; (b) H. L. M. van Gaal and F. L. A. van Bekeron, *J. Organometallic Chem.* (1977) **134**, 237; (c) C. Busetto, A. D'Alfonso, E. Maspero, G. Perego and A. Zazzetta, *J. C. S.* (*Dalton*) (1977) 1828; (d) M. Aresta, C. F. Nobile, *Inorg. Chim. Acta* (1977) **24**, L49; (e) Y. Iwashita and A. Hayata, *J. Amer. Chem. Soc.* (1969) **91**, 2525.
28. (a) L. S. Pu, A. Yamamoto and S. Ikeda, *J. Amer. Chem. Soc.* (1968) **90**, 3896; (b) A. Misono, Y. Uchida, M. Hidai and T. Kuse, *Chem. Commun.* (1968) 981; (c) A. Yamamoto, S. Kitazume, L. S. Pu and S. Ikeda, *J. Amer. Chem. Soc.* (1971) **93**, 371; (d) V. D. Bianco, S. Doronzo and M. Rossi, *J. Organometallic Chem.* (1972) **35**, 337; (e) I. S. Kolomnikov, A. I. Gusev, G. G. Aleksandrov, T. S. Lobeeva, Yu. T. Struchkov, M. E. Vol'pin, *J. Organometallic Chem.* (1973) **59**, 349; (f) S. Komiya, A. Yamamoto, *Bull. Chem. Soc. Japan*, (1976) **49**, 784.

29. (a) B. Jezowska-Trzebiatowska and P. Sobota, *J. Organometallic Chem.* (1974) **80**, C27; (b) *J. Organometallic Chem.* (1974) **76**, 43; (c) G. Fachinetti, C. Floriani, A. Roselli and S. Pucci, *J. C. S. Chem. Commun.* (1978) 269; (d) I. S. Kolomnikov, T. S. Lobeeva and M. E. Vol'pin, *Izvest. Akad. Nauk S.S.S.R., Ser. Khim.* (1970) 2650; (e) Y. Inoue, Y. Sasaki, H. Hashimoto, *J. C. S. Chem. Commun.* (1975) 718; (f) G. O. Evans, C. J. Newell, *Inorg. Chim. Acta* (1978) **31**, L387; (g) P. Haynes, L. H. Slaugh and J. F. Kohnle, *Tetrahedron Letters* (1970) 365.
30. (a) I. S. Kolomnikov, T. S. Lobeeva, V. V. Gorbachevskaya, G. G. Aleksandrov, Yu. T. Struckov and M. E. Vol'pin, *Chem. Commun.* (1971) 972; (b) I. S. Kolomnikov, A. O. Gusev, T. S. Belopotapova, M. Kh. Grigoryan, T. V. Lysyak, Yu. T. Struchkov and M. E. Vol'pin, *J. Organometallic Chem.* (1974) **69**, C10; (c) Y. Inoue, Y. Sasaki and H. Hashimoto, *Bull. Chem. Soc. Japan* (1978) **51**, 2375; (d) A. Musco, C. Perego and V. Tartiari, *Inorg. Chim. Acta* (1978) **28**, L147; (e) R. J. De Pasquale, *J. C. S. Chem. Commun.* (1973) 157; (f) T. V. Ashworth and E. Singleton, *J. C. S. Chem. Commun.*, 1976, 204; (g) B. R. Flynn and L. Vaska, *J. Amer. Chem. Soc.* (1973) **95**, 5081; (h) E. Chaffee, T. P. Dasgupta and C. M. Harris, *J. Amer. Chem. Soc.* (1973) **95**, 4169; (i) R. J. Crutchley, J. Powell, R. Faggiani and C. J. L. Lock, *Inorg. Chim. Acta* (1977) **24**, L15.
31. (a) A. Yamamoto, S. Go, M. Ookawa, M. Takahashi, S. Ikeda and T. Keii, *Bull. Chem. Soc. Japan* (1972) **45**, 3110; (b) P. Sobota, B. Jezowska-Trzebiatowska and Z. Janas, *J. Organometallic Chem.* (1976) **118**, 253; (c) T. V. Ashworth, M. Nolte and E. Singleton, *J. Organometallic Chem.* (1976) **121**, C57; (d) M. H. Chisholm and M. Extine, *J. Amer. Chem. Soc.* (1974) **96**, 6214; (e) (1975) **97**, 1623; (f) (1975) **97**, 5625; (g) (1977) **99**, 782; (h) T. Tsuda, H. Washita and T. Saegusa, *J. C. S. Chem. Commun.* (1977) 468; (i) T. Tsuda, H. Washita, K. Watanabe, M. Miwa and T. Saegusa, *J. C. S. Chem. Commun.* (1978) 815.
32. J. W. Byrne, H. U. Blaser, and J. A. Osborn, *J. Amer. Chem. Soc.* (1975) 3871.
33. R. D. W. Kemmitt and D. I. Nichols, *Inorg. Nuclear Chem. Letters* (1968) **4**, 739.
34. (a) J. W. McDonald, J. L. Corbin and W. E. Newton, *J. Amer. Chem. Soc.* (1975) **97**, 1970; (b) P. Sobota, B. Jezowska-Trzebiatowska, *J. Organometallic Chem.* (1977) **131**, 341.
35. J. L. Thomas, *J. Amer. Chem. Soc.* (1973) **95**, 1838.
36. (a) G. N. Schrauzer and T. D. Guth, *J. Amer. Chem. Soc.* (1977) **99**, 7189; (b) A. H. Boonstra and C. A. H. A. Mutsaers, *J. Phys. Chem.* (1975) **79**, 2025.
37. (a) D. Evans, J. A. Osborn and G. Wilkinson, *J. Chem. Soc.* (*A*) (1968) 3133 and references therein; (b) C. K. Brown and G. Wilkinson, *J. Chem. Soc.* (*A*) (1970) 2753; (c) C. O'Connor and G. Wilkinson, *J. Chem. Soc.* (*A*) (1968) 2665; (d) M. Yagupsky and G. Wilkinson, *J. Chem. Soc.* (*A*) (1970) 941.
38. E. E. van Tamelen, W. Cretney, N. Klaentschi and J. S. Miller, *J. C. S. Chem. Commun.* (1972) 481.
39. (a) G. N. Schrauzer, *Angew. Chem.* (*Internat. Edn.*) (1975) **14**, 514; (b) G. N. Schrauzer and D. A. Doemeny, *J. Amer. Chem. Soc.* (1971) **93**, 1608; (c) G. N. Schrauzer and G. S. Schlesinger, *J. Amer. Chem. Soc.* (1970) **92**, 1808; (d) G. N. Schrauzer, G. W. Kiefer, K. Tano and P. R. Robinson, *J. Amer. Chem. Soc.* (1975) **97**, 6088.
40. T. A. Vorontsova and A. E. Shilov, *Kinet. Katal.* (1973) **14**, 1326 (English translation, p. 1166).

41. K. Tano and G. N. Schrauzer, *J. Amer. Chem. Soc.* (1975) **97**, 5404.
42. D. A. Ledwith and F. A. Schultz, *J. Amer. Chem. Soc.* (1975) **97**, 6591.
43. (a) A. Shilov, N. Denisov, O. Efimov, N. Shuvalov, N. Shuvalova and A. Shilova, *Nature* (*London*) (1971) **231**, 460; (b) A. E. Shilov, A. K. Shilova and T. A. Vorontsova, "Reaction Kinetics and Catalysis Letters", (1975) **3**, 143.
44. (a) S. I. Zones, T. M. Vickrey, J. G. Palmer and G. N. Schrauzer, *J. Amer. Chem. Soc.* (1976) **98**, 7289; (b) S. I. Zones, M. R. Palmer, J. G. Palmer, J. M. Doemeny and G. N. Schrauzer, *J. Amer. Chem. Soc.* (1978) **100**, 2113.
45. (a) N. T. Denisov, N. I. Shuvalova and A. E. Shilov, *Kinet. Katal.* (1973) **14**, 1325 (English translation, p. 1164); (b) A. E. Shilov, "Biological Aspects of Inorganic Chemistry" (Ed. A. W. Addison, W. R. Cullen, D. Dolphin and B. R. Jones). Wiley, New York (1977); (c) L. A. Nikonova, S. A. Isaeva, N. I. Pershikova and A. E. Shilov, *J. Mol. Cat.* (1975/76) **1**, 367 and references therein.
46. (a) E. B. Fleischer and M. Krishnamurthy, *J. Amer. Chem. Soc.* (1972) **94**, 1382; (b) J. Chatt, C. M. Elson and G. J. Leigh, *J. Amer. Chem. Soc.* (1973) **95**, 2408.
47. (a) C. Dudley and G. Read, *Tetrahedron Letters* (1972) **52**, 5273; (b) C. Dudley, G. Read and P. J. C. Walker, *J. C. S.* (*Dalton*) (1974) 1926.
48. S. D. Pell and J. N. Armor, *J. Amer. Chem. Soc.* (1975) **97**, 5012.
49. (a) J. A. Dineen and P. L. Pauson, *J. Organometallic Chem.* (1974) **71**, 91; (b) E. O. Fischer and R. J. J. Schneider, *J. Organometallic Chem.* (1968) **12**, P 27.
50. (a) G. N. Schrauzer, P. R. Robinson, E. L. Moorehead and T. M. Vickrey, *J. Amer. Chem. Soc.* (1975) **97**, 7069; (b) *J. Amer. Chem. Soc.* (1976) **98**, 2815.
51. (a) P. Rigo and A. Turco, *Coord. Chem. Rev.* (1974) **13**, 133; (b) W. P. Griffith, *Coord. Chem. Rev.* (1975) **17**, 177.
52. G. Rouschias and G. Wilkinson, *J. Chem. Soc.* (*A*) (1968) 489.
53. (a) T. Tatsumi, M. Hidai and Y. Uchida, *Inorg. Chem.* (1975) **14**, 2530; (b) A. Misono, Y. Uchida, M. Hidai and T. Kuse, *Chem. Commun.* (1969) 208.
54. (a) R. E. Clarke and P. C. Ford, *Inorg. Chem.* (1970) **9**, 227; (b) 495; (c) P. C. Ford, *Coord. Chem. Rev.* (1970) **5**, 75; (d) R. D. Froust, Jr. and P. C. Ford, *J. Amer. Chem. Soc.* (1972) **94**, 5686.
55. J. L. Thomas, *J. Amer. Chem. Soc.* (1975) **97**, 5943.
56. G. N. Schrauzer, P. A. Doemeny, R. H. Frazier, Jr. and G. W. Kiefer, *J. Amer. Chem. Soc.* (1972) **94**, 7378.
57. S. Tyrlik, *J. Organometallic Chem.* (1972) **39**, 371.
58. J. N. Armor, *Inorg. Chem.* (1973) **12**, 1959.
59. (a) S. Pell and J. Armor, *J. Amer. Chem. Soc.* (1972) **94**, 686; (b) *J. Amer. Chem. Soc.* (1973) **95**, 7625; (c) F. Bottomley, E. M. R. Kiremire and S. G. Clarkson, *J. C. S.* (*Dalton*) (1975) 1909.
60. (a) K. R. Grundy, C. A. Reed and W. R. Roper, *Chem. Commun.* (1970) 1501; (b) C. A. Reed and W. R. Roper, *J. Chem. Soc.* (*A*) (1970) 3054; (c) J. P. Collman, N. W. Hoffman and D. E. Morris, *J. Amer. Chem. Soc.* (1969) **91**, 5659.
61. (a) J. N. Armor and M. Buchbinder, *Inorg. Chem.* (1973) **12**, 1086; (b) J. N. Armor, M. Buchbinder and R. Cheney, *Inorg. Chem.* (1974) **13**, 2990.
62. (a) J. H. Enemark and R. D. Feltham, *Coord. Chem. Rev.* (1974) **13**, 339; (b) R. D. Harcourt and J. A. Bowden, *J. Inorg. Nuclear Chem.* (1974) **36**, 1115.
63. B. L. Haymore and J. A. Ibers, *Inorg. Chem.* (1975) **14**, 3060.
64. T. M. Vickrey and G. L. Blackmer, *J. Inorg. Nuclear Chem.* (1976) **38**, 1519.
65. B. Folkesson, *Acta Chem. Scand. A.* (1974) **28**, 491.

66. (a) J. H. Swinehart, *Coordination Chem. Rev.* (1967) **2**, 385; (b) N. G. Connelly, *Inorg. Chim. Acta Rev.* (1972) **6**, 47; (c) F. Bottomley, S. G. Clarkson and S. B. Tong, *J. C. S.* (*Dalton*) (1974) 2344.
67. (a) P. G. Douglas, R. D. Feltham and H. G. Metzger, *J. Amer. Chem. Soc.* (1971) **93**, 84; (b) P. G. Douglas and R. D. Feltham, *J. Amer. Chem. Soc.* (1972) **94**, 5254; (c) F. J. Miller and T. J. Meyer, *J. Amer. Chem. Soc.* (1971) **93**, 1294; (d) F. Bottomley and E. M. R. Kiremire, *J. C. S.* (*Dalton*) (1977) 1125.
68. (a) M. Rossi and A. Sacco, *Chem. Commun.* (1971) 694; (b) W. B. Hughes, *Chem. Commun.* (1969) 1126.
69. D. B. Brown, *Inorg. Chem.* (1975) **14**, 2582 and references therein.
70. (a) L. A. P. Kane-Maguire, F. Basolo and R. G. Pearson, *J. Amer. Chem. Soc.* (1969) **91**, 4609; (b) L. A. P. Kane-Maguire, P. S. Sheridan, F. Basolo and R. G. Pearson, *J. Amer. Chem. Soc.* (1970) **92**, 5865.
71. F. Monacelli, G. Mattogno, D. Gattegno and M. Maltese, *Inorg. Chem.* (1970) **9**, 686.
72. J. A. Broomhead, J. R. Budge, W. D. Grumley, T. R. Norman and M. Sterns, *Austral. J. Chem.* (1976) **29**, 275.
73. B. C. Lane, J. W. McDonald, F. Basolo, and R. G. Pearson, *J. Amer. Chem. Soc.* (1972) **94**, 3786.
74. (a) T. W. Swaddle and E. L. King, *Inorg. Chem.* (1964) **3**, 234; (b) G. C. Lalor and A. Moelwyn-Hughes, *J. Chem. Soc.* (1963) 1560; (c) G. C. Lalor, *J. Chem. Soc.* (*A*) (1966) 1.
75. R. D. Bereman, *Inorg. Chem.* (1972) **11**, 1149.

11

Perspectives

JOSEPH CHATT

A.R.C. Unit of Nitrogen Fixation

THE contributors to this book have considered the main lines of current research into the improvement or development of new chemical processes for nitrogen fixation. Now we must ask ourselves what are the prospects for improved chemical methods of fixing nitrogen. These we shall consider in the order of the book.

I. Heterogeneous Catalysts

In the short term, improvement in the Haber-type catalysts is probably the most likely to be rewarding. The chemical plant using such catalysts has now undergone some seventy years of development and reached a very advanced stage of design and operation. The methods of producing molecular nitrogen and hydrogen, the raw materials for the Haber–Bosch process, and of isolating the product are also well established. Every large chemical company concerned with ammonia manufacture must have put considerable effort into the improvement of the Haber catalyst, and so it seems unlikely that there is room for any but minor improvements by empirical methods, but the scale of operation is so large than even minor improvements can be profitable.

Despite the importance of the Haber process and the vast amount of empirical and fundamental research which has gone into the ammonia synthesis reaction, it is still imperfectly understood. Even the nature of the most abundant intermediate on the catalyst surface is not yet certain. Is it chemisorbed dinitrogen, mononitrogen, or some nitrogen hydride? What exactly is the role of the promoter? Does the alumina play a chemical role or does it merely prevent the sintering of the iron crystallites on which the reaction occurs? Do all the surfaces of the iron crystallites catalyse the reaction equally, or is one, say the 111 face as has been suggested, more active than others? These and many more partially answered or unanswered questions still need further investigation. Only a very fundamental study of the ammonia synthesis reaction can lead to new concepts which are likely to produce significantly more effective catalysts.

The materials for the manufacture of promoted iron–alumina catalysts are all cheap. Better catalysts based on ruthenium and osmium are known, but they cannot compete with iron unless the catalyst is more active by some orders of magnitude. Nevertheless their existence shows that the ultimate in Haber-type catalysts has not necessarily been reached. The fundamental studies presented in the first chapter of this book are essential steps towards that difficult goal.

The reaction of molecular nitrogen with molecular hydrogen is a remarkable reaction in view of the very high stability of the nitrogen molecule. The fundamental study of its activation on metal surfaces can have valuable "spin-off" into areas of catalysis other than the vital one of ammonia synthesis. A better understanding of the fundamental chemistry involved is the only way forward from the present position towards better synthesis catalysts. It might also lead to the direct production of organo-nitrogen compounds from molecular nitrogen and common hydrocarbon molecules when a heterogeneous catalyst capable of activating molecular nitrogen at temperatures much lower than those at present used in ammonia synthesis has been found.

II. New Nitrogen-Fixing Reactions

All newly discovered reactions of molecular nitrogen have involved the element under moderate conditions of temperature and pressure, but they involve somewhat sophisticated metal complexes and/or very vicious reducing agents such as metallic sodium, metallic magnesium, Grignard, or lithium reagents. These are themselves much too expensive to form the basis of any ammonia process.

A. Nitriding reactions

In the nitriding reactions described in Chapter 3 the final product of reduction appears to be a nitride complex. The most studied nitriding system involves the reduction of biscyclopentadienyl titanium chlorides and the nitrido-complex so obtained gives ammonia on hydrolysis, in good yield based on the amount of titanium present. It will also react with certain organic compounds which, followed by hydrolysis, produce amines. These reactions are generally stoichiometric and their chemistry is still not completely understood. Useful general chemical knowledge might be obtained from their further study when more sensitive and sophisticated spectroscopic means of following the reactions become available. However nitriding reactions depend on the use of relatively expensive reducing agents and in no way could they provide a viable process for the production of commercially important nitrogen compounds from molecular nitrogen.

B. Nitrogen fixation through dinitrogen complexes

It is believed that the biological process of nitrogen fixation proceeds through the formation of a dinitrogen complex on one or two atoms of molybdenum in the enzyme and that this is reduced to ammonia by feeding electrons through the molybdenum atom and protons from the solution to the dinitrogen molecule until reduction to ammonia is complete. The many attempts to emulate such reactions have been described in Parts 2 and 3. There is no question that molecular nitrogen can be reduced in the presence of vanadium or molybdenum complexes in a protic medium as described in Part 2. Moreover, as described in Part 3, when dinitrogen is bonded to certain tungsten(0) complexes a simple treatment with acid leads to 90% yields of ammonia based upon the reducing capacity of the tungsten(0) atom. In the latter case about half of the remaining 10% of the reducing power appears as hydrazine. This demonstrates that when the nitrogen bonding site is correct a highly efficient reduction of dinitrogen to ammonia on a single metal atom is possible, even in an acid medium where strong reducing agents normally produce dihydrogen. However, no method of reducing ammonia in aqueous or alcoholic solution is ever likely to produce a manufacturing process to compete with the Haber process which produces anhydrous ammonia.

The importance of the work described in Parts 2 and 3 lies in its demonstration that it is possible for dinitrogen, modified by attachment to an element such as molybdenum, tungsten, or vanadium to be reduced to ammonia in high yield in a protic medium. It opens up the possibility of preparing an electrode with complex metal sites designed to take up dinitrogen and feed electrons into it in a protic electrolyte (e.g. water) to produce ammonia. If ammonia were produced in this manner it would require at 100% efficiency slightly less energy than that needed to produce dihydrogen from water. A viable industrial process for ammonia could never be based on such a reaction, but it might be possible for the specialized needs of agriculture or horticulture.

If a highly efficient electrode for the reduction of dinitrogen at atmospheric pressure could be devised, it would produce ammonia in a simple and relatively cheap plant such as is used for electroplating. It would open up the possibility of converting atmospheric nitrogen to ammonia in irrigation channels at the point where it is needed. The ammonia supply need not be constant and the plant could use intermittent energy such as provided by wind, tide, sun, or off-peak electrical supply. Such a solution did not appear possible only ten years ago. Now the chemistry of dinitrogen shows that it is possible, but it is still very far from being realized.

Chapter 10 describes the study of nitrogen analogues on dinitrogen binding sites. Such studies have given important information about the chemical characteristics of these sites. It could eventually lead to some interesting new synthetic chemistry and point the way to more interesting reactions of ligating dinitrogen itself.

The problem of the production of ammonia for agricultural purposes is an important one. When scientists towards the end of the last century saw the end of Chilean saltpetre as heralding the beginning of famine in the Western world, the problem of producing nitrogenous fertilizer was placed fairly and squarely on the shoulders of the chemists. It was solved around the end of the first decade of this century by the Haber process. Now when the problem is arising again as the world's oil and gas reserves are being depleted it has returned to the scientists, but it is not necessarily a problem for the chemists.

The problem is no easier to solve and no less acute than before, but it may be that a biological solution will be the right one this time. Considerable effort is being expended in microbial genetics, to map the genes responsible for nitrogen fixation and for its regulation in bacteria with the eventual goal that nitrogen fixation genes may be transferred to other bacteria to create more nitrogen fixers or even be transferred to plants or animals themselves. Also the genetic manipulation of bacteria may create new symbioses so that nitrogen-fixing bacteria symbiotic with the main cereal crops such as wheat or rice may be developed. Although these possible solutions present enormous difficulties, they are no greater than those which faced the chemists who had the problem of finding an ammonia process at the beginning of this century. There appears at present to be no theoretical reason why gene transfer as discussed above cannot be done. The energy required for nitrogen fixation would then be collected through the leaves of the plants and the ammonia provided in the plant where and when it is needed. It would provide an ideal solution to the nitrogen problem. In the meantime, the study of dinitrogen complexes and their reactions holds out the best hope for chemists to find some alternative to the Haber process for the production of ammonia for agricultural and horticultural purposes.

Subject Index